## UNITS (S.I. AND ENGLISH)

| Parameter | S.I. | English |
| --- | --- | --- |
| mass | kilogram (kg) | lbm |
| length | meter (m) | foot (ft) |
| force | newton (N) | lbf |
| time | second (s) | second (s) |
| density | $kg/m^3$ | $lbm/ft^3$ |
| angular displacement | radian (rad) | radian (rad) |
| moment of force | N·m | pound·foot, lbf·ft |
| mass moment of inertia | $kg·m^2$ | $lbm·ft^2$ |
| stress, elastic modulus | pascal (P), $N/m^2$ | psi, $lbf/in^2$ |
| velocity | m/s | ft/s |
| acceleration | $m/s^2$ | $ft/s^2$ |
| energy (work) | joule (J), N·m | ft·lbf |
| power | watt (J/s) | ft·lbf/s |
| spring coefficient | | |
|   translation | N/m | lb/ft |
|   rotation | N·m/rad | lbf·ft/rad |
| damping coefficient | | |
|   translation | N·s/m | lbf·s/ft |
|   rotation | N·m·s/rad | lbf·ft·s/rad |

Donated by:

Matt Meyer

# Principles of Vibration

# Principles of Vibration

Benson H. Tongue
*Department of Mechanical Engineering*
*University of California at Berkeley*

*New York    Oxford*
OXFORD UNIVERSITY PRESS
1996

Oxford University Press

Oxford   New York
Athens   Auckland   Bangkok   Bombay
Calcutta   Cape Town   Dar es Salaam   Delhi
Florence   Hong Kong   Istanbul   Karachi
Kuala Lumpur   Madras   Madrid   Melbourne
Mexico City   Nairobi   Paris   Singapore
Taipei   Tokyo   Toronto

and associated companies in

Berlin   Ibadan

Published by Oxford University Press, Inc.
198 Madison Avenue, New York, New York, 10016-4314

http://www.oup-usa.org

**Library of Congress Cataloging-in-Publication Data**
Tongue, Benson H.
Principles of vibration / Benson H. Tongue.
p. cm.      Includes bibliographical references and index.
ISBN 0-19-510661-X (cl)
1. Vibration.      I. Title
TA355.T64      1996
620.2--dc20      95-45031

Printing (last digit): 9 8 7 6 5 4 3 2 1

Printed in the United States of America
on acid-free paper

To my wife Claire, without whom this
book never would have been written

A subject is difficult only so long as you don't
understand it. Once you do, it becomes intuitively
obvious.

*Benson Tongue (1957–present)*

# Contents

Contents                                                                    **ix**

# Foreword

It is a distinct pleasure to contribute a word of appreciation for this new book. The subject material is a standard part of a modern engineer's vocabulary and education. Vibration theory has wide applications and a deep intellectual foundation. Professor Benson Tongue has struck an attractive balance between the intriguing applications that permeate this topic and the rigorous basis for its study.

His engaging writing style, with the student very much in mind (including practicing engineers as well as undergraduate and graduate students), will find a wide and receptive audience, I am sure. The emphasis on understanding and the development of intuition, most fully addressed in his delightful chapter on "seat of the pants" engineering, will be welcomed by those current and future engineers who discover it desirable or necessary to acquire a working knowledge of vibrations. Vibration is ubiquitous in many venues of engineering practice. The subject and the author are well matched in this highly recommended book.

Professor Earl H. Dowell

*Dean of Engineering*
*Duke University*

# Preface

Authors take the time to write prefaces in order to tell readers their hopes and aspirations, why they wrote their books, and what they set out to accomplish. Unfortunately, the result usually is a boring list of topics covered and how they differ from coverage in existing books, something that rarely makes a good read. And am I going to be any different? That's up to you to decide. However, you'll never know if you don't read the next few paragraphs.

A question for anyone who's thinking of writing a book is *why*, when there are so many books already on the market. Writing a book takes an enormous amount of time that could be spent on other things. So why did I do it? Was it just to keep myself out of trouble? Well, that was part of it, but there were other reasons as well. For one thing (and this is definitely addressed to the instructors out there), the other books dealing with this subject usually range from a trifle dry to downright dehydrated. The reason (and we're all guilty of it) is that professors just love to get into the details of their subject. They've mastered the material themselves and want to lay out all the fascinating bits of information that make the subject so interesting to someone who's really digging into it. And once you get started, it's easy to go deeper and deeper, covering all the possible variations on a theme that the subject supports. Of course the book, by necessity, becomes denser and denser. The end product is something that contains a great deal of highly correct and useful material. The difficulty is that, for someone just learning the subject, it's in too concentrated a form for easy digestion. It's incredibly easy for an author to forget how hard this material is when it's being encountered for the first time.

Learning new things is always difficult. Always. No matter what the subject, be it karate or differential calculus, forcing your brain to accept new material is never easy; it takes disciplined attention and hard work. Given this fact, it seems reasonable to try to present the material in as accessible a form as possible. This is what I tried to do in this book. The presentation is informal, more informal than is standard. My hope is that when reading it, the readers will feel as though I'm talking with them, offering the information for their perusal rather than forcing them to search out the facts and techniques from the pages like a prospector digs gold from the rocks. Usually, a technique is used first without any supporting derivations, in order to motivate anyone who might care about it in the first place. Only then, once its use has been demonstrated, will a more in-depth examination take place. Be assured that the detailed examination will indeed occur, just after the fact rather than before.

The book was written to be self-contained. However, the reader who has access to MATLAB will find it very helpful. MATLAB is a great piece of software for doing matrix manipulations. Since vibration analysis is pretty much all matrix operations, MATLAB is obviously an excellent tool to have. It's not necessary, but access to it will really speed up the learning process. I know when I was learning vibrations I had to take a lot of the information on faith since it was too much of a chore to verify some of the assertions through calculation. Having a convenient software package running on a PC or a larger machine makes it easy to check the book and also easy to try different cases—to poke around with the mathematics a bit.

It was a temptation to include a disk with presolved problems that students could fire up at will. But, as you can see, I didn't. There were a couple of reasons. First, some instructors have no interest in MATLAB and some schools don't have it resident on their machines. Thus they'd be subsidizing the software for those schools that do use it. More importantly, I feel that MATLAB is very user friendly and that the more a student looks into it, the more that student will be inclined to use it and to learn more about it. None of the problems in the book require more than a few lines of commands to run, and writing them will be good for the student. Thus I've indicated what commands will enable the reader to solve the problems in the book and I've included some explicit MATLAB commands to start those readers who plan to use this package.

The material in the book grew out of lectures to junior, senior, and first-year graduate students at Georgia Tech and at Berkeley. The presumption is that the student using the text encountered linear algebra as a sophomore (but has likely forgotten much of it), has had a course in deformable bodies, and has taken a first course in dynamics. The text starts with the classical material (single-degree-of-freedom systems) and then branches out into more modern material. A good deal of emphasis is given to multiple-degree-of-freedom systems, which is probably the most important subject for the practicing engineer. Whether obtained through a modal approach or a lumped discretization, most practical vibrations problems ultimately lead to multiple-degree-of-freedom systems.

The material in the book reflects my own opinions as to what can and cannot reasonably be taught in a quarter or semester course. Many topics could have been included but have not been. For instance, there is no attempt to cover finite elements. I have found that there is never time to do anything beyond mentioning the topic in class. Given that the subject deserves an entire course of study (and almost all schools offer such a course), no purpose seemed to be served by including a brief overview. The aim of this book is to give a real understanding of the basic principles of vibrations—to present the core ideas and concepts that make up the field.

Many of the problems are meant to challenge the student to think deeply about a particular problem, to really analyze why a certain thing happens and what can be accomplished with a given technique. The instructor should definitely look at the solution book before assigning a problem; some have very short solutions, whereas others are much more involved. Many absolutely require the use of a computer and a good set of eigenanalysis algorithms.

One unique aspect of the book is Chapter 7, Seat-of-the-Pants Engineering. Many ways to get a quick answer or to verify an analysis are known to experienced engineers but are only picked up slowly, after years of working on vibration problems. And yet these approaches are useful for everyone to know right from the start. Chapter 7 brings together a few of these approaches and applies them to all the previously discussed systems. I felt that this material deserved the recognition of a separate chapter as it essentially carries the idea of approximate analyses (discussed in Chapter 6) to another level. Rather than being an approximate analysis that, with more effort, approaches the actual solution, these techniques are inherently approximate approaches that can't be made to emulate the exact solution but can give the user a fast and accurate estimation of the real solution.

I hope that this book strikes the right balance between detail and accessibility and that it makes learning vibrations just a bit more fun than might otherwise be the case. I'd really enjoy hearing from you and finding out if the book achieved its aims. If you want to pass on your compliments or criticisms, I can be reached at bhtongue-vibrations@euler.me.berkeley.edu. It may well be that some instructors have interesting homework problems or examples that they'd like to see included in this book. I encourage you to send these to me. If I agree that they'd be an interesting addition to the book, I'll add them and include an acknowledgment of the contributor in the next edition.

As a final note, I must cheerfully admit that there are undoubtedly some errors (simply typos, of course!) still floating around the book, in spite of the many proofreadings. As an aid in clearing this up, I'll be maintaining a net-accessible errata file. Whenever I discover or am informed about a problem, I'll include the information in this errata file. You can then access this file via the internet and download it to your machine. The procedure is as follows (regular text indicates what you type and computer font text represents computer prompts):

ftp ftp.me.berkeley.edu
`login`: anonymous
`password`: (your e-mail address goes here)
`ftp>` bin
`ftp>` cd pub/bhtongue/vibrations
`ftp>` get errata.vib
`ftp>` quit

Once you've gotten the file vib.err you can read it with pretty much any editor you have, since the file is a purely ASCII one. As an encouragement to letting me know of errors you might find, the contributors of the corrections will be included in the file.

*Berkeley, California*
*January 1996*

B. H. T.

# Acknowledgments

This book wouldn't be what it is without contributions from many people over the years. My views on the subject were strongly influenced by my Ph.D. advisor, Earl H. Dowell, then a professor at Princeton and now Dean of Engineering at Duke University. His approach to learning and teaching has made an indelible mark on me.

Many semesters' worth of students have listened patiently as I presented this material to them. Their input led to changes in presentation that, I hope, have led to a better end product. I'm most especially grateful to those that suffered through preliminary versions of this book without storming my office and demanding a "real" book.

Professor Aldo Ferri of Georgia Tech gave me much appreciated feedback and participated in useful discussions throughout the genesis of the book. He's a superb teacher and his insights were most welcomed. Bonnie Korpi deserves particular thanks for typing the homework problems. Professor L. van Niekerk, once a Ph.D. student of mine and now a professor at the University of Pretoria, graciously provided the nonlinear beam data included at the end of Chapter 8.

My entire family was very supportive during the long preparation period of this book. A special thanks to my father, Ben H. Tongue, for asking me every week, "Is it done yet?" Yes, it's finally done.

# 1

# Free Vibration of Single-Degree-of-Freedom Systems

## 1.1 INTRODUCTION

In this chapter we will examine the responses of single-degree-of-freedom systems. This is one of the most important topics to master, since the more complicated cases (multiple-degree-of-freedom (MDOF) and continuous systems) can often be treated as if they are simply collections of several, individual, single-degree-of-freedom systems. Once the responses of these simpler systems are well understood, branching out to the more complicated situations becomes relatively easy. Initially, we will ignore damping, since for many common structures the presence of damping changes the overall responses only slightly. Once we have a good grasp of how undamped problems behave, we'll look at the effects of damping.

Before we dive in, it makes sense to explain what single degree of freedom actually means. This idea is pretty simple—if an object is free to move in a particular direction, then it has a degree of freedom in that direction. Thus, a bead that slides on a wire of an abacus has one degree of freedom. Figure 1.1 shows another example of a single-degree-of-freedom system, a mass in a channel. Since the mass is only free to move up or down the channel, it has one degree of freedom. We've used the coordinate $x$ to describe the position of the mass. With this representation it is clear

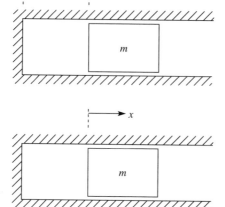

**Figure 1.1** Single-degree-of-freedom system, single mass

**Figure 1.2** Single-degree-of-freedom system, single mass

what $x$ stands for; it gives us the displacement of the mass from the left end. The more common way that we'll be illustrating coordinates is shown in Figure 1.2. Here it looks likes the $x$ coordinate is attached to the mass, but it isn't. $x$ motions are still referenced to the fixed channel; the starting point of the $x$ coordinate is aligned with the mass because the initial (zero displacement) reference occurs exactly at the mass's initial position. Any variations of the mass from the position it is in currently will produce a finite $x$ reading.

What about more than one degree of freedom? A point mass on a flat plane has two degrees of freedom (it can move both up or down and horizontally, for instance, as shown in Figure 1.3). A penny on a flat plane has three degrees of freedom, two to define the position of the penny's center and a third (an angular rotation) to describe at what angle the penny is oriented. Rigid

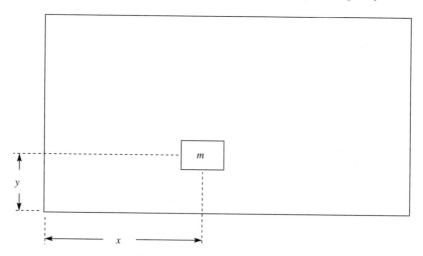

**Figure 1.3** Two-degree-of-freedom system, single mass

bodies in three space have six degrees of freedom, three translational (telling the location of the center of mass) and three rotational (telling the angular orientation of the body). In this text you'll sometimes see *degree of freedom* written out, but most often it will be abbreviated as *DOF*. Thus 2-DOF should be read as *two degrees of freedom*.

## 1.2 TRANSLATIONAL VIBRATIONS—UNDAMPED

The first vibrational device that we will discuss is a spring-mass system, something that you can make very easily. In fact, I strongly encourage you to take a minute right now to make one, since doing so will help you understand several basic vibrations concepts. Although you can follow along without making an actual device, using one will greatly aid you in developing your physical intuition. Not only that, you've already paid for half the system by buying this book and you might as well get your money's worth. This half is, of course, the spring that was shrink-wrapped with the book. (If you didn't buy a new book, check around at a hardware store or use some rubber bands.) Now, attach one end of the spring to a handy weight (a ceramic coffee cup will do nicely) and hold the free end of the spring in your hand. Make sure that the weight you've chosen isn't so heavy that the spring is stretched to near its limit, that is, there must be a good bit of stretch remaining in the spring beyond its extension due to gravity. When you're done, you'll have a high-class vibrational test device like the one shown in this chapter opener.

A cautionary note is needed here. If you use rubber bands because you bought a used book and can't find a spring, you should be aware that since the rubber dissipates much more energy than a spring, the vibrations will die away quickly and it will take more effort to work the forced tests that come in the next chapter.

These two elements, the mass and spring, are the most basic building blocks for vibrational analyses. Even though we know that our mass has some degree of internal flexibility, our model is considered to be infinitely rigid. Similarly, although the physical spring clearly has a mass of its own (it's not weightless, after all), our model will have none. Thus we are already approximating the real world. This idealization is necessary if we are to come up with reasonable mathematical models, but you should keep in mind that we are using approximations that never exactly match reality. The fact that actual objects combine mass and stiffness (and damping) will become much more important when we look at the vibration of continuous systems, such as strings, rods, beams, and such. Happily, even though our system models will not match reality exactly, they can come very close to doing so, thus enabling us to use the approximations and to feel confident that our results have physical meaning.

Having introduced masses and springs with words, it is appropriate to pin down exactly what we mean. Mass is a fundamental quantity, possessed by all macroscopic objects. Since this is a vibrations text, we won't get into the details of how physicists currently view matter, but instead use the well-known Newtonian view that mass times acceleration equals force and that the force generated by a mass in a gravitational field is called its weight.

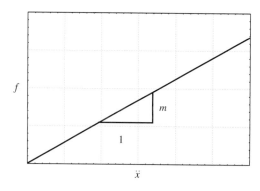

**Figure 1.4**  Force versus acceleration for a mass

Although we won't get any vibrations out of a mass by itself, it is useful to analyze the behavior of simple masses, especially since complicated objects such as airplanes and spacecraft can behave in a manner similar to a lumped mass under normal operating conditions. If we write out Newton's second law for a constant mass, we have

$$m\ddot{x} = f \tag{1.2.1}$$

Note that an overdot indicates differentiation with respect to time. If the applied force $f$ is zero, then the solution to this equation is simply

$$x(t) = a + bt \tag{1.2.2}$$

where $a$ and $b$ are constants. Thus the mass can be stationary (if $b = 0$) or can move with a constant speed $b$, in what we call rigid body motion. In our case, there really isn't any body to speak of, since we're dealing with a mass particle. But when we move on to more realistic extended bodies, such as aircraft, they'll show the same behavior. If they aren't restrained, then they can remain in one spot or, depending on the initial conditions, translate and/or rotate at a constant speed. Those of you who've been in a jet when it hits a sudden downdraft can certainly attest to the fact that the whole plane can sharply move down without pitching or rolling to any great extent.

If we plot the relation between acceleration and force we obtain Figure 1.4. The plot is linear and the slope varies with the mass $m$. Thus the larger the mass, the less it accelerates for a given force.

Springs are a little more complicated than masses, but not by much. As we've already mentioned, to accelerate a given mass $m$ at an acceleration $a$, we must apply a force $f$, where $f = ma$. In the case of a spring, an applied force doesn't produce an acceleration, but rather a displacement. Figure 1.5 shows the relationship between force and displacement for the illustrated spring. As we can see from the figure, the graph is linear, indicating that an increase in the applied force will produce a proportional change in the displacement. The constant of proportionality is commonly denoted by $k$ and is called the spring constant. Thus

$$f = kx \tag{1.2.3}$$

When we increase $k$ we're increasing the stiffness of the spring, making it harder to deflect.

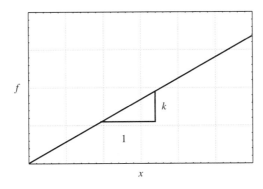

**Figure 1.5**   Force versus displacement for a spring

## Example 1.1

**Problem**   Find the local spring constant for motions about a given displacement for the nonlinear spring characteristic shown in Figure 1.6.

**Solution**   Figure 1.6 shows a general force versus deflection curve for a nonlinear spring. In order to determine the linear spring characteristic, we must examine the system for motions about some equilibrium. If we assume that the equilibrium position corresponds to $x = x_0$, then the local linear spring characteristic is given by

$$k = \frac{df}{dx} \tag{1.2.4}$$

evaluated at $x = x_0$. This is simply equal to the slope of the force-deflection curve, as shown in the figure. If we take the specific example of

$$f(x) = x + x^3 \tag{1.2.5}$$

and want to find the local spring constant around $x = 1$, then we have

$$k = 1 + 3x^2|_{x=1} = 4 \tag{1.2.6}$$

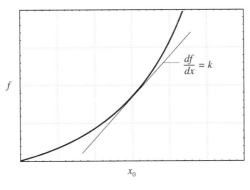

**Figure 1.6**   Local spring constant

Throughout the rest of the book we will use $m_1$, $m_2$, $k_1$, $k_2$, etc., to indicate both the particular object we're examining, as well as its associated physical quantity. Thus we might call a particular spring $k_1$ and also indicate its stiffness by $k_1$. Otherwise we'd have to say "consider the spring $s_1$, with spring constant $k_1$ and the mass $n_1$, with mass equal to $m_1$," which is long-winded and also uses up too many symbols. It will be obvious from the sentence whether $m_1$ refers to the object itself or the mass of the object and whether $k_1$ refers to the spring or its spring constant.

The first experimental test we'll run will be a free vibration experiment. Keeping the free end of the spring fixed with the mass suspended beneath it, pull the mass down a bit and then release it. You'll note that the mass oscillates up and down at a fairly regular rate. Furthermore, it seems to oscillate with close to a constant amplitude. We can easily demonstrate that the mathematical equations of motion of this system exhibit these same properties. From the free-body diagram of Figure 1.7 we see that the equation of motion for the system is

$$m\ddot{y}(t) = -k\left(y(t) - l\right) + mg \tag{1.2.7}$$

The spring's unstretched length is equal to $l$ (thus the spring force is proportional to the stretch in the spring beyond this length) and $g$ is the acceleration due to gravity. Although this representation is perfectly valid, it doesn't use an important fact, namely that the oscillations will occur about the static equilibrium position of the mass due to gravity. This position is the one at which the restoring force due to the spring just balances the force due to gravity:

$$k(y_0 - l) = mg \tag{1.2.8}$$

If we let $y(t) = y_0 + x(t)$, that is, $x$ represents the motions about the equilibrium position, then we can rewrite (1.2.7) as

$$m\ddot{x}(t) = -k(y_0 + x(t) - l) + mg \tag{1.2.9}$$

or, using our equilibrium equation (1.2.8) to eliminate $mg$,

$$m\ddot{x}(t) = -k(y_0 + x(t) - l) + k(y_0 - l) = -kx(t) \tag{1.2.10}$$

Moving all the dependent variables to the left side of the equation leaves us with

$$m\ddot{x}(t) + kx(t) = 0 \tag{1.2.11}$$

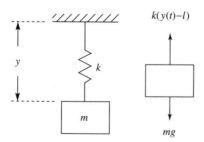

**Figure 1.7**   Free-body diagram of a spring-mass system

Make sure that this is clear in your mind since it is often a point of confusion. Gravity simply serves to set up an equilibrium configuration for the system and the system's vibrations occur about this equilibrium. You'll examine this more fully in some of the homework problems.

At this point, we'll stop showing the time dependency as $(t)$ when it is clear from the context which of the variables are time dependent and which are just constants:

$$m\ddot{x} + kx = 0 \qquad (1.2.12)$$

Equation (1.2.12) is an autonomous (no forcing), constant coefficient, ordinary linear differential equation. All constant coefficient, ordinary linear differential equations have solutions of the form

$$x = ae^{\lambda t} \qquad (1.2.13)$$

that is, they exhibit exponential behavior. This is worth repeating. *All* autonomous, constant coefficient, ordinary linear differential equations have exponential solutions. Burn this into your brain; it underlies all the solution procedures we'll be meeting later in the book.

Using this assumed solution, (1.2.12) becomes

$$(m\lambda^2 + k)ae^{\lambda t} = 0 \qquad (1.2.14)$$

Since $a$ can't be equal to 0 (unless $x(t) = 0$ seems interesting to you) and $e^{\lambda t}$ isn't equal to 0 for any finite time $t$, we're left requiring that

$$m\lambda^2 + k = 0 \qquad (1.2.15)$$

Therefore $\lambda^2 = -\frac{k}{m}$ or $\lambda_{1,2} = \pm i\sqrt{\frac{k}{m}}$. Thus the full solution can be written out as

$$x(t) = a_1 e^{it\sqrt{\frac{k}{m}}} + b_1 e^{-it\sqrt{\frac{k}{m}}} \qquad (1.2.16)$$

Using the fact that $e^{\pm i\omega t} = \cos(\omega t) \pm i \sin(\omega t)$, we can just as easily express $x(t)$ as

$$x(t) = a_2 \sin\left(\sqrt{\frac{k}{m}}t\right) + b_2 \cos\left(\sqrt{\frac{k}{m}}t\right) \qquad (1.2.17)$$

You can check that both (1.2.16) and (1.2.17) are valid solutions by using them in (1.2.12) and observing that they satisfy the equation of motion.

We see from (1.2.17) that $\sqrt{\frac{k}{m}}$ has the units of $\frac{1}{\text{seconds}}$ (since the argument of a sine or cosine is nondimensional) and thus has the dimensions of a frequency. We commonly refer to this particular frequency as the system's natural frequency, $\omega_n$:

$$\omega_n \equiv \sqrt{\frac{k}{m}} \qquad (1.2.18)$$

We'll see when we start looking at forced vibrations that this frequency plays an important role in a system's vibrational response.

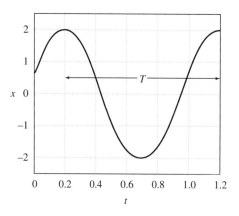

**Figure 1.8**   Sinusoidal waveform

If we plot (1.2.17) versus time for particular values of $a_2$, $b_2$, $k$, and $m$, we obtain Figure 1.8. You can see that the response is a periodic signal, meaning that the response repeats itself over and over. The time necessary for the signal to repeat ($T$) is the fundamental period of the oscillation, which is related to the natural frequency by

$$\omega_n = \frac{2\pi}{T} \tag{1.2.19}$$

In the absence of damping, the amplitude of the solution stays constant. When you performed the free-vibration experiment with the spring and mass, you saw the mass undergo several oscillations without much decay in the response. This was due to the fact that the system didn't have much damping.

The details of what a particular response will look like depend upon the system's initial conditions. Given the right start, we could see a pure sine wave, a cosine, or a combination of the two. To see how only the cosine term would appear, let $\omega_n = 2\pi$ and let the initial conditions be

$$x_0 \equiv x(0) = 2$$

$$\dot{x}_0 \equiv \dot{x}(0) = 0$$

To match these initial conditions we would need $a_2 = 0$ and $b_2 = 2$, leading to

$$x(t) = 2\cos(2\pi t) \tag{1.2.20}$$

shown in Figure 1.9. If, on the other hand, the initial conditions were

$$x_0 = 1.0$$

$$\dot{x}_0 = 5.0$$

then we will find that $a_2 = \frac{5}{2\pi}$ and $b_2 = 1$ or

$$x(t) = \frac{5}{2\pi}\sin(2\pi t) + 1\cos(2\pi t) \tag{1.2.21}$$

which is illustrated in Figure 1.10.

$x(0) = 2, \ \dot{x}(0) = 0$

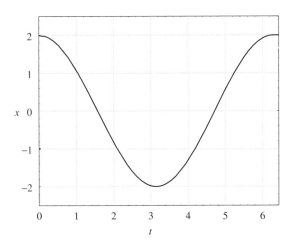

**Figure 1.9** Free response of a spring-mass system

Although (1.2.20) and (1.2.21) are both perfectly valid ways to write the solutions, there are actually other forms that are just as correct. For instance, say we don't like to use cosines. Would it be possible to express (1.2.20) in terms of a sine? The answer is "yes." There's no problem with doing this. All we need to do is include the appropriate phase shift. Our original solution was $x(t) = 2\cos(\omega_n t)$. If we want to use a sine, then we'd say $x(t) = 2\sin(\omega_n t + \frac{\pi}{2})$. Check it out. At every time $t$, both these expressions produce the same answer. The $\frac{\pi}{2}$ term is called a "phase shift," and does what the name implies—it shifts the phase of the sine wave by $\frac{\pi}{2}$. This can also be viewed as shifting the effective origin of the time axis by $t = \frac{2\pi}{\omega n}$.

$x(0) = 1, \ \dot{x}(0) = 5$

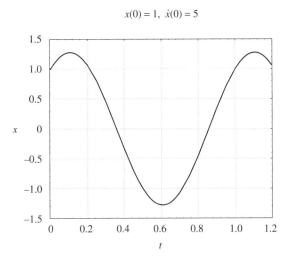

**Figure 1.10** Free response of a spring-mass system

Since the idea of phase shifts will be of fundamental importance to understanding vibrational responses, we'll spend some time on the concept now. You'll recall from trigonometry that sines (or cosines) of the sum of two quantities can be expanded in terms of sines and cosines of those quantities through the formulas

$$\sin(x \pm y) = \sin(x)\cos(y) \pm \cos(x)\sin(y) \tag{1.2.22}$$

$$\cos(x \pm y) = \cos(x)\cos(y) \mp \sin(x)\sin(y) \tag{1.2.23}$$

Thus, if we wished to express

$$a_1 \sin(\omega_n t) + a_2 \cos(\omega_n t) \tag{1.2.24}$$

as a cosine with a phase shift $\phi$, we would say

$$d \cos(\omega_n t + \phi) = d \cos(\omega_n t)\cos(\phi) - d \sin(\omega_n t)\sin(\phi) \tag{1.2.25}$$

and, comparing the coefficients of $\cos(\omega t)$ and $\sin(\omega t)$ in (1.2.24) and (1.2.25), we would see that

$$d \cos(\phi) = a_2 \tag{1.2.26}$$

and

$$d \sin(\phi) = -a_1 \tag{1.2.27}$$

Squaring both sides of (1.2.26) and (1.2.27) and adding them together yields

$$d^2 = a_1^2 + a_2^2 \tag{1.2.28}$$

and dividing $d \sin(\phi)$ by $d \cos(\phi)$ yields

$$\tan(\phi) = -\frac{a_1}{a_2} \tag{1.2.29}$$

or

$$-\phi = \tan^{-1}\left(\frac{a_1}{a_2}\right) \tag{1.2.30}$$

Thus we can always express the overall vibration as a cosine function with a phase shift and, in an exactly parallel manner, also express it as a sine function with a phase shift.

**Example 1.2**

    **Problem**    Put $2 \sin(3t) + 4 \cos(3t)$ into the form $a \cos(3t + \phi)$.

    **Solution**    For this problem, $a_1 = 2$ and $a_2 = 4$. Thus, using (1.2.28), we have

$$d = \sqrt{2^2 + 4^2} = \sqrt{20} \tag{1.2.31}$$

and (1.2.30) gives us

$$\phi = -\tan^{-1}\left(\tfrac{2}{4}\right) = -.464 \tag{1.2.32}$$

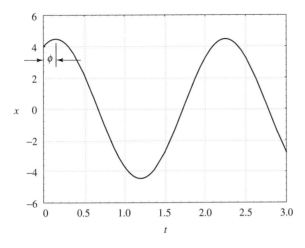

**Figure 1.11**   $4.47\cos(3t - .464)$

Therefore

$$2\sin(3t) + 4\cos(3t) = \sqrt{20}\cos(3t - .464) \tag{1.2.33}$$

Figure 1.11 shows what the final waveform looks like and shows how the phase shift acts to move the cosine wave over to the right. The effective time origin of the cosine wave is at $t = .464/3 = .155$.

Okay, now we know how to find the free response of a system that consists of one mass and one spring. But what if our system is more complex than this; what if it involves several springs? What do we do? Well, let's consider the problem. Say we're examining a mass that's supported by four isolators (springs) as shown in Figure 1.12 If we're concerned only with vertical motions (one degree of freedom) then we can redraw the system as shown in Figure 1.13. A free body diagram (Figure 1.13*b*) indicates that the relevant equation of motion will be

$$m\ddot{x} = -k_1 x - k_2 x - k_3 x - k_4 x \tag{1.2.34}$$

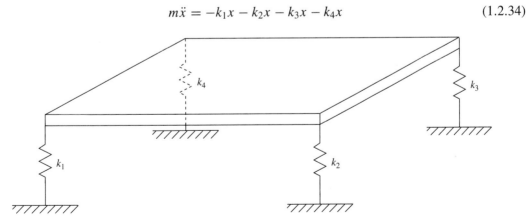

**Figure 1.12**   System supported by four suspension springs

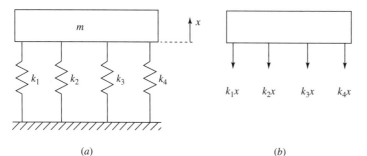

Figure 1.13 Free-body diagram of a multi-spring system

or

$$m\ddot{x} + k_1 x + k_2 x + k_3 x + k_4 x = 0 \tag{1.2.35}$$

This can be written more simply as

$$m\ddot{x} + k_{eq} x = 0 \tag{1.2.36}$$

where

$$k_{eq} = k_1 + k_2 + k_3 + k_4 \tag{1.2.37}$$

Thus we see that springs that are in *parallel* simply can be added together to find the equivalent spring constant.

More complicated than this would be to have springs in *series*, as shown in Figure 1.14a. This is a very common situation. For instance, if you were modeling an automobile, you'd want to take account of the spring stiffness of the tire first (our $k_1$). Next, you'd consider the actual suspension springs of the vehicle (our $k_2$). Finally, you'd have the mass of the car body (our $m$).

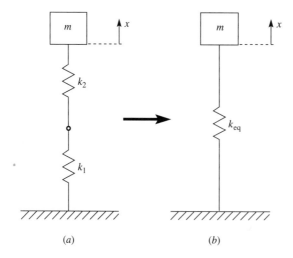

Figure 1.14 Representing two springs in series as a single equivalent spring

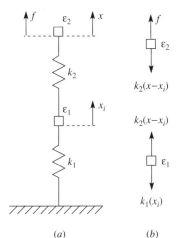

**Figure 1.15**  Free-body diagram with two springs in series

$(a)$                    $(b)$

How can we determine the equivalent $k$ (Figure 1.14$b$) for this system? Well, the approach we'll use is one that you'll find helpful in the future when you need to determine the boundary conditions for continuous problems (and, of course, it's helpful here). What we'll do is create a new problem, shown in Figure 1.15. For this problem we've taken the two springs that are in series and put a tiny mass $\epsilon_1$ between them and also put another tiny mass $\epsilon_2$ at the end of $k_2$. Finally, we've applied a force $f$ to the top mass.

As indicated in Figure 1.14, what we eventually want is an expression such as

$$f = k_{eq}x \tag{1.2.38}$$

where $k_{eq}$ is our equivalent spring constant. The reason for introducing the tiny masses is to make it easier to apply Newton's laws to this problem in a way that is familiar to you. However, once we get the equations of motion, we'll set $\epsilon_1$ and $\epsilon_2$ to zero, thus reducing the problem to our original one. The $\epsilon$'s are just a conceptual aid.

Going ahead and determining the equations of motion for this system (using the free body diagrams from Figure 1.15$b$) gives us

$$\epsilon_2 \ddot{x} = -k_1(x - x_i) + f \tag{1.2.39}$$

and

$$\epsilon_1 \ddot{x}_i = k_2(x - x_i) - k_1 x_i \tag{1.2.40}$$

As mentioned, we'll now set the $\epsilon$'s to zero, thus giving us the correct static equations

$$k_2(x - x_i) = f \tag{1.2.41}$$

$$k_2(x - x_i) - k_1 x_i = 0 \tag{1.2.42}$$

If we use (1.2) to eliminate $k_2(x - x_i)$ from (1.2.42), we'll obtain

$$f = k_1 x_i \tag{1.2.43}$$

**Figure 1.16**   Electrical resistors in series

or

$$x_i = \frac{f}{k_1} \tag{1.2.44}$$

Substituting this value for $x_i$ into (1.2.42) gives us

$$k_2 \left( x - \frac{f}{k_1} \right) - f = 0 \tag{1.2.45}$$

which, when rearranged, yields

$$\left( \frac{k_1 k_2}{k_1 + k_2} \right) x = f \tag{1.2.46}$$

Comparing (1.2.46) with (1.2.38), we see that we've found our equivalent spring constant

$$k_{\text{eq}} = \frac{k_1 k_2}{k_1 + k_2} \tag{1.2.47}$$

If you've taken a circuits course, this will remind you of the formula for the equivalent resistance of two resistors in parallel, *not* two resistors in series. This is a point that sometimes causes confusion, since the schematic respresentation of two springs in series looks very much like two resistors in series (Figure 1.16). The relevant equations for this electrical example are

$$v - v_i = r_2 i \tag{1.2.48}$$

and

$$v_i = r_1 i \tag{1.2.49}$$

Combining (1.2.48) and (1.2.49) yields

$$v = (r_1 + r_2) i \tag{1.2.50}$$

thus the equivalent resistance is $r_{\text{eq}} = r_1 + r_2$.

Calculating the equivalent spring constant for springs in parallel is easier than the previous case. Figure 1.17 shows this case and the equivalent formulation we're trying for. Figure 1.18 illustrates the relatively straightforward free body diagram for this problem. Applying $f = ma$ in the vertical direction gives us

$$m\ddot{x} = -k_1 x - k_2 x \tag{1.2.51}$$

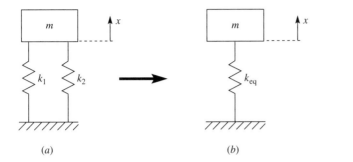

(a)                                      (b)

**Figure 1.17** Representing two springs in parallel as single equivalent spring

$k_1 x$       $k_2 x$

**Figure 1.18** Free-body diagram for a mass with two springs in parallel

or

$$m\ddot{x} + (k_1 + k_2)x = 0 \tag{1.2.52}$$

Thus the equivalent spring constant for two parallel springs is given by

$$k_{eq} = k_1 + k_2 \tag{1.2.53}$$

Note that this looks just like the equivalent resistance formula we derived a moment ago. If you take a moment to look at the equivalent resistance of resistors in parallel, you'll have the completed picture—namely that series arrangements of springs are like parallel arrangements of resistors and parallel arrangements of springs are like series arrangements of resistors. Actually there is a logical way to look at both systems with a method that predicts this behavior, but we won't go into it here. The interested reader can check [8] for more information. Now that we have (1.2.47) and (1.2.53), we can determine the equivalent spring constant for any particular arrangement of springs simply by breaking the more complicated problem down into smaller pieces.

Even though we really never have pure springs and masses in practice, we often approximate our systems as if the lumped elements exhibited only a single characteristic. For instance, examine Figure 1.19, which shows an extensible bar with a lumped mass at the end. We know that the bar itself has mass, but if the lumped mass is much greater than the bar's mass, we might decide to

$l$

$x$

$m_1$

$\rho, EA(x)$

**Figure 1.19** Schematic of a bar with a lumped mass

model the bar as a pure spring. We'll revisit this problem in Chapter 7, to see how more exact solutions compare to our approximate ones.

If we're going to ignore the bar's own mass, then the system can be represented as a lumped mass $m_1$ at the end of a spring. All we need to do is determine the spring constant, then we can solve for the system's natural frequency. The governing static equation for this case is

$$EA\xi_x = f \tag{1.2.54}$$

where $f$ is a constant force applied at the end of the bar, $\xi$ is the bar's spatial deflection, and $EA$ is the bar's stiffness. Integrating this gives us

$$EA\xi(x) = fx + b \tag{1.2.55}$$

Our boundary condition of zero displacement at $x = 0$ causes $b$ to equal zero and, evaluating this expression at $x = l$, we find

$$EA\xi(l) = fl \tag{1.2.56}$$

This can be rewritten as

$$f = \frac{EA}{l}\xi(l) \tag{1.2.57}$$

If we draw a parallel with our well-known spring-force equation ($f = kx$), we can identify $\frac{EA}{l}$ as our equivalent spring constant $k_b$. Thus our natural frequency is given by

$$\omega_n = \sqrt{\frac{k_b}{m_1}} = \sqrt{\frac{EA}{lm_1}} \tag{1.2.58}$$

As we've mentioned, the accuracy of this approximation depends upon the end mass $m_1$ being substantially greater than the bar's mass. To see this, we can look at two cases. First, we'll consider the case for which the bar's linear density $\rho$ is equal to 1 kg/m. The other dimensions will be $l = 1$ m, $m_1 = 10$ kg, and $EA = 10$ N. Thus the total mass of the bar is one-tenth that of the end mass. For these values we'll obtain $\omega_n = 1.00$ rad/s. If we solve this problem exactly (using the techniques discussed in Chapter 5), we will find that the exact answer is equal to .9836 rad/s. Thus our error is only around 2 percent.

If we now increase the bar's linear density by a factor of 10, to 10 kg/m, our estimate doesn't change at all (since it ignores the bar's mass). However, the mass certainly affects the actual answer and lowers the natural frequency to $\omega_n = .8603$ rad/s, giving us an error of 16 percent. This really isn't too bad if you consider the fact that the bar's mass is now equal to the end mass. It's actually possible to correct our approximation by using some engineering insight; we'll go into this in Chapter 7.

As an aid to the homework problems, Appendix B contains lumped spring constants for a variety of common situations. In addition, Appendix C contains physical constants for several common structural materials.

## 1.3 ROTATIONAL VIBRATIONS—UNDAMPED

Although many systems involve only translational motion of the sort we've been studying, many also include rotational motions. As long as the rotation angles are small, these systems can be analyzed just like the systems we've already seen. Consider, for example, a simple pendulum (Figure 1.20). This consists of a point mass, suspended from a massless rod. The generalized coordinate is $\theta$, the angle by which the rod deviates from the vertical position.

It's important to realize that there are actually two equilibium positions for this problem. The first corresponds to the $\theta = 0$. The pendulum will oscillate nicely about this equilibrium. However, there is another equilibrium, namely when $\theta$ is equal to 180 degrees. In this case the pendulum points directly upwards. This is what we call an unstable equilibrium, since it is precarious at best. If we very, very carefully place the pendulum in this upright position, it should remain upright. At least that's what the mathematics will tell us. But the mathematics will also tell us that any disturbance, no matter how small, will cause the pendulum to fall away from this equilibrium (and end up moving about the stable equilibrium at $\theta = 0$). Thus, in this and other examples, you'll want to be sure you're analyzing the system about the appropriate equilibrium. We'll mention how to do this after finding the equations of motion.

Constructing a free-body diagram of this system and summing moments about the attachment point 0 leads to

$$\sum \text{moments}_O = -mgl\sin(\theta) \tag{1.3.1}$$

This is equal to the time rate of change of the angular momentum about $O$:

$$ml^2\ddot{\theta} = -mgl\sin(\theta) \tag{1.3.2}$$

or

$$ml^2\ddot{\theta} + mgl\sin(\theta) = 0 \tag{1.3.3}$$

Keep in mind that in this example the point $O$ is stationary, thus the equation used is correct. If the attachment was moving we'd have additional terms in our moment equation.

All right, at this point we'll look at equilibria. Simply put, an equilibrium position is one for which the system is happy to remain at rest. No motion. Not even a little. A straightforward way

**Figure 1.20**   Simple pendulum

to find an equilibrium position is to take the equations of motion ((1.3.3) in our case) and cross off the terms that imply motion. For us, this means crossing off the $\ddot{\theta}$ term. Obviously, if there's no motion, then all derivatives with respect to time must be equal to zero. This leaves us with

$$mgl\sin(\theta) = 0 \qquad (1.3.4)$$

Since the $mgl$ term isn't equal to zero, we need to find when $\sin(\theta)$ is equal to zero. This occurs at $\theta = 0$ and at $\theta = \pi$. Thus our two equilibrium positions involve the pendulum hanging straight down or standing straight up, as mentioned before. We'll examine the case of motions about the $\theta = 0$ equilibrium here and leave the $\theta = \pi$ case to the homework section.

We can't solve (1.3.3) in the same manner as we solved the previous translational problems because this equation is nonlinear (i.e., the equation doesn't involve just the first power of the dependent variable $\theta$ and its time derivatives). In order to solve this problem we need to *linearize* the equation. Simply put, this means that we expand out all terms involving the dependent variables and their time derivatives, then throw away all that don't consist solely of the dependent variables (and their time derivatives) raised to the first power. Thus, any terms like $\theta\dot{\theta}$, $\theta^2$, $\theta^2\ddot{\theta}$, etc., will be discarded, while terms like $\theta$, $\dot{\theta}$, $\ddot{\theta}$, etc., will be kept. When equations involve only terms like $\theta$, $\dot{\theta}$, $\ddot{\theta}$, etc. (dependent variables and their derivatives raised to the first power), we call them linear. If other terms, such as $\theta^2$, $\dot{\theta}\theta$, $\theta^3$, etc., are present, then the equation is nonlinear.

Since we know that $\sin(\theta)$ can be expanded in a power series

$$\sin(\theta) = \theta - \tfrac{1}{3}\theta^3 + \cdots \qquad (1.3.5)$$

we can discard everything except $\theta$ to obtain

$$\ddot{\theta} + \frac{g}{l}\theta = 0 \qquad (1.3.6)$$

We discarded the higher-order terms because they become vanishingly small as $\theta$ is made small, that is, as $\theta$ becomes small, all powers of $\theta$ go to zero.

Note that we've divided through by $ml^2$. Also note that when we expanded out, we were doing the expanding about an equilibium position (in this case $\theta = 0$). This is always the case. You get your nonlinear equations of motion, find the equilibria, then linearize about a particular equilibrium.

Equation (1.3.6) is a linear equation and can be solved in the same manner as the translational problems. Since we've thrown away several terms in order to get to this linearized form, the question arises as to whether the resulting equation has any validity. Clearly, tossing out the terms as we did will be sensible only if, during the motion of the actual system, these terms are negligibly small. This condition is equivalent to saying that $\sin(\theta)$ is very close to equaling $\theta$ itself. We know from trigonometry that this is true only if $\theta$ is small. How small? You can easily verify with your calculator that, if $\theta$ is less than .24 rad, then the error in approximating $\sin(\theta)$ by $\theta$ is less than 1 percent and that the error for angles up to .53 rad is under 5 percent. These are pretty fair-sized angles, so we see that the approximation is a good one for realistic amplitudes of oscillation.

**Example 1.3**

**Problem**   Find the natural frequency of the streetlight shown in Figure 1.21 under the influence of gravity. Compare the result to the case in which gravity is neglected. View the pole as being infinitely rigid. The mass of the light is the dominant mass and is represented by $m$. The rotational stiffness of the arm is represented by a lumped rotational spring, $k_\theta$. The values of the various parameters are given by $m = 50$ kg, $g = 9.81$ m/s$^2$, $k_\theta = 29,393$ Nm, and $l = 3$ m.

**Solution**   When the arm is horizontal, the spring is unstretched. Gravity is going to work to push the end mass down while the torsional spring will resist this motion. Thus the equilibrium position will be somewhere between $\theta = 0$ and $\theta = \frac{\pi}{2}$. If we sum moments about 0, we'll find

$$ml^2\ddot{\theta} = -k_\theta\theta + mgl\cos(\theta) \tag{1.3.7}$$

Setting the time derivatives to zero gives us

$$-k_\theta\theta + mgl\cos(\theta) = 0 \tag{1.3.8}$$

or

$$k_\theta\theta = mgl\cos(\theta) \tag{1.3.9}$$

Using the given parameter values, this can be solved to give

$$\theta_{\text{eq}} = .05 \text{ rad} \tag{1.3.10}$$

If we expand about the equilibrium, $\theta_{\text{eq}}$, by representing $\theta$ by

$$\theta = .05 + \eta \tag{1.3.11}$$

(where $\eta$ is small), we'll have

$$ml^2(\ddot{.05 + \eta}) + k_\theta(.05 + \eta) - mgl\cos(.05 + \eta) = 0 \tag{1.3.12}$$

We now need to expand $\cos(.05 + \eta)$ into $\cos(.05)\cos(\eta) - \sin(.05)\sin(\eta)$. Since $\eta$ is assumed to be small, this can be approximated by $\cos(.05) - \sin(.05)\eta$ (where we've used $\sin(\eta) \approx \eta$ and $\cos(\eta) \approx 1$).

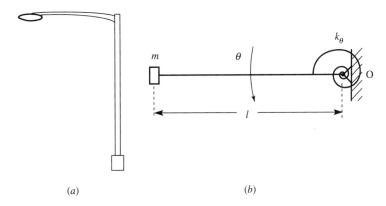

(a)                                                (b)

**Figure 1.21**   Streetlight and lumped approximation

Substituting this approximation into our equation of motion gives us

$$ml^2\ddot{\eta} + k_\theta(.05 + \eta) - mgl\Big(\cos(.05) - \sin(.05)\eta\Big) = 0 \tag{1.3.13}$$

Expanded and simplified, this will finally yield

$$ml^2\ddot{\eta} + \Big(k_\theta + mgl\sin(.05)\Big)\eta = 0 \tag{1.3.14}$$

If we go ahead and substitute the appropriate values for $m$, $l$, etc., we'll have

$$450\ddot{\eta} + 29,467\eta = 0 \tag{1.3.15}$$

giving us a value of 8.09 rad/s as the system's natural frequency. Note that in the absence of gravity, the natural frequency would be equal to $\sqrt{\frac{29,393}{450}} = 8.08$ rad/s. Thus the effect on the slight steady-state rotation due to gravity had a negligible effect on the system's natural frequency.

**Example 1.4**

**Problem**   Find the equations of motion for the system shown in Figure 1.22.

**Solution**   The system we'll be examining is shown in Figure 1.22 and contains both rotational and translational components. The horizontal bar can rotate about its hinge and the angle of rotation is given by $\theta$. Rotations are resisted by the spring $k_1$ and $k_2$. As we've done in the previous problems, we'll assume that all rotations are small, thus the displacement of the right end is given by $l_2\theta$. A mass is connected by a spring to the right end of the bar and an external vertical force is applied to the same end.

The first step in solving this problem is to construct a free-body diagram that shows what forces are acting on the system (Figure 1.23). Note that no reaction forces are drawn for the left end of the bar because the bar is physically pivoting about that point, thus the reaction forces will apply no moments there.

Summing moments about the pivot gives us

$$I_A\ddot{\theta} = -k_1(l_1\theta)l_1 - k_2(l_2\theta - x)l_2 + fl_2 \tag{1.3.16}$$

**Figure 1.22**   Spring-restrained/spring-mass system

**Figure 1.23**   Free-body diagram for spring-restrained/spring-mass system

while equating force and acceleration in the vertical direction for the lumped mass yields

$$m_2\ddot{x} = k_2(l_2\theta - x) \tag{1.3.17}$$

Putting all the dependent variables on the left-hand side then gives us the final equations of motion

$$I_A\ddot{\theta} + (k_1l_1^2 + k_2l_2^2)\theta - k_2l_2x = fl_2$$
$$m_2\ddot{x} + k_2x - k_2l_2\theta = 0 \tag{1.3.18}$$

## 1.4  VISCOUS DAMPING

The difficulty with the analyses so far is that our systems never stop moving; they oscillate forever, something that clearly doesn't happen in the real world. When we pull down our coffee cup and release it, the magnitude of the resulting oscillations becomes less and less until the oscillations die down completely. The energy of the system is slowly dissipated. Can we modify our equations so that they demonstrate this behavior? We have two ways to go if we wish to add dissipation (damping) to our system model. First, we can try to accurately model the processes by which dissipation occurs in real life, and add the appropriate terms to our equation of motion. In real life, damping is a very intricate and involved process, and in order to correctly model it we would need a complex set of equations that are very system dependent.

   The other approach would be to ignore the "real" damping behavior and try to convince ourselves that a simple linear term is a good enough approximation. The advantage to this approach is that it's much simpler than doing it the "correct" way, and we can obtain closed-form solutions for our problems rather than resort to numerical (computer-based) solutions.

   Given these two choices, the scientist in us would opt for the first approach, while our engineering half would probably vote for the simple solution. Since this is an engineering text, the second approach wins! Before those with a more scientific bent cry foul, we should state that the simple linear approach, although not exactly correct, is usually very close. When the level of dissipation is small, as it almost always is for engineering structures, it really doesn't make a great deal of difference in the final results if we use a linear damping approximation or a more realistic one. Only when the damping is large do we find sizable discrepancies creeping in between the different approaches. We'll see in a moment what is really meant by "large" and "small" values of damping.

Thus far, our formulations have involved terms for which the force is proportional to acceleration ($m\ddot{x}$) and displacement ($kx$). What we will do now is add a force that's proportional to velocity,

$$f = c\dot{x} \tag{1.4.1}$$

This force is called a *linear damping force*. It makes physical sense to let our damping term scale with velocity since we're very familiar with the fact that drag on cars and boats increases with velocity; thus the damping force in our equations should increase with velocity. (By the way, the drag on cars actually doesn't increase linearly, it scales more closely with the square of the velocity.) Besides being called linear, this type of damping is also referred to as viscous damping.

Putting this new term into our equation of motion yields

$$m\ddot{x} + c\dot{x} + kx = 0 \tag{1.4.2}$$

Although we just threw this damping term in, we could just as well have postulated the existence of a physical damper (shown in Figure 1.24) that is attached in parallel with the spring of our system. If we assume that the force generated is proportional to velocity, then a free-body force balance would give us

$$m\ddot{x} = -c\dot{x} - kx \tag{1.4.3}$$

which is simply (1.4.2) arranged in a different order. The nice thing about visualizing the damper as a physical component is that it allows us to see why the damper will act to reduce the motions in the system. Since the force is directed toward the system's equilibrium position (just as the spring's force is) and is proportional to velocity, it generates a force that moves the mass toward the equilibrium point, when the mass is moving away from that point. The faster the mass is moving, the greater the force acting on it.

**Figure 1.24**  Spring-mass damper system

Having added what we think will give us a damping effect, the next step is to solve the equations. To start, we'll deal with the unforced case, just as we did for the undamped case. Since our equation is still linear, and we remember that all unforced, linear, constant coefficient, ordinary differential equations have exponential solutions, we'll solve the problem in exactly the same manner as for the undamped case, by assuming a solution of the form

$$x = ae^{\lambda t} \tag{1.4.4}$$

Using (1.4.4) in (1.4.2) we obtain

$$m\lambda^2 ae^{\lambda t} + c\lambda ae^{\lambda t} + kae^{\lambda t} = 0 \tag{1.4.5}$$

Grouping terms gives us

$$ae^{\lambda t}(m\lambda^2 + c\lambda + k) = 0 \tag{1.4.6}$$

Once again, $e^{\lambda t}$ is never 0 for finite $t$ and $a = 0$ is uninteresting, so we are left with

$$m\lambda^2 + c\lambda + k = 0 \tag{1.4.7}$$

Solving this gives us

$$\lambda_{1,2} = \frac{1}{2m}(-c \pm \sqrt{c^2 - 4mk}) \tag{1.4.8}$$

Although we could leave the solution in the above form, we've already seen that dividing our undamped equation by $m$ led us to the natural frequency $\omega_n$ and put the equation into a very neat form. If we carry through the division by $m$ shown above, we'll obtain

$$\lambda_{1,2} = -\frac{c}{2m} \pm \sqrt{\frac{1}{4}\left(\frac{c}{m}\right)^2 - \frac{k}{m}} = -\frac{c}{2m} \pm \sqrt{\frac{1}{4}\left(\frac{c}{m}\right)^2 - \omega_n^2} \tag{1.4.9}$$

Note that all we really need to do now is to specify the ratio of $c$ to $m$. The absolute values of these quantities don't really matter, which is one of the reasons why we divided by $m$ in the first place. Although it isn't clear why, at the moment, it is convenient to replace $\frac{c}{m}$ with $2\zeta\omega_n$. If you'll withhold your disbelief for just a second, you'll see why.

Using this definition, our equation of motion becomes

$$\ddot{x} + 2\zeta\omega_n\dot{x} + \omega_n^2 x = 0 \tag{1.4.10}$$

Substituting $2\zeta\omega_n$ for $\frac{c}{2m}$ in (1.4.9), we obtain

$$\lambda_{1,2} = -\zeta\omega_n \pm \omega_n\sqrt{\zeta^2 - 1} \tag{1.4.11}$$

This form is great as long as $\zeta$ is larger than one. But for $\zeta < 1$ we have the square root of a negative number. In this case we'll get an imaginary value and the roots are most conveniently expressed as

$$\lambda_{1,2} = -\zeta\omega_n \pm i\omega_n\sqrt{1 - \zeta^2} \tag{1.4.12}$$

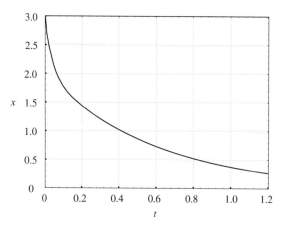

**Figure 1.25**   Overdamped response

Note that we get two forms that depend upon $\zeta$'s value. For (1.4.11), both roots are real, negative numbers. This means that the overall solution to the free-vibration problem will include two decaying exponentials. A typical response to an initial displacement is shown in Figure 1.25. For this example $\omega_n = 6.28$ and $\zeta = 2$. The mass starts off displaced from the origin, then smoothly approaches it as time increases.

This is different from the behavior we will find for the kinds of roots expressed by (1.4.12). For this case we no longer have two negative, real roots, but rather have complex roots. The real part of each root is the same, while the imaginary part differs in sign. Roots like these are called complex conjugates and are very common solutions for vibrations problems. To see what sort of overall solutions these roots support, we'll use the trigonometric identity

$$e^x = \cos(x) + i\,\sin(x) \tag{1.4.13}$$

Thus, instead of solutions that look like

$$x(t) = a_1 e^{-\zeta\omega_n t} e^{i\omega_n t\sqrt{1-\zeta^2}} + a_2 e^{-\zeta\omega_n t} e^{-i\omega_n t\sqrt{1-\zeta^2}} \tag{1.4.14}$$

we can express the solution as

$$x(t) = e^{-\zeta\omega_n t}\left[b_1 \cos(\omega_d t) + b_2 \sin(\omega_d t)\right] \tag{1.4.15}$$

where $\omega_d \equiv \omega_n\sqrt{1-\zeta^2}$.

Thus we have a solution that is made up of a sinusoidal part multiplied by a decaying exponential. $\omega_d$ is known as the "damped natural frequency." Since the exponential is equal to 1.0 for $t = 0$ and decays from there, the effect of multiplying the sine or cosine by it is to cause the oscillations to decay within an exponential envelope. This is graphically shown in Figure 1.26 for the response

$$x(t) = e^{-1.256t}\cos(6.153t) \tag{1.4.16}$$

which corresponds to $\omega_n = 6.28$ and $\zeta = .2$.

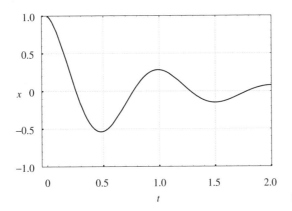

**Figure 1.26**   Underdamped response

Note that the general form of this type of solution is $x(t) = e^{b_1 t} \cos(b_2 t + \phi)$. If changes to our system cause $b_1$ to increase, then the speed at which the oscillations decay is increased. Also, increasing $b_2$ causes the frequency of the oscillations to increase. Since we can control either of these by varying $\omega_n$ and $\zeta$, we have the tools necessary for modifying the decay characteristics of our system.

The critical parameter that divides the responses between those that oscillate and decay (1.4.12) and those that are purely decaying (1.4.11) is $\zeta$. For $\zeta > 1.0$ we'll have exponential decay (because the number under the square root of (1.4.11) is positive), and for $\zeta < 1.0$ we'll have decaying oscillations. This is the main reason for the introduction of $\zeta$; it characterizes these two response regimes. $\zeta$ is called different things in different books (damping coefficient, damping ratio, damping factor). In this book we'll refer to it as the damping factor. If our system has a $\zeta < 1.0$, we call the system *underdamped*. Systems with $\zeta > 1.0$ are *overdamped* while systems having $\zeta$'s of exactly 1.0 are *critically damped*.

Looking at our definition of $\omega_d$ ($\omega_d = \omega_n \sqrt{1 - \xi^2}$), we see that the frequency of our damped oscillations depends upon both $\omega_n$ and $\zeta$. For small values of $\zeta$ (lightly damped) the oscillation rate is essentially equal to $\omega_n$. Increasing the damping factor serves to reduce the oscillation rate. This makes physical sense since the mass will have more difficulty in moving if the damping is raised, thus lowering the overall rate of oscillation.

Although $\zeta$ is the damping factor for our system, keep in mind that it is not the sole determiner of the response's decay rate. Equation (1.4.15) shows that the decay rate is determined by the product of $\zeta$ and $\omega_n$. Thus the decay rate is increased by increasing the damping factor or by increasing the system's natural frequency. If we think about what's going on with our system physically, this makes sense. First of all, it certainly seems obvious that increasing the amount of damping in the system will increase the rate of decay, and $\zeta$ is linearly proportional to the physical damping coefficient $c$. But it is also true that a larger $\omega_n$ means a higher rate of oscillation, leading to higher velocities and consequently more dissipation (since the damping is proportional to velocity). Thus both the damping factor and natural frequency play a part in determining how fast an unforced oscillation will die away.

The general responses we've just examined are absolutely central to understanding and describing the responses of complex systems. The speed at which responses decay and the frequency at which they do so are determining factors in how well the systems will perform the tasks they were designed for.

**Example 1.5**

**Problem** Consider a system like that shown in Figure 1.24. For this example $m = 10$ kg, $c = 100$ N/m/s, and $k = 30,000$ N/m. If the initial conditions are $x(0) = .01$ m and $\dot{x}(0) = 0$, what will the response look like for $t > 0$?

**Solution** Knowing $m$, $c$, and $k$, we can immediately form

$$\omega_n^2 = \frac{30,000}{10}$$

and so

$$\omega_n = 54.77 \text{ rad/s}$$

Knowing that $\frac{c}{m} = 2\zeta\omega_n$, we have

$$\zeta = .0913$$

Since

$$x(t) = e^{-\zeta\omega_n t} \left[ b_1 \cos(\omega_d t) + b_2 \sin(\omega_d t) \right]$$

we can apply the initial conditions and obtain

$$.01 = b_1$$

and

$$0 = -\zeta\omega_n b_1 + \omega_d b_2$$

Solving both these equations simultaneously gives us

$$b_1 = .01$$

and

$$b_2 = b_1 \frac{\zeta}{\sqrt{1 - \zeta^2}} = .0921$$

Thus

$$x(t) = .01 \left( \cos(54.5t) + .0921 \sin(54.5t) \right) e^{-5t}$$

## 1.5 LAGRANGE'S EQUATIONS (OPTIONAL)

No vibrations book seems to be complete without a discussion of Lagrange's equations. All of you have already taken a dynamics class or two, which, in the vast majority of cases, has involved Newtonian mechanics, that is, finding the equations of motion by applying Newton's laws. The fundamental distinguishing feature of Newtonian mechanics is that it treats the force interactions within a system as the fundamental quantities of interest. Once you've determined all the relevant forces, you simply use $f = ma$ (for constant mass) and you're done. Sounds simple. However, as you probably discovered in class, actually applying these laws is a bit trickier than one might at first think. Correctly analyzing a system using Newton's laws requires us to form free-body diagrams and demands that we account for all interaction forces exactly. For instance, let's consider the problem shown in Figure 1.27a, that of a cart-pendulum. There is more than one way to solve this problem with Newton's laws, and one approach to accounting for the applied and interaction forces is shown in Figure 1.27b. Note that we *must* solve for the interaction forces between the two masses, even if we don't really care about them. As vibration engineers, probably all we want to know is the governing equations of motion, something that doesn't explicitly involve these forces at all, as can be seen from the solution.

The fact that we must spend our time getting intermediate results that we might have no interest in bothered several researchers in the past, Lagrange being one of them. Lagrange wanted to see if he could come up with an alternate way of finding the equations of motion, a way that didn't necessarily involve interaction forces at all. The notion he came up with was to look, not at the forces in a system, but at the *energies*. This viewpoint is quite different from Newton's, and there is a huge body of literature detailing different approaches to mechanics from an energy viewpoint. Since this isn't a dynamics course, we'll look only at one of these approaches, namely Lagrange's equations.

It is important at the outset to tell you that we'll be indulging in some major handwaving in the next few pages. The "standard" derivation of Lagrange's equations involves the notions of virtual displacements, something that brings a shudder to some dynamicists, even today. Others couldn't care less. In any event, we'll present this approach, point out the places that you might question it, and leave the rest to you. If you're really interested in this area, there are other, more involved approaches that some people find more appealing.

Okay, enough with the preliminaries. In order to apply Lagrange's equations, we first need to define what are called *generalized coordinates*. As you know, we can't really do much in the way of dynamics without having some coordinates. At the very least, we want our coordinates to pin down exactly the physical configuration of our system, at any point in time. It is possible for different people to come up with different, and equally valid, sets of generalized coordinates for a given problem. One example is shown in Figure 1.28. Here we have a bar that's supported by two springs. We could decide to use the displacement of the bar at each end as our coordinate set; $x_1(t)$ and $x_2(t)$ are all that we need to accurately describe the bar's position (we're assuming small motions only). Alternatively, we could use $x(t)$ and $\theta(t)$ to describe the vertical motion of the bar's midpoint and the rotation about this point. This would also work just fine.

Actually, you can come up with any number of coordinate systems; all that matters is that they successfully describe the system's configuration. Most commonly we refer to these coordinates by the letter $q$. Thus, for our second set of coordinates in Figure 1.28 we'd have $q_1 = x$ and $q_2 = \theta$.

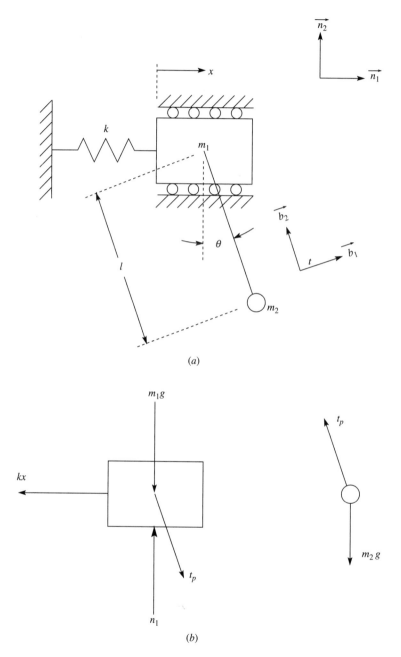

(a)

(b)

**Figure 1.27**   Mass/pendulum: physical system and free-body diagram

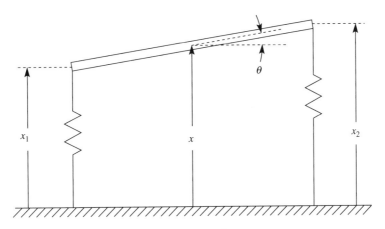

**Figure 1.28**   Two coordinate systems

If we have no externally applied forces and no dissipation, then Lagrange's equations are given by

$$\frac{d}{dt}\frac{\partial L}{\partial \dot{q}_i} - \frac{\partial L}{\partial q_i} = 0, \qquad i = 1\cdots n \tag{1.5.1}$$

in which

$$L = k_e - p_e \tag{1.5.2}$$

where $k_e$ is the kinetic energy, $p_e$ is the potential energy of our system, and $L$ is called the Lagrangian.

Where did this result come from? I told you, we weren't going to go into details. You'll just have to believe that it is correct. Or, if you prefer, you can read more about it in [4, 5]. To demonstrate that it actually works, let's consider the problem shown in Figure 1.27. What's shown is a mass that can only move horizontally with a pendulum suspended from it. There is no friction in the problem and no external forces. Because gravity will cause the pendulum to oscillate, and the spring $k$ will induce oscillations in $m_1$, we should expect equations of motion that will involve periodic solutions, presumably ones in which both masses are moving. If we were going to approach this problem from a Newtonian standpoint, we'd first break the system up into two parts and then apply Newton's laws to them individually, as shown in Figure 1.27b. The unknown force $t_p$ is the force of interaction between $m_1$ and $m_2$; it's what keeps the pendulum attached to $m_1$. We'll need to set up two force-balance equations and one moment-balance equation to solve this problem. The two force balances in the $x$ direction (for $m_1$ and $m_2$) are

$$m_1\ddot{x} = t_p \sin(\theta) - kx \tag{1.5.3}$$

and

$$m_2\left(\ddot{x} + l\cos(\theta)\ddot{\theta} - \dot{\theta}^2 l \sin(\theta)\right) = -t_p \sin(\theta) \tag{1.5.4}$$

Using both (1.5.3) and (1.5.4) together (and eliminating $t_p$) gives us

$$(m_1 + m_2)\ddot{x} + m_2 l\ddot{\theta} \cos(\theta) - m_2 \dot{\theta}^2 l \sin(\theta) + kx = 0 \qquad (1.5.5)$$

Next, we must sum moments about some convenient point. If we use the point of attachment of the pendulum to $m_1$, our equation will be

$$-m_2 gl \sin(\theta) = m_2 \left(l^2 \ddot{\theta} + \ddot{x} l \cos(\theta)\right) \qquad (1.5.6)$$

or, rearranging slightly and canceling out the $m_2$ term,

$$l^2 \ddot{\theta} + \ddot{x} l \cos(\theta) + gl \sin(\theta) = 0 \qquad (1.5.7)$$

Our equations of motion are given by (1.5.5) and (1.5.7). You saw that the interaction force $t_p$ showed up in the initial work and dropped out in the final equations, just as we expected since we were using a Newtonian approach.

Now let's use Lagrange's equations. To find the kinetic energy, we need to know the velocities of the two masses. Finding $m_1$'s velocity ($v_1$) is the easiest, it's just $\dot{x} \vec{n}_1$. However, $m_2$ is a bit trickier since it rotates about $m_1$. If we use the coordinate vectors shown in the figure, we see that $m_2$'s velocity is

$$v_2 = \left(\dot{x} + l\dot{\theta} \cos(\theta)\right)\vec{n}_1 + l\dot{\theta} \sin(\theta)\vec{b}_1 \qquad (1.5.8)$$

In order to find the square of the velocities that we need for the kinetic energy, we must express the velocity in one coordinate system. If we choose the fixed reference frame as the one to use, $v_2$ becomes

$$v_2 = \left(\dot{x} + l\dot{\theta} \cos(\theta)\right)\vec{n}_1 + l\dot{\theta} \sin(\theta)\vec{n}_2 \qquad (1.5.9)$$

The kinetic energy is simply

$$k_e = \tfrac{1}{2}m_1 v_1^2 + \tfrac{1}{2}m_2 v_2^2 \qquad (1.5.10)$$

Using our knowledge of what $v_1$ and $v_2$ are, we can reexpress this as

$$k_e = \tfrac{1}{2}m_1 \dot{x}^2 + \tfrac{1}{2}m_2\left((\dot{x} + l\dot{\theta} \cos(\theta))^2 + (l\dot{\theta} \sin(\theta))^2\right) \qquad (1.5.11)$$

The potential energy comes from the translational spring and the gravitational potential that results when the pendulum swings up:

$$p_e = \tfrac{1}{2}kx^2 + m_2 gl\left(1 - \cos(\theta)\right) \qquad (1.5.12)$$

Using (1.5.11) and (1.5.12), we can form the Lagrangian ($L = k_e - p_e$) and apply Lagrange's equations. The two dependent variables (the $q_i$'s) are $x$ and $\theta$. Applying Lagrange's equations with respect to $x$ first, we find

$$\frac{d}{dt}(m_1 \dot{x}) + m_2 \frac{d}{dt}\left(\dot{x} + \dot{\theta} l \cos(\theta)\right) + k_1 x = 0 \qquad (1.5.13)$$

or, performing the differentiations and grouping terms

$$(m_1 + m_2)\ddot{x} + m_2 l\ddot{\theta} \cos(\theta) - m_2 \dot{\theta}^2 l \sin(\theta) + kx = 0 \qquad (1.5.14)$$

If we next apply Lagrange's equations with respect to $\theta$, we'll obtain

$$m_2 \frac{d}{dt}(\dot{x}l\cos(\theta) + l^2\dot{\theta}) - m_2\dot{x}\dot{\theta}^2 l\sin(\theta) + m_2 g l \sin(\theta) \qquad (1.5.15)$$

Note that the $-m_2\dot{x}\dot{\theta}^2 l\sin(\theta)$ term came about when we differentiated the kinetic energy with respect to $\theta$. Since the kinetic energy has a $\theta$ dependence, we obtained this additional term. This almost always happens when you're dealing with rotational systems and rarely occurs for translational ones, since a translational system's kinetic energy usually depends upon $\dot{x}$ type terms, not terms proportional to $x$. It's easy to forget this fact if most of your examples are of the translational type, and you might begin to differentiate just the kinetic energy in the $\frac{d}{dt}\left(\frac{\partial}{\partial \dot{q}}\right)$ term, forgetting about the $\frac{\partial}{\partial q}$ part of the equations. So now you've been warned! Make sure to take account of all the relevant differentiations.

We can differentiate and group the terms of (1.5.15) to get

$$l^2\ddot{\theta} + \ddot{x}l\cos(\theta) + gl\sin(\theta) = 0 \qquad (1.5.16)$$

If you compare (1.5.14) and (1.5.16) with (1.5.5) and (1.5.7), you'll quickly notice that they are exactly the same. Thus it appears that Lagrange's equations do what was advertised; they come up with fully valid equations of motion.

Interestingly, you can get into some real battles with certain researchers over which is better to use, Lagrange's equations or Newton's equations (or even other approaches). However, the simple fact of the matter is that you can almost always get your answer with either approach. In some problems a Newtonian approach might be less work, and in others you might find Lagrange's equations to be preferable. It's not a bad idea to simply view both as different tools with which to get the equations of motion and not to get too caught up with which is "better." As a general guide, Lagrange's equations are "easier," in that you don't need to use any physical insight in modeling the force interactions. The method simply asks for the energies. Thus you need find only the relevant velocities with which to form the kinetic energy and account for any potential energies. Then you can merely "plug and chug in" Lagrange's equations. The downside is that after you get the velocities, you must square them, then differentiate. If you've got a bunch of reference frames, because your system has several moving parts, then your velocity terms will be long and involved. Therefore you'll have a lot of terms to track. For complicated systems, this can be a big drawback, since errors can easily creep in when you're trying to keep track of so many terms. Also, computer simulations will tend to run slower because the equations are so complex. But for relatively simple problems, Lagrange's equations are just fine.

Newton's equations shine when you need interaction forces, since you can't solve the problem without considering them. Often you will need these interaction terms if you're evaluating the forces in a joint or a slider. The equations are also more physical since they involve moments and forces, things we're comfortable dealing with as engineers. In order to apply Newton's equations, you need to find accelerations, something that is generally simpler than the manipulations needed for Lagrange's equations. Also, you need to think about the system in physical terms in order to construct the necessary free-body diagrams, something that isn't necessary with Lagrange's equations. Finally, you must include all the interaction forces, even if you don't care about them, something that Lagrange's equations don't require.

This short discussion has shown that either approach will work. Although for the problems we'll be dealing with most people find Lagrange's equations to be the most convenient, you should probably become familiar with both approaches; this will help make your life simpler in the long run.

All right, now we've got the unforced, undamped case under control. But what about damping? Can we add a viscous damper to our problem? Yes we can, but it is going to look like a Band-Aid fix. If our damper has a known force characteristic, we need to integrate with respect to the time derivative of the generalized coordinates that apply to it. Thus, if we wanted to add a damper to our cart-pendulum problem, we'll get the system shown in Figure 1.29. From our previous discussions, we recognize that the force of the damper is given by

$$f_d = c\dot{x} \tag{1.5.17}$$

and so, integrating with respect to $\dot{x}$, we obtain

$$\int f_d d(\dot{x}) = \tfrac{1}{2}c\dot{x}^2 \tag{1.5.18}$$

We define this quantity as the Rayleigh dissipation function, $r_d$. Thus

$$r_d = \tfrac{1}{2}c\dot{x}^2 \tag{1.5.19}$$

In order to use this in Rayleigh's equations, we need to modify the equation a bit, to

$$\frac{d}{dt}\frac{\partial L}{\partial \dot{q}_i} + \frac{\partial r_d}{\partial \dot{q}_i} - \frac{\partial L}{\partial q_i} = 0, \qquad i = 1 \cdots n \tag{1.5.20}$$

This modification will allow us to deal with any linear, unforced problem we might encounter.

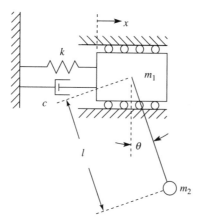

**Figure 1.29**   Damped mass/pendulum

In order to add forcing to our equations, we must determine what work would have to be done to change the coordinates by a small amount. If the work is equal to $fd(q_i)$, for instance, then the forcing associated with the equation of motion involving $q_i$ (called the *generalized force*) is simply $f$ (and usually represented by $Q_i$).

Thus the forced form of Lagrange's equations is given by

$$\frac{d}{dt}\frac{\partial L}{\partial \dot{q}_i} - \frac{\partial r_d}{\partial q_i} = Q_i, \qquad i = 1 \cdots n \qquad (1.5.21)$$

This is most clearly shown by an example. Figure 1.30 shows the example we've already seen (Figure 1.29) but with the addition of external force $\vec{F}_{ext} = f_{ext}\vec{n}_1$. All we need to determine is what the $Q_i$ are for this problem (since we've solved all the rest of it earlier in the chapter). The differential work due to $\vec{F}_{ext}$ is given by $f_{ext}\vec{n}_1$ dotted into the differential displacement of the tip mass. Our differential displacement is given by

$$d(x)\vec{n}_1 + ld(\theta)\vec{b}_1 \qquad (1.5.22)$$

If we express this purely in terms of the $\vec{n}_i$ coordinate frame, we'll see that the differential displacement is equal to

$$(d(x) + ld(\theta)\cos(\theta))\,\vec{n}_1 + ld(\theta)\sin(\theta)\vec{n}_2 \qquad (1.5.23)$$

The total differential work is, therefore, found from

$$f_{ext}\vec{n}_1 \cdot ((d(x) + ld(\theta)\cos(\theta))\,\vec{n}_1 + ld(\theta)\sin(\theta)\vec{n}_2) = f_{ext}(d(x) + l\cos(\theta)d(\theta)) \qquad (1.5.24)$$

Therefore our generalized force with respect to $x$ is given by $f_{ext}$ and our generalized force with respect to $\theta$ is given by $f_{ext}l\cos(\theta)$.

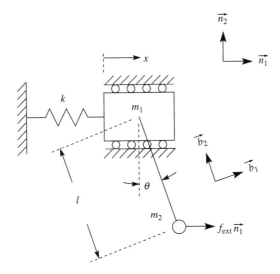

**Figure 1.30**  Mass/pendulum with external force

Knowing the generalized forces, we can write our equations of motion as

$$(m_1 + m_2)\ddot{x} + m_2 l \ddot{\theta} \cos(\theta) - m_2 \dot{\theta}^2 l \sin(\theta) + kx = f_{ext} \tag{1.5.25}$$

and

$$l^2 \ddot{\theta} + \ddot{x} l \cos(\theta) + gl \sin(\theta) = f_{ext} l \cos(\theta) \tag{1.5.26}$$

## 1.6 HOMEWORK PROBLEMS

### Section 1.2

1. If $x = a_1 e^{i\omega t} + a_2 e^{-i\omega t}$, $x(0) = 2$, and $\dot{x}(0) = 3$, what are $a_1$ and $a_2$ equal to?

2. If $x = b_1 \cos(\omega t) + b_2 \sin(\omega t)$, with the same initial conditions as in problem 1, what are $b_1$ and $b_2$ equal to?

3. A sinusoid, $x(t) = a \cos(7t + \phi)$, has initial conditions $x(0) = 3$, $\dot{x}(0) = 20$. Determine $a$ and $\phi$.

4. A mass moves with the illustrated acceleration profile. Determine the position of the mass at $t = 8$ s. $x(0) = \dot{x}(0) = 0$.

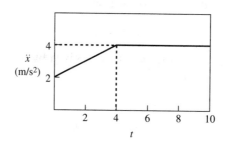

5. Plot the velocity and displacement versus time graphs for the illustrated acceleration profile. $x(0) = 4$ m/s and $\dot{x}(0) = -2$ m/s.

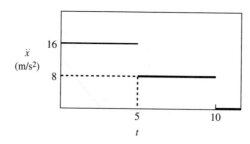

**6.** We know that the spring constant of two identical springs, used in series, is one-half the individual spring's spring constant. Show from the formula for the stiffness of a coil spring (Appendix B) that doubling the number of coils (which is equivalent to putting two coil springs in series) will give the same result.

**7.** What is the amplitude, frequency, and period of oscillation for the illustrated signal?

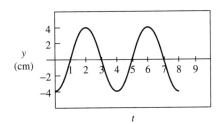

**8.** Determine the phase difference between the two illustrated signals. Which signal is lagging?

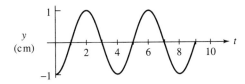

**9.** What is $\omega_n$ for the illustrated system in terms of $m$, $k_1$, $k_2$, and $k_3$?

**10.** The figure below shows a Ferris wheel at a carnival. Assume that the sun is directly overhead and that the wheel has a radius of 20 m. It takes 2 min to complete a single revolution. Draw a plot of what the shadow cast by a single car on the Ferris wheel would look like as a function of time. Identify the amplitude, period, and frequency of the resulting waveform.

**11.** Consider a physical system governed by

$$m\ddot{x} + kx = 0$$

Is it possible, given some $m, k$, and initial conditions, for the free vibration solution to be given by

$$x(t) = 2e^{3it}?$$

**12.** Determine the natural frequency of oscillation, $\omega_n$, for small oscillations about the equilibrium position of the following spring-mass system. Note that the spring is not linear, and thus you'll have to linearize about the system's equilibrium position in order to find the equivalent linear stiffness. $m_1 = 1$ kg and $g = 9.8$ m/s$^2$.

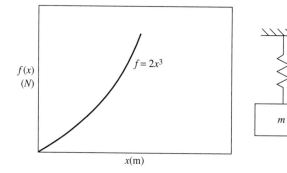

**13.** Consider the illustrated system having two linear springs. The unstretched length of $k_1$ is .5 m and the unstretched length of $k_2$ is .25 m. $k_1 = 1000$ N/m, $k_2 = 2000$ N/m, $m = 2$ kg, and $l = .5$ m. Find the equilibrium position of the mass and determine the natural frequency of the system. Compare this natural frequency to that associated with $l = .75$ m (i.e., no precompression in the spring).

14. Consider the system in which the linear spring of Problem 13 is replaced with a nonlinear spring, as shown below. The unstretched length of $k_1$ is .5 m and the unstretched length of the nonlinear spring is .25 m. $k_1 = 1000$ N/m, $f(x) = 24{,}000x^2$, $m = 2$ kg, and $l = .5$ m. Find the equilibrium position of the mass and find the nonlinear spring's equivalent linear stiffness at the calculated deflection. (Hint: The linear stiffness is found by evaluating $\dfrac{df}{dx}$ at a given calculated deflection. This tells you the linear change of force per unit deflection). Calculate the natural frequency. Does the natural frequency change if $l$ is set equal to .75 m? How does this situation differ from that of Problem 13?

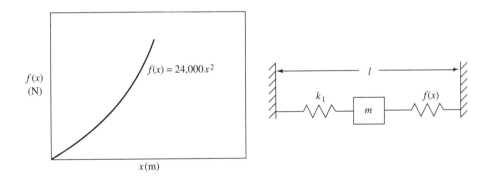

15. In an effort to add realism to cartoons, you've been asked to dynamically analyze a cartoon gag (assume that you're working during the summer as an animator). In this gag, Dopey Dog is falling from an airplane and, as he passes a ledge, grabs an elastic rope he tied there earlier. The unstretched length of the rope is 2 m and he falls 50 m before grabbing it. His mass is 35 kg and the rope can be considered massless. The spring constant of the rope is 420 N/m. Calculate Dopey's position as a function of time after he grabs the rope, and determine how long it takes for his velocity to reach zero. Assume he doesn't hit the canyon floor. (Note: Don't neglect the static offset he'd have due to gravity.) Calculate a trajectory only for motions in which the rope is under tension.

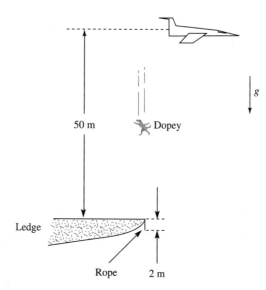

**16.** Determine the natural frequencies for the illustrated systems. What is the difference between the two cases? Which springs are most active and which are relatively uninvolved in the oscillations?

**17.** A static horizontal load of 50 N, applied to the lamp housing of the illustrated street lamp, caused a horizontal deflection of 4 cm. Assume that the motion of an earthquake can be modeled as a step change in position. Thus the initial conditions are $x(0) = 10$ cm and, $\dot{x}(0) = 0$ cm/s. Determine the response of the lamp housing. Consider the lamppole to be massless and the mass of the housing to be 20 kg.

18. Determine the free vibration of the illustrated system (water in a U-shaped tube). The tube has an inner area of $s$, the water has density $\rho$, and the length of the water-filled section is $l$.

19. What is the difference between system ($a$) and system ($b$) (if any)?

20. Consider two masses joined by a spring and sliding on a frictionless surface. If the distance between them is given by $z$, determine the equation of motion of the system that involves only $z$, $m_1$, and $m_2$. (Note: The motion will be a "breathing" action in which the masses move together and apart.)

21. With use, the end conditions of the illustrated system have gone from cantilevered-cantilevered to pinned-pinned. What is the percentage change in the system's natural frequency? Consider the beams to be massless. (Use Appendix B.)

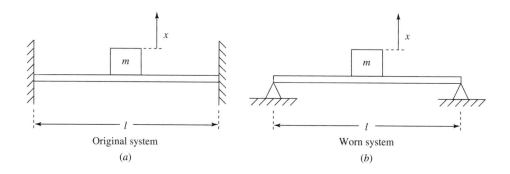

Original system

(a)

Worn system

(b)

22. A simple model of a building is given by a horizontal floor supported by two vertical beams. An even more simplified model would consist of a single beam and mass. If the stiffness of each original beam (for which both ends remain vertical) is $k$, determine what $m'$ must equal in the single-mass, single-beam case if the natural frequencies are to agree. (For the single-mass, single-beam case the beam end doesn't remain vertical.)

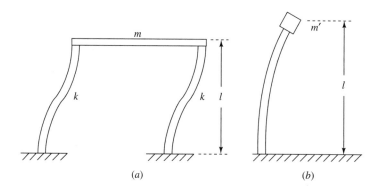

(a)

(b)

23. Demonstrate that the stiffness of system (a) is equal to $\dfrac{192EI}{l^3}$, using the knowledge that the stiffness of system (b) is $\dfrac{12EI}{l^3}$.

$$f = k_1 x, \quad k_1 = \frac{192EI}{l^3}$$

(a)

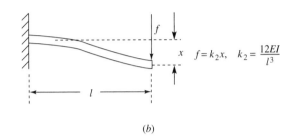

$$f = k_2 x, \quad k_2 = \frac{12EI}{l^3}$$

(b)

**24.** Demonstrate that the stiffness of system (*a*) is equal to $\dfrac{48EI}{l^3}$, using the knowledge that the stiffness of system (*b*) is $\dfrac{3EI}{l^3}$.

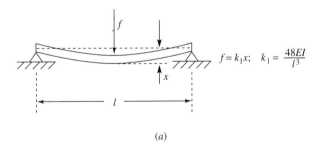

$$f = k_1 x; \quad k_1 = \frac{48EI}{l^3}$$

(a)

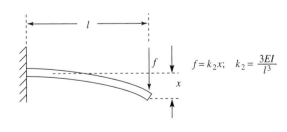

$$f = k_2 x; \quad k_2 = \frac{3EI}{l^3}$$

(b)

**25.** A SDOF spring-mass system has a natural frequency equal to 20 rad/s. How will this change if the suspension-spring constant is doubled?

**26.** When a 1000-kg mass is placed on a suspension, the springs deflect 1 cm. The spring force relation is linear. What is the natural frequency of oscillation for the system?

**27.** What is the natural frequency for the system illustrated here? (Assume that it oscillates about its equilibrium position.) The radius of each pulley is .2 m, $k = 400$ N/m, and $m = 10$ kg. The pulleys are massless.

**28.** What is the natural frequency for the system illustrated? Keep in mind that we're looking at *small* oscillations about equilibrium, so you needn't worry about geometric effects from the springs due to the mass's motion. (Hint: Consider polar coordinates.)

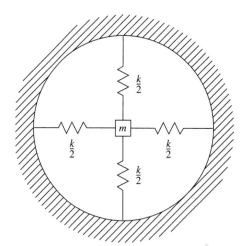

**29.** Consider a spring that has an unstretched length of .5 m and a spring constant equal to 500 N/m. Calculate the natural frequency if a 1 kg mass is suspended from the spring. Then calculate how the natural frequency will change if the spring is cut in half and the mass is suspended from just half the spring.

**30.** The natural frequency of a one-story building can be calculated by determining the stiffness of the supporting beams and assuming that the roof moves horizontally. If the width of the building is $l_1$, the depth is $l_2$, the stiffness of each beam (4 total) is $EI$, and the length is $l_3$, calculate the natural frequency, assuming a roof mass of $m$. Treat the beams as massless.

**31.** As a practical joke, you've put some fast-acting glue in your friend's slingshot. Assume your friend puts a mass of .05 kg in the slingshot, pulls it back 1 m, and then releases it. If the mass travels out 1 m beyond the slingshot and then returns to strike his hand .6 s after the release, what must the spring constant be for the massless elastic band? Assume that the elastic band has an unstretched length equal to 0 m.

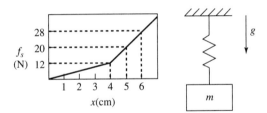

Slingshot

**32.** Given the illustrated spring-force characteristics, determine the natural frequency of oscillation of the system illustrated. Assume the system executes small motions about the static equilibrium due to gravity. Take $g$ to be 10 m/s$^2$ and $m = 2$ kg.

**33.** In an effort to add realism to cartoons, you've been asked to dynamically analyze a new episode. The action of interest is when Dopey Dog, having attached himself by means of an elastic band to a rocky outcropping, is dismayed to see the outcropping come apart from the cliff and fall. After a drop of 49 m, the outcropping lands on a ledge. The part that Dopey is attached to projects outward from the ledge. Thus Dopey continues to fall. The unstretched length of the elastic is 1 m. Just before the fall, Dopey was in static equilibrium and was 2 m from the outcropping (i.e., the total length of the stretched elastic was 2 m). Assume that the separation between Dopey and the outcropping stayed constant during the fall.

Determine the value of the linear spring constant $k$ necessary to keep Dopey from hitting the ground (which is 3.1 m below his position at the instant that the outcropping hits the second ledge). Considering the actual spring constant (as found from the static equilibrium conditions), will he hit the ground? Finally, what piece of information in the above description doesn't make physical sense?

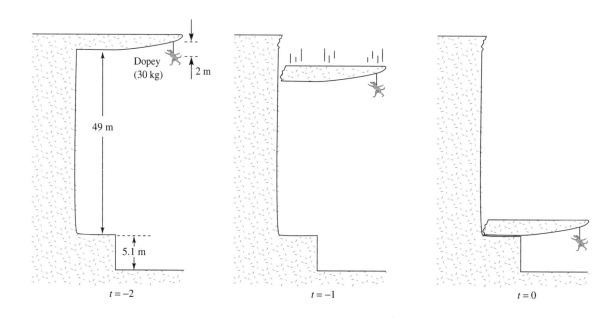

**34.** The coupling between two railroad boxcars can be viewed as a lumped spring. If the masses of the two cars are $m_1 = 1.1 \times 10^5$ kg and $m_2 = 1.3 \times 10^5$ kg, and the spring constant $k$ is equal to $4 \times 10^5$ N/m, determine the natural frequency of the system's oscillations.

**35.** Determine the frequency of oscillation of a cube of wood bobbing up and down in the water, in terms of the water's density, the wood's density, $g$ and $l$.

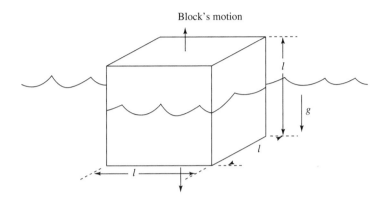

Block's motion

## Section 1.3

**36.** Determine the natural frequency of oscillation for the illustrated half disc. Assume no slip conditions.

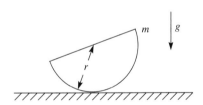

**37.** In this problem Dopey Dog is rolling a large cylindrical drum up a slope. The actual shape of the ground he's walking on is described by a circle, with a radius equal to $10r_1$. When Dopey reaches point A, he slips and the cylinder rolls over him. Assuming pure rolling, how long will it be before the cylinder returns to Dopey? The cylinder's mass is equal to $m$ and it has a radius of $r_1$. Assume that linearization about $\theta = 0$ degrees is a valid approximation.

**38.** Determine how long it will take the drum of Problem 37 to return to Dopey if, instead of pure rolling, we have a frictionless interface between the drum and the slope.

**39.** For the system shown, the mass is constrained to ride in a tube that is rotating about the point 0 at the constant rate $\omega$. The mass is connected to the spring $k$. Determine the response of the system about 0 for the following cases:

(**a**) $\omega^2 < \frac{k}{m}$

(**b**) $\omega^2 = \frac{k}{m}$

(**c**) $\omega^2 > \frac{k}{m}$

**40.** Find the linearized equation of motion about the illustrated system's equilibrium position. You don't have to solve for the equilibrium position itself—just show what you'd need to do to obtain it. The torsional spring is uncompressed when $\theta = 0$. What is the linearized natural frequency equal to?

**41.** The system shown below consists of a uniform bar (mass $m$, length $2l$) suspended by two inextensible wires (massless) of length $h$. What is the frequency of oscillation for rotary oscillations in the $xy$ plane?

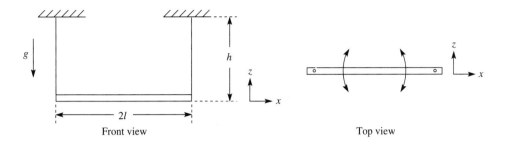

Front view                                                    Top view

**42.** Find the equilibrium position for the illustrated system and compute the natural frequency of oscillations about this equilibrium position. $m = 1$ kg, $l_1 = .7$ m, $l_2 = .1$ m, and $k = 683.4$ N/m.

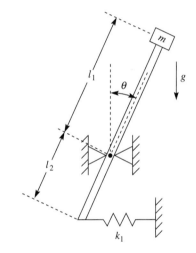

**43.** What is the natural frequency of oscillation of a point mass sliding without friction on a circular track? (Use a small angle approximation.)

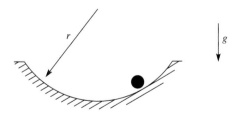

**44.** What is the natural frequency of the illustrated system? (Neglect gravity.) At static equilibrium the two arms are horizontal.

**45.** Find the natural frequency of oscillation for the illustrated system.

**46.** What are the equations of motion for the system shown? Consider motions about the vertical position.

**47.** Find the solution for a mass particle sliding on a semicircle. Consider the system to be frictionless and the equilibrium position to be $\theta = 0$. Linearize the equations of motion ($\theta$ small), find the solution, and explain what each solution physically represents. (Use $\theta(t) = ae^{\lambda t}$ as the assumed form for the solution.)

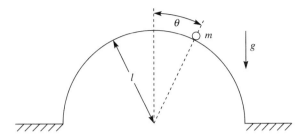

**48.** What is the natural frequency for the system shown?

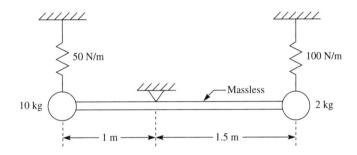

**49.** What is the natural frequency for the system shown?

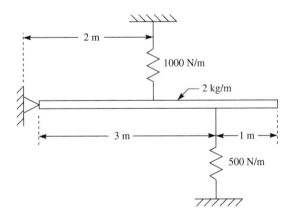

**50.** The system shown consists of a mass $m$ that rides along a horizontal track and a tensioned string that connects the mass to the point 0. Determine the natural frequency of the system under the assumption that the tension doesn't change as the mass changes position. $m = 1$ kg, $\rho = .2$ kg/m, $l = 1$ m, and $T = 50$ N.

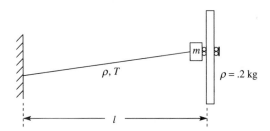

**51.** A band saw can be modeled as a continuous loop that runs around two pulleys, as shown. The steel has some flexibility, represented by the springs in the figure. Neglect the mass of the steel loop and determine the natural frequency of the pulley/band. Assume that the oscillation mode involves the pulleys moving in opposition to each other. The rotational inertia of each pulley about its shaft is equal to $.001$ kg·m$^2$, the spring stiffness is 10,000 N/m, and the radius of the pulleys is .4 cm.

**52.** To analyze the motions of an automobile on its tires, we approximate the auto by a bar of uniform mass (1000 kg) with mass moment of inertia equal to 1333.3 kg·m$^2$. The bar is 4 m long. The tire/suspension is approximated by a pair of springs, with spring constant equal to 11,697 N/m for each one. Considering motions in which the car oscillates about its center of mass (no translation), what is the natural frequency of motion for the system?

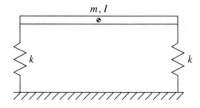

**53.** One way to determine the mass moment of inertia, $I_{\overline{AO}}$, of a body is to attach it to a torsional spring, observe the free vibrations, and deduce $I_{\overline{AO}}$ from the frequency of the oscillation. If the torsional stiffness of a vertical rod is 81 N·m/rad and the free vibrations have a frequency of 13 rad/s, what must $I_{\overline{AO}}$ be equal to?

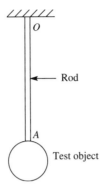

**54.** You need to determine the torsional stiffness of a vertical rod, rigidly attached to the ceiling at 0, and attached to a bar of length 1 m and mass 3 kg (as shown). If the free vibrations occur at 5 rad/s, what is the rod's torsional stiffness?

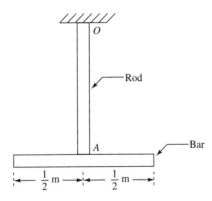

**55.** Shown here is a lumped mass in the middle of two tensioned strings. Treat the strings as massless and neglect gravity. Determine the period of the free vibrations if the tension in the string is equal to 100 N, $l = 2$ m, and the mass is equal to .3 kg. (Note: Only small amplitude motions of the mass take place.)

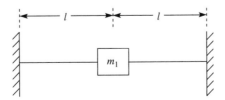

**56.** A circular disk is supported by three massless wires, as shown. The dish has a density $\rho$. Determine the frequency of oscillation if the dish is given an initial twist and released. Assume that the initial angular displacement of the disk is small.

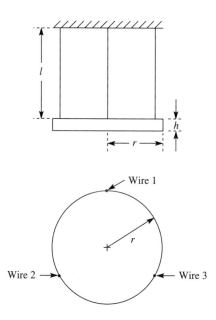

**57.** Find $\omega_n$ for the illustrated system. Consider the beams to be massless. The mass of the restrained dish is equal to $m$.

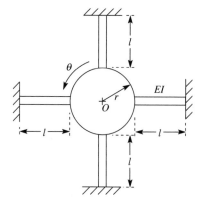

**58.** Find $\omega_n$ for the illustrated system. Consider the tensioned restraining wires to be massless and have uniform tensions $T$. All wires are of length $l$.

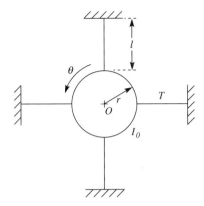

**59.** Determine the period of oscillation for the illustrated system. (Consider small oscillations about $\theta = 0$.)

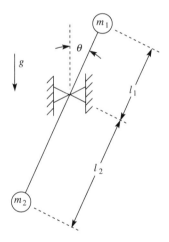

## Section 1.4

**60.** Does the addition of damping increase or decrease the frequency of the system oscillations?

**61.** As long as a SDOF spring-mass damper is underdamped, its vibrational response will cross zero an infinite number of times, just as in the undamped case. By how much will the "period" of the oscillations be decreased from that of the undamped case if $\zeta = .05$? What if $\zeta = .8$? Comment on the results.

**62.** What effect on the decay rate does $m$ have for a viscously damped SDOF system?

**63.** Is it more effective to alter $\omega_n$ or $\zeta$ if we wish to increase the decay rate for a viscously damped SDOF?

**64.** Can the frequency of oscillation for a viscously damped spring-mass damper be altered without changing the oscillation decay rate by varying the system parameters?

**65.** Is $\zeta$ more strongly affected by $c$, $m$, or $k$?

**66.** If the mass of a spring-mass damper is doubled in value, while the spring stiffness is halved, how will the decay rate and frequency of subsequent oscillations be affected?

**67.** How many oscillation cycles will elapse if you're waiting for the oscillation amplitude of a spring-mass damper to drop by 20 percent and $\zeta = .05$? Let $\omega_n = 100$. What if $\zeta$ is decreased to .0005?

**68.** Oscillation amplitudes are reduced by 50 percent in 1 s for a given spring-mass damper system. Say you want to have the amplitude reduced to 25 percent of the original value in 1 s. How effective would it be to vary the value of $k$?

**69.** What does simultaneously doubling the mass and stiffness of a spring-mass damper system do to the free vibration's decay rate?

**70.** Consider a variation of Problem 31. This time, when the mass returns it fails to reach the friend's hand, coming up short by 27 cm. Determine the damping factor of the elastic band/mass system. The time at which the mass came closest to the friend's hand was .3 s after release.

**71.** Say you have only a little over one-half the time response for a linear, damped oscillator, as shown. Can you use this information to determine $\zeta$ for the system?

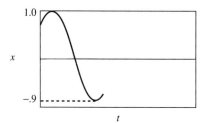

**72.** What can you say about the damping in the system that produced the following free response?

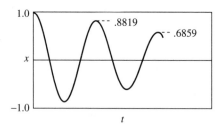

**73.** The dampers on automobiles are supposed to work in the same manner during compression and extension. Due to a manufacturing defect, one that you are examining is operative only during one-half the cycle (the compression stroke). If you know that the damping is viscous during compression, and you also know that the amplitude drops from 1.0 to .2 over one cycle of free vibration, calculate the value of $\zeta$ that would be produced by a properly operating damper. The time elapsed between the two amplitude measurements is equal to .5 s.

**74.** The illustrated mass vibrates in a viscous flow. The force experienced by the mass as a function of velocity is shown on the accompanying graph. Determine the damping factor for this system. $k = 1058$ N/m and $m = 2$ kg.

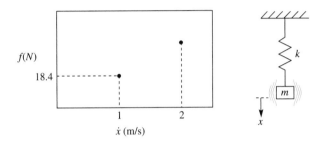

## Section 1.5

(Note: Use Lagrange's equations to find the equations of motion for all the problems in this section.)

**75.** Find the equation of motion for the illustrated system.

**76.** Find the equation of motion for the illustrated system.

**77.** Find the equation of motion for the illustrated system.

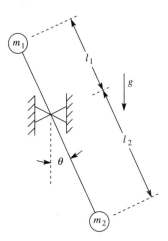

**78.** Find the kinetic and potential energies for the illustrated system and determine the full equations of motion. After you've determined the equations of motion, linearize them (retaining terms like $x$, $\dot{x}$, $\dot{\theta}$, etc.).

**79.** What are the equations of motion for the illustrated system?

**80.** Find the equation of motion for the following system and linearize it.

**81.** Find the linearized equation of motion for the following system. Assume that the cylinder is of mass $m$ and rolls without slipping on the illustrated circular path.

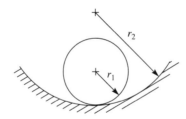

**82.** Find the equation of motion for the following system. There is no interface friction.

**83.** Find the equation of motion for the following system. There is no interface friction.

**84.** Find the equations of motion for the following system and linearize them. The bar has length $l$.

**85.** Find the equations of motion for the following system and linearize them. $m_1 = 1$ kg, $m_2 = 3$ kg, $l_1 = 2$ m, $l_2 = 1$ m, $g = 9.81$ m/s$^2$.

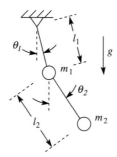

# 2

## Forced Vibration of Single-Degree-of-Freedom Systems

Don't you think it's about time to balance your tires, Dad?

## 2.1 INTRODUCTION

In the previous chapter we saw how to determine the equations of motion for systems that aren't experiencing any external forcing. Somehow they started into some motion, but for the time under consideration they were simply reacting to that initial condition, oscillating with or without some amplitude variation. Now we're going to see what happens when we continually force the system. We'll begin our forced analyses by considering the most widely encountered excitation, that of a sinusoidal disturbance. Since our interest in this chapter is in steady-state systems, we're not going to be concerned about how the system reaches this steady state, but only about the specifics of the motion once all the transient effects have died down. Thus this kind of approach would identify the long-term vibrational forces that act on a passenger in an automobile due to sinusoidal vibrations of the engine. It would also let us determine the forces transmitted to the ground due to an unbalanced shaft in a steam turbine. In fact, the number of examples you can come up with is almost unlimited. Any time there is a periodic disturbance, you're going to get a periodic response. Of course, the question of a system's transient response is an important one, and we'll tackle this particular subject in Chapter 3.

## 2.2 SEISMIC EXCITATION

To introduce you to the behavior of forced, single-degree-of-freedom systems, we will again use our physical mass-spring experiment. As before, hold the free end of the spring and let the mass hang down freely. However, instead of giving the mass an initial displacement and then releasing it, this time we will force the system in a periodic manner. There are two obvious ways to apply a force to a system of this sort. The first, and the one discussed at length in most texts, is to apply a sinusoidally varying force to the mass itself while keeping the free end of the spring fixed. The problem with this approach is that there is no way you can replicate this sort of forcing experimentally unless you have access to some noncontacting force generator. Therefore we'll start with the second approach, that of sinusoidally moving your hand. This is called base or seismic excitation. Although our system hangs down, it is clear that we could just as well have a system supported by springs (as shown in Figure 2.1), and in this case movements of the spring support are obviously base motions. The term seismic excitation comes from the observation that this is precisely the sort of excitation acting on buildings during earthquakes. Since you are moving one end of the spring, its subsequent elongation or compression impresses a force on the mass that is proportional to the amount of stretch.

   To perform our first experiment, we must let the mass hang down freely from our hand. Once it is hanging quietly, we can quickly raise our hand to a different height and keep it there. As a result, the mass will start oscillating about a new equilibrium (one that's exactly the same distance up from the old one as the amount your hand moved). As time goes on, the oscillations die away and the steady-state position of the mass will alter by the amount you moved your hand. Although this might not seem to be a big deal, it actually is pretty significant. For instance, say you're designing a positioning device, such as the one shown in Figure 2.2. We'll assume that a

**Figure 2.1**   Seismically excited, spring-mass system

**Figure 2.2**   Positioner arm

**Figure 2.3**   Seismically excited, spring-mass damper

very powerful motor can move the left end to a specified position. The positioner arm is flexible, and at the right end there is a gripper that can hold onto a payload. If the arm is much lighter than the payload, we can neglect its mass and model it as a pure spring, with spring constant $k$. The mass of the payload and gripper is $m$. The simplified model for such a system would look like that shown in Figure 2.3 and the equations of motion for this system would be given by

$$m\ddot{x} + c\dot{x} + kx = ky + c\dot{y} \qquad (2.2.1)$$

where $y$ is the displacement of the controlled end and $x$ is the displacement of the gripper. The damping $c\dot{x}$ is included to model energy losses in the overall system. If we divide (2.2.1) by $m$ we can then reexpress it in terms of $\omega_n$ and $\zeta$,

$$\ddot{x} + 2\zeta\omega_n\dot{x} + \omega_n^2 x = \omega_n^2 y + 2\zeta\omega_n\dot{y} \qquad (2.2.2)$$

You'll note that this is exactly the same situation we had with our physical experiment, a mass-spring combination and a controlled end. We even had the damping; it's certain you noticed that the oscillations didn't go on forever. Situations like this are extremely common in machine design. Very often we'll design a structure that's meant to translate or rotate. Since the structure has mass and flexibility, it will allow vibrations, vibrations that are often undesirable. To see this, let's analyze (2.2.2). Our input is illustrated in Figure 2.4 and is simply a step change in position. This

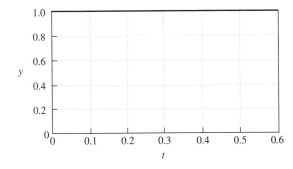

**Figure 2.4**   Unit step displacement input

is one of the simplest forced examples, since all the transient excitation occurs at $t = 0$; after that the forcing is constant. Since the steady-state solution is just a constant displacement (as you saw from your experiment), we'll assume a solution of the form

$$x(t) = a + be^{\zeta \omega_n t} \cos(\omega_d t + \phi) \tag{2.2.3}$$

Here, we've explicitly assumed that the system is underdamped (which is by far the most common situation). Note that we're using $\omega_d$, which is equal to the damped frequency ($\omega_d = \omega_n\sqrt{1 - \zeta^2}$). The constant $a$ is meant to account for the steady-state displacement of the mass due to the base movement and $b$ and $\phi$ are needed to account for initial conditions.

If the system was initially at rest, then $x(0) = \dot{x}(0) = 0$. If the base displacement is equal to 1 cm, then our forced solution must match our steady-state conditions for large $t$:

$$a = 1 \text{ cm} \tag{2.2.4}$$

Thus our solution becomes

$$x(t) = 1 \text{ cm} + e^{-\zeta \omega_n t}\left(a \cos(\omega_d t) + b \sin(\omega_d t)\right) \text{ cm} \tag{2.2.5}$$

Finally, we account for the zero initial conditions

$$x(0) = 1 + a = 0 \tag{2.2.6}$$

and

$$\dot{x}(0) = \zeta \omega_n + b\omega_d = 0 \tag{2.2.7}$$

The solutions to this are $a = -1$ and $b = -\dfrac{\zeta}{\sqrt{1-\zeta^2}}$. Thus

$$x(t) = 1 + e^{-\zeta \omega_n}\left(-\cos(\omega_d t) - \frac{\zeta}{\sqrt{1 - \zeta^2}} \sin(\omega_d t)\right) \text{ cm} \tag{2.2.8}$$

Now let's consider a couple of cases. First, let's assume that our system has very little damping, say a damping factor of .05. We'll let $m = 1$ kg and $k = 900$ N/m. The response for a step input to this system is shown in Figure 2.5a. You'll note that there's some good and some bad with this response. If we continue to use our analogy to a positioner (Figure 2.2), the point of the device is to position the end at the steady-state position quickly and accurately. Although the payload is brought to the correct position (1 cm) quickly, it then overshoots and rings (or oscillates) for a long time. Only after many oscillation cycles does the final response start to look like a constant offset. Obviously, this ringing is not generally considered desirable. If our positioning arm was part of a robotic welder in an automotive plant, the robot would have to wait for a good while before making the weld, since the oscillations would degrade the welding process if they were appreciable in magnitude.

One way to reduce the amount of ringing is to increase the damping. If we increase $\zeta$ from .05 to .8, then our reponse will look that of Figure 2.5b. Note that the ringing is greatly reduced, a good characteristic for our welder. However, the response time has now increased a great deal.

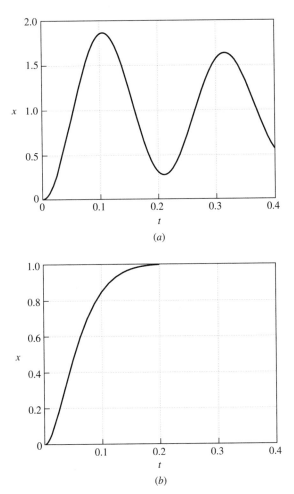

(a)

(b)

**Figure 2.5**   Effect of damping on system's time response

Instead of reaching the .9 cm position in .05 (as it did for the previous case), it now requires almost .12 sec. This is a not-so-good result. Whereas before, we had lots of oscillations to deal with, now we have the problem of a relatively slow response.

This illustrates the sort of trade-off you'll be facing when attempting any design problem. Changing a parameter to make some aspect of the problem better invariably makes another aspect worse. Your job will be to find the optimum trade-off that gives you the behavior you need at an acceptable cost.

The general response shown in Figure 2.6 is a good one, since it allows us to illustrate how transient responses are used to characterize a system. If a step input is our input, we'd like the response to also be a step. Unfortunately, this is never the case, and thus we characterize the actual response by how much it varies from the ideal. One commonly used quantity is the *rise time*, shown in Figure 2.6 as $\tau_r$. This can be defined in different ways; the way we've chosen to define

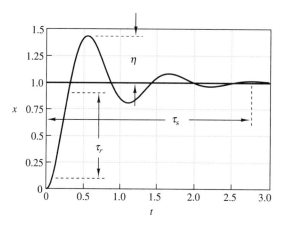

**Figure 2.6**   Step response for an underdamped system

it is as the time from when the response is at 10 percent of the steady-state value to when it's at 90 percent of this value. We can also talk about *settling time*, shown in Figure 2.6 as $\tau_s$. This is the time necessary for the oscillation to die away almost completely. How much oscillation to allow is arbitrary; in Figure 2.6 we've set it to be equal to the time needed for the oscillations to die down to 2 percent of the steady-state value. The last quantity we'll introduce is the overshoot, that is, the amount by which the response exceeds the steady-state response, indicated by $\eta$.

The type of analysis we've been going over here is also central to work in the area of controls, and you'll be seeing the same material if you take a controls course. These transient measures of system response are very easy to visualize and play a large part in our intuition about how systems respond. Although we won't be going into any great detail in this book, these concepts also form the basis of several methods of system identification. When you first start to analyze a system, you may not know its mass, stiffness, or damping properties. In fact, you probably won't know them. Thus, a central concern is simply figuring out how to mathematically represent the system. This problem is the motivation behind system identification methods. Simply put, these methods allow you to run tests on a system and to determine what the mass, stiffness, and damping properties are. You can intuitively understand how this could be done just from our example. If you know how SDOF systems respond to a step input, then you can apply a step to the system, record the output, then calculate $\tau_s$, $\tau_r$, etc. By comparing these values to those you obtained from (2.2.1), you can figure out what the physical parameters of your system must be. We'll talk a bit more about identification in Section 2.9, but for now let's move on to more involved types of forcing inputs.

To start our next experiment, hold the free end of the spring, begin to *slowly* move your hand up and down, and observe the mass. It moves up and down in the same direction as your hand and also seems to move the same amount. If your hand moves up one inch, the mass moves up around 1 in. This is our first observation about vibrational systems. When you excite a vibrational system at a *low* frequency (which is what we were doing), the spring in the system acts like a rigid rod. The motion is the same as if the spring were actually replaced by a rigid rod. This is a limiting case response, a topic we'll look at more closely in Chapter 7.

Now start moving your hand up and down as quickly as you can. This is a *high* frequency excitation. You'll notice that the mass stays essentially fixed in space. Even though your hand is moving up and down, the mass doesn't move (at least not much). Observation number two! In the limiting case of high-frequency vibrations, it is as if the spring disappears; almost no motion is transmitted through the spring to the mass.

At this point we're ready to move on to a more dramatic effect. If you experiment a very little bit, you'll find that there is a frequency of excitation at which, with almost no motion of your hand, the mass exhibits a very large response. In fact, you'll probably find that the motion keeps growing and that soon the mass is flying out of control. This phenomenon is called *resonance*.

Let's recap for a moment. At very high frequencies the mass didn't move at all. At resonance it moved a tremendous amount, while at low frequencies the mass moved only as much as your hand moved. The question is, what happens for other frequencies of excitation? Now that you know the resonant frequency, this will be easy to answer. Start oscillating your hand again, but at a frequency below the resonance frequency. You'll see that the mass again moves in the direction of your hand (up when your hand goes up, down when your hand goes down), but that the amount of motion is now *greater* than your hand's motion. Next, try moving your hand at a frequency above the resonant frequency. This time we see something new. Depending on the precise frequency with which you move your hand, the mass's motion can be greater than or less than your hand's motion. In either case, the motion of the mass is in opposition to that of your hand. When your hand goes up, the mass goes down, and vice versa. This is called *out-of-phase* motion. Similarly, for the case in which the mass moved in the same direction as your hand, we had *in-phase* motion. When damping is added the situation will become a bit more complicated, but for now, the mass can only be completely in phase or out of phase with the excitation. We'll explain why we use the word *phase* once we've examined the problem from a more mathematical viewpoint.

Now that we have some physical feel for the system, we can analyze it mathematically, and then correlate the results to our physical experiments. Looking at Figure 2.1, we can see that the governing equation is

$$m\ddot{x} + kx = ky \qquad (2.2.9)$$

We've neglected damping here, a reasonable assumption for the kinds of responses we're going to be examining. Dividing by $m$, we obtain

$$\ddot{x} + \frac{k}{m}x = \frac{k}{m}y \qquad (2.2.10)$$

or, using our knowledge that $\frac{k}{m} = \omega_n^2$,

$$\ddot{x} + \omega_n^2 x = \omega_n^2 y \qquad (2.2.11)$$

We can model the excitation $y$ as

$$y = \bar{y}\sin(\omega t) \qquad (2.2.12)$$

that is, a purely sinusoidal forcing. You'll recall from your course in ODEs that the general solution for such a system is

$$x = a \sin(\omega t) + b \cos(\omega t) \tag{2.2.13}$$

where $\omega$ is not equal to $\omega_n$. If $\omega$ was equal to $\omega_n$ then our assumed solution would be the same as the forcing itself, in which case (2.2.13) would no longer be adequate [1].

Note that (2.2.13) represents the solution for the steady-state problem. We are ignoring the effect of initial conditions (for which we'd need to add the homogeneous solution $c \sin(\omega_n t) + d \cos(\omega_n t)$). Although we can obviously deal with both the homogeneous and nonhomogeneous solutions simultaneously, it is vastly more common in vibration analyses to be concerned only with the steady-state responses of a system. Only if we cared about the transient response would we retain both solutions. Problems for which we'd care about the transient motion would include the response of an aircraft to a sudden wind gust or to a landing, the response of a building to an earthquake, the motion of an offshore oil rig to a large wave, etc.

Just this once we'll work the problem out, keeping both the $\sin(\omega t)$ and $\cos(\omega t)$ terms in order to illustrate why only one is actually needed for the undamped case. Differentiating (2.2.13) twice to obtain an expression for $\ddot{x}$ and substituting for $x$, $\ddot{x}$, and $y$ in (2.2.11), we obtain

$$a(\omega_n^2 - \omega^2) \sin(\omega t) + b(\omega_n^2 - \omega^2) \cos(\omega t) = \bar{y} \omega_n^2 \sin(\omega t) \tag{2.2.14}$$

Now $\sin(\omega t)$ and $\cos(\omega t)$ are linearly independent functions (i.e., you cannot combine multiples of them to get zero). Therefore, the only way that the left-hand side of the equation can equal the right-hand side is for all the terms multiplying $\sin(\omega t)$ on the left to equal the terms multiplying $\sin(\omega t)$ on the right (and the same for the $\cos(\omega t)$ terms).

Thus we must have

$$a(\omega_n^2 - \omega^2) = \omega_n^2 \bar{y} \tag{2.2.15}$$

and

$$b(\omega_n^2 - \omega^2) = 0 \tag{2.2.16}$$

It's pretty clear from (2.2.16) that, unless $\omega = \omega_n$, the only way to satisfy the equation is for $b$ to equal zero. And we've already mentioned that $\omega$ can't be equal to $\omega_n$. Thus we can drop the $\cos(\omega t)$ part of the solution.

What we've seen is a general result. When dealing with a sinusoidally forced oscillator that has only mass and spring terms, the assumed solution need include only the trigonometric terms found in the forcing. As we can see, this means that the response can be either motion that moves in the same direction as the forcing (in-phase motion) or motion that moves in opposition to the forcing (out-of-phase motion). At this point we can clear up what we mean by the word *phase*. Instead of calling the response $a \sin(\omega t)$, we can write it as $|a| \sin(\omega t - \phi)$. $|a|$ represents the absolute value of the amplitude of the response, and it will not switch sign as the frequency goes above resonance. Simply the phase angle, $\phi$, will change. When the mass moves in the same direction as the excitation, the phase angle is equal to zero. After resonance, the phase angle will abruptly switch to $\pi$ radians (180 degrees), indicating that the response lags the input by $\pi$ radians. This kind of representation will be essential once we add damping to the problem, since we'll then have values of $\phi$ between 0 and 180 degrees.

Now that (2.2.16) is out of the way, it is easy to use (2.2.15) to solve for $a$, the unknown response amplitude

$$a = \frac{\omega_n^2 \bar{y}}{(\omega_n^2 - \omega^2)} \tag{2.2.17}$$

Thus the final solution is

$$x(t) = \frac{\omega_n^2 \bar{y}}{(\omega_n^2 - \omega^2)} \sin(\omega t) \tag{2.2.18}$$

This is a good point at which to compare our experimental findings with the mathematical result we've just obtained. Figure 2.7 shows a plot of $\dfrac{\omega_n^2}{\omega_n^2 - \omega^2}$ versus $\omega$. The dots labeled 1–5 correspond to the test cases of our physical experiments. At point 1 we're forcing the system at a low frequency and find that the response amplitude is almost equal to the forcing amplitude. Furthermore, the sign of $a$ is the same as the sign of $\bar{y}$. We call this in-phase motion since, if we had used $a \sin(\omega t - \phi)$ as our guess for the response, $\phi$ would equal zero, thus the input and output would be locked together. As we increase in frequency to point 2, the response stays in phase with the forcing, but the amplitude grows. At point 3, the resonance condition, the amplitude of the response has gone to infinity. Increasing the forcing yet further (point 4), causes the response to go out of phase with the input (indicated by the negative values for the response). Finally, as the frequency becomes very large (point 5), the response becomes very small.

Happily, all the mathematical predictions match the experimental results. Even at this early stage, we are able to do a good bit of engineering. We can determine the frequencies at which the

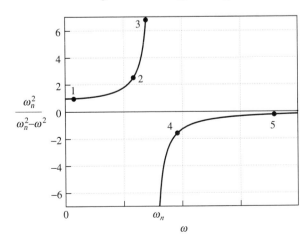

**Figure 2.7**  Displacement response of a seismically excited spring-mass system

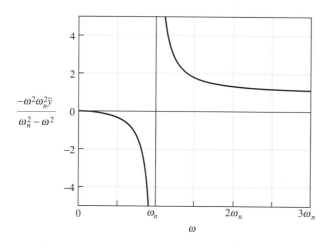

$$\frac{-\omega^2 \omega_n^2 \bar{y}}{\omega_n^2 - \omega^2}$$

$\omega$

**Figure 2.8**   Acceleration response of a seismically excited, spring-mass system

response amplitude of the mass is less than the forcing amplitude and those for which it is greater. Furthermore, it is a simple matter to consider the accelerative loadings encountered by the mass (an item of interest when packaging fragile goods). All we need do is differentiate (2.2.18) twice to obtain $\ddot{x}$, the acceleration of the mass

$$\ddot{x} = \frac{-\omega^2 \omega_n^2 \bar{y}}{(\omega_n^2 - \omega^2)} \sin(\omega t) \tag{2.2.19}$$

Figure 2.8 shows a plot of the acceleration, $\dfrac{-\omega^2 \omega_n^2 \bar{y}}{\omega_n^2 - \omega^2}$ versus the frequency of excitation. (Again, $\bar{y}$ is set to 1.0 for simplicity.) Although the overall results are somewhat similar to those shown in Figure 2.7, we see that the acceleration goes to zero at low frequencies (instead of going to a nonzero constant) and it becomes constant at higher frequencies (instead of going to zero). Once again, the response switches phase at the system's resonant frequency.

From the foregoing we can see that the system's natural frequency is a very special frequency. With this in mind, we might well want to consider introducing the following, nondimensional, frequency:

$$\Omega \equiv \frac{\omega}{\omega_n} \tag{2.2.20}$$

In this case, resonance occurs when $\Omega = 1.0$. Dividing both numerator and denominator of (2.2.17) by $\omega_n^2$ and using (2.2.20), we find

$$a = \frac{\bar{y}}{1 - \Omega^2} \tag{2.2.21}$$

This is certainly a clean and compact form for the displacement response. It is also a very general solution. If we plot (2.2.21) (with $\bar{y} = 1.0$) versus $\Omega$, as is shown in Figure 2.9, we have a response plot that is valid for any seismically excited, single-degree-of-freedom, spring-mass system. In order

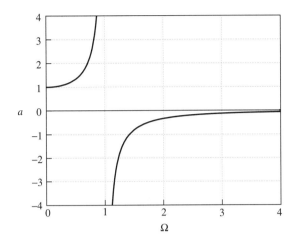

**Figure 2.9**   Response of seismically excited spring-mass system, nondimensional frequency

to relate this figure to a particular system, we simply need to be given the actual mass and stiffness values (in order to convert the nondimensional $\Omega$ into an actual frequency).

Just as we could put (2.2.17) in a very clear form by using $\Omega$, we can also reexpress (2.2.19) if we wish

$$\ddot{x} = \frac{-\omega_n^2 \Omega \bar{y}}{1 - \Omega^2} \sin(\omega t) \tag{2.2.22}$$

## Example 2.1

**Problem**   Consider the system shown in Figure 2.1. Determine the absolute value of the response if the base is oscillated at 25 rad/s with an amplitude of 1 mm and if the system has a spring constant equal to 1000 N/m and a mass of 2.0 kg.

**Solution**   The governing equation is given by

$$m\ddot{x} + kx = ky \tag{2.2.23}$$

Putting this into the form of (2.2.11) gives us

$$\ddot{x} + \omega_n^2 x = \omega_n^2 y \tag{2.2.24}$$

Since $\omega_n^2 = \frac{k}{m} = 500$, we have

$$\ddot{x} + 500x = 500y \tag{2.2.25}$$

Since the particular phase of the forcing isn't important, we'll assume a cosine for the forcing

$$\ddot{x} + 500x = .001 \cos(25t) \tag{2.2.26}$$

where the particular amplitude and frequency of the base excitation has been used.

The solution amplitude is given by (2.2.17) as

$$a = \frac{(500)(.001)}{500 - 25^2} = -.004 \tag{2.2.27}$$

Thus the absolute value of the response is equal to 4 mm.

## 2.3 DIRECT FORCE EXCITATION

Now that we've examined the case of unforced seismic excitation, we can consider the problem in which the free end of the spring is fixed and a force is directly applied to the mass (Figure 2.10). Clearly, we won't be able to experimentally investigate this particular case as we did for seismic excitation, but, as we'll soon see, the mathematics of the two problems are almost identical. Thus we can have some confidence in the analysis, even though we don't perform a supporting experiment. Keep in mind that the two problems are almost identical only when there is no damping in the systems. The presence of damping will cause some complications and will be examined in Sections 2.5 and 2.7.

As you can easily determine from Figure 2.10, the equation of motion for this system is simply

$$m\ddot{x} = f(t) - kx \tag{2.3.1}$$

or, dividing by $m$ and using the definitions already introduced

$$\ddot{x} + \omega_n^2 x = \frac{f(t)}{m} \tag{2.3.2}$$

Assuming that the excitation can be expressed as

$$f(t) = \bar{f} \sin(\omega t) \tag{2.3.3}$$

and using the general solution

$$x(t) = \bar{x} \sin(\omega t) \tag{2.3.4}$$

leads to

$$-\omega^2 \bar{x} \sin(\omega t) + \omega_n^2 \bar{x} \sin(\omega t) = \frac{\bar{f}}{m} \sin(\omega t) \tag{2.3.5}$$

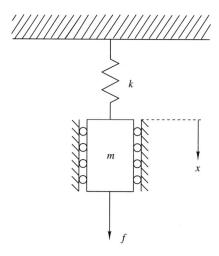

**Figure 2.10**  Direct-force-excited, spring-mass system

Cancelling out the $\sin(\omega t)$ factor and solving for $\bar{x}$, we find that

$$\bar{x} = \frac{\bar{f}}{m(\omega_n^2 - \omega^2)} \tag{2.3.6}$$

If we wish to express this in terms of the nondimensional frequency, $\Omega$, we have

$$\bar{x} = \frac{\bar{f}}{m\omega_n^2(1 - \Omega^2)} \tag{2.3.7}$$

Some authors prefer to clean up this solution by letting $f(t)$ equal $f'k \sin(\omega t)$, which would lead to

$$\bar{x} = \frac{f'}{(1 - \Omega^2)} \tag{2.3.8}$$

just the same form as we had in (2.2.21). However, this often leads to confusion, so we'll express the forcing as just $f \sin(\omega t)$.

The nondimensional response $\frac{m\omega_n^2 \bar{x}}{\bar{f}}$ (from (2.3.6)) is plotted versus $\Omega$ in Figure 2.11. Clearly, the forms of (2.2.21) and (2.3.7) are the same; the only difference is in their magnitude at a particular frequency.

## Example 2.2

**Problem**   For the system shown in Figure 2.10, determine the response if the input is a cosine excitation with a magnitude of 2 N. The mass of the system is equal to 5 kg, the spring constant is equal to 20,000 N/m, and the frequency of excitation is equal to 60 rad/s.

**Solution**   The system's natural frequency is found from

$$\omega_n^2 = \frac{20,000}{5} = 4000 \tag{2.3.9}$$

**Figure 2.11**   Displacement response of direct-force-excited, spring-mass system, nondimensional frequency

Using (2.3.7) we'll find

$$\bar{x} = \frac{2}{5(4000 - 60^2)} = .001 \qquad (2.3.10)$$

Thus $x(t) = .001 \cos(60t)$, where the units of displacement are in meters.

## 2.4 TRANSFER FUNCTIONS

A very useful characteristic of linear systems is that they support a phenomenon known as super-position. This means if an input $i_1$ produces on output $o_1$ for a particular linear system, and another input $i_2$ produces the output $o_2$, then the output of the same system for the combined input $\alpha i_1 + \beta i_2$ (where $\alpha$ and $\beta$ are arbitrary constants) is simply $\alpha o_1 + \beta o_2$. Thus, if you already know the output of the system when the forcing is $\sin(\omega t)$, you don't have to recalculate the response if the input amplitude is changed to, say, $5 \sin(\omega t)$. It is simply the original output amplitude multiplied by five.

We've already seen, both experimentally and theoretically, that a seismically excited spring-mass system, when acted on by an input sinusoid, reacts by producing a sinusoidal motion of the mass. The only thing that can change is the degree to which the mass responds and whether it is in or out of phase with the input. Thus we can view the system as a sort of amplifier, one that takes in sinusoidal signals and produces sinusoidal signals. The degree of amplification, or gain, is dependent upon frequency. Therefore, what we really should be concerning ourselves with is not the actual response for a particular sinusoidal input, but the gain characteristics of the system as a function of frequency. This information is contained in what we shall call the *transfer function* of the system.

A quick examination of (2.2.18) reveals that the response actually contains the entire input excitation $\left( \bar{y} \sin(\omega t) \right)$, along with other terms. We are now in a position to recognize these other terms as the transfer function of the system. Dividing (2.2.18) by $\bar{y} \sin(\omega t)$, we find that the transfer function $g(\omega)$ is given by

$$g(\omega) \equiv \frac{x(t)}{y(t)} = \frac{\omega_n^2}{\omega_n^2 - \omega^2} \qquad (2.4.1)$$

or, in terms of the nondimensionalized frequency $\Omega$

$$g(\Omega) = \frac{1}{1 - \Omega^2} \qquad (2.4.2)$$

found by dividing the numerator and denominator by $\omega_n^2$. A plot of $g(\Omega)$ versus frequency is shown in Figure 2.12. Figures like this are called frequency-response plots, for the obvious reason that they show us the response amplitude of the system to an input sinusoid as a function of frequency.

Although you will sometimes encounter frequency responses in the form shown, the more common way to present this information is to express the transfer function in the form $|g|e^{i\phi}$, where $|g|$ is the absolute value of the transfer function and $e^{i\phi}$ represents any complex phase shifts.

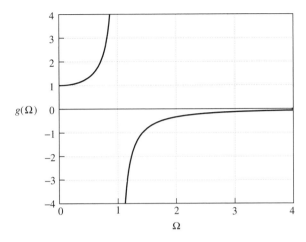

**Figure 2.12**   Transfer function of a seismically excited spring-mass system, nondimensional frequency

This is especially important if the transfer function is complex (as they usually are), since we'd then need to plot the real part versus frequency, as well as the imaginary part versus frequency, in order to adequately express the complete response. Looking at the magnitude and phase is usually more physically meaningful.

Since our transfer function is real (for now), $\phi$ will be either $0$ or $-\pi$ rad. The first plot simply presents the absolute value of the amplitude response as a function of frequency ($|g(\Omega)|$) and is shown in Figure 2.13a. In this case the resonant response at $\omega = \omega_n$ shows up as an infinitely tall spike. The other plot (Figure 2.13b) gives us the phase information that is lost when taking the absolute value of $g(\Omega)$.

The reason that $\phi$ is chosen to be $-\pi$ rather than $\pi$ is that the output will always lag the input. Thus we'd expect a negative phase shift. This will be much clearer in a few pages, when we introduce damping.

**Example 2.3**

**Problem**   Determine the transfer function for the system of Example 2.1.

**Solution**   From (2.4.1) we have

$$g(\omega) = \frac{\omega_n^2}{\omega_n^2 - \omega^2} \tag{2.4.3}$$

Thus for our system we'll have

$$g(\omega) = \frac{500}{500 - \omega^2} \tag{2.4.4}$$

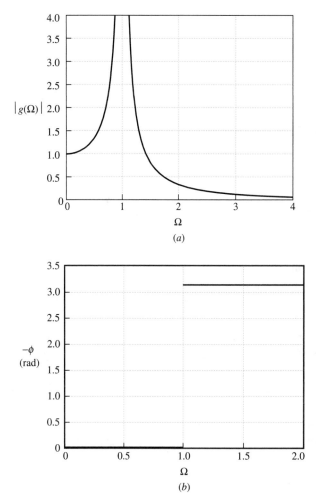

**Figure 2.13**  Magnitude and phase transfer function plots of a seismically excited, spring-mass system, nondimensional frequency

## 2.5 VISCOUS DAMPING

All the work we've done so far has involved undamped systems. If no damping is present, then once a motion has been started in the system it will persist forever. Clearly, undamped systems are not highly accurate models of the real world since we know that the unforced oscillations of real systems always decay away eventually. Thus we're motivated to add damping to our system description. We've already described the most common type of damper, the viscous damper, in Chapter 1. Section 2.10 of this chapter will introduce other types of damping, but, for the vast majority of vibrational analyses, viscous damping will be the damping model of choice.

If we add a viscous damper to our system (as shown in Figure 2.14), then the mass will have three forces acting on it: the applied external force $f(t)$, the force due to the spring $(kx)$, and the

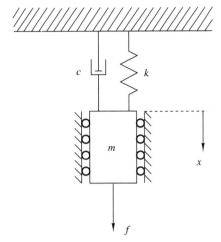

**Figure 2.14**  Direct-force-excited, spring-mass damper system

new force due to the damper ($c\dot{x}$). Thus our equation of motion becomes

$$m\ddot{x} + c\dot{x} + kx = f(t) \tag{2.5.1}$$

Dividing by $m$, as we did for the undamped case, allows us to write this as

$$\ddot{x} + \frac{c}{m}\dot{x} + \frac{k}{m}x = \frac{f(t)}{m} \tag{2.5.2}$$

If we go ahead and use the identities we derived when looking at free vibrations, namely $\omega_n^2 = \frac{k}{m}$ and $\zeta = \frac{c}{2\sqrt{mk}}$, we'll obtain

$$\ddot{x} + 2\zeta\omega_n\dot{x} + \omega_n^2 x = \frac{f(t)}{m} \tag{2.5.3}$$

If we next assume that our forcing is sinusoidal (say a sine), then we'll have

$$\ddot{x} + 2\zeta\omega_n\dot{x} + \omega_n^2 x = \frac{\bar{f}}{m}\sin(\omega t) \tag{2.5.4}$$

Now that we're including damping, our approach to solving the problem becomes important. The obvious method is to assume that the response is of the form

$$x(t) = a\cos(\omega t) + b\sin(\omega t) \tag{2.5.5}$$

just as we did with the undamped case. Although the additional harmonic wasn't needed in that case (the response was exactly in or exactly out of phase with the forcing), in this case it definitely is needed. The advantage to this assumed form is that it is quite physical; the values of $a$ and $b$ will indicate how much of the cosine and sine solutions make up the total response. The disadvantage, which will become clear in a moment, is that the solution is a bit messy. Since this approach is the most physically obvious, we'll start with it, completely characterize our solution, then introduce a more compact way of getting the same answer.

If we use (2.5.5) in (2.5.3), we'll obtain

$$-\omega^2\big(a\cos(\omega t)+b\sin(\omega t)\big)+2\omega_n\zeta\big(-\omega a\sin(\omega t)+\omega b\cos(\omega t)\big)$$

$$+\omega_n^2\big(a\cos(\omega t)+b\sin(\omega t)\big)=\frac{\bar{f}}{m}\sin(\omega t) \qquad (2.5.6)$$

If we group the sines and cosines, we'll have

$$\big(-\omega^2 a+2\zeta\omega_n\omega b+\omega_n^2 a\big)\cos(\omega t)+\big(-\omega^2 b-2\zeta\omega_n\omega a+\omega_n^2 b\big)\sin(\omega t)=\frac{\bar{f}}{m}\sin(\omega t) \qquad (2.5.7)$$

We now need to realize that sines and cosines are linearly independent functions; there's no way to get one from the other. Therefore, the coefficient of the $\sin(\omega t)$ term on the left side of the equation must equal the coefficient of the $\sin(\omega t)$ term on the right side and the coefficients of the cosines must match also. Matching the coefficients of the cosine terms gives us

$$-\omega^2 a+2\zeta\omega_n\omega b+\omega_n^2 a=0 \qquad (2.5.8)$$

and matching the sine coefficients yields

$$-\omega^2 b-2\zeta\omega_n\omega a+\omega_n^2 b=\frac{\bar{f}}{m} \qquad (2.5.9)$$

Equations (2.5.8) and (2.5.9) are two equations in two unknowns. If we group the $a$ and $b$ terms we'll have

$$(\omega_n^2-\omega^2)a+2\zeta\omega_n\omega b=0 \qquad (2.5.10)$$

and

$$-2\zeta\omega_n\omega a+(\omega_n^2-\omega^2)b=\frac{\bar{f}}{m} \qquad (2.5.11)$$

If we go ahead and solve for $a$ and $b$ we'll obtain

$$a=-\frac{(2\zeta\omega\omega_n)\frac{\bar{f}}{m}}{(\omega_n^2-\omega^2)^2+(2\zeta\omega\omega_n)^2} \qquad (2.5.12)$$

and

$$b=\frac{(\omega_n^2-\omega^2)\frac{\bar{f}}{m}}{(\omega_n^2-\omega^2)^2+(2\zeta\omega\omega_n)^2} \qquad (2.5.13)$$

Only $b$ can ever equal zero and this occurs only if the forcing frequency is equal to the system's natural frequency. At all frequencies $a$ is nonzero, and thus we always will have some finite response. Therefore the addition of damping has eliminated the infinite resonant response we saw in the undamped case when the system was forced at its natural frequency. In complete form our response looks like

$$x(t)=-\frac{(2\zeta\omega\omega_n)\frac{\bar{f}}{m}}{(\omega_n^2-\omega^2)^2+(2\zeta\omega\omega_n)^2}\cos(\omega t)+\frac{(\omega_n^2-\omega^2)\frac{\bar{f}}{m}}{(\omega_n^2-\omega^2)^2+(2\zeta\omega\omega_n)^2}\sin(\omega t) \qquad (2.5.14)$$

The new behavior due to the damping results in the addition of another harmonic component in the response (in addition to the one displayed by the forcing). When we had no damping, a sine input would give us a purely sine output. Now we've got both a cosine and a sine in the response. Therefore the question of the day is: What do we get when we add a sine and a cosine? Luckily, we've already answered this, way back in Chapter 1. If you check back, you'll see that a sine and cosine of the same frequency add together to yield either a sine with a phase shift or a cosine with a different phase shift.

Just to make sure all is clear, we'll work the problem out again, this time using the particular forms we've been dealing with. We're given

$$x(t) = a\cos(\omega t) + b\sin(\omega t) \tag{2.5.15}$$

Next, we'll write (2.5.15) as either

$$a\cos(\omega t) + b\sin(\omega t) = |\bar{x}|\cos(\omega t - \phi) \tag{2.5.16}$$

or as

$$a\cos(\omega t) + b\sin(\omega t) = |\bar{x}|\sin(\omega t - \phi) \tag{2.5.17}$$

Both forms for the solutions are equally possible; which we use will simply depend on which is convenient to use. Generally, the problem you're working on will suggest the appropriate form. For instance, if the forcing is a cosine, then it makes sense to assume that the solution is of the form $\cos(\omega t - \phi)$. In this case $\phi$ has a physical meaning, that is, the amount that the output lags the input. For illustration purposes we'll choose (2.5.17), and you can derive the appropriate results for (2.5.16) for yourself, if you're interested in doing so.

The first step in showing that (2.5.17) is a reasonable equation is to expand $\sin(\omega t - \phi)$. To do this, we'll use the trigonometric formula

$$\sin(a - b) = \sin(a)\cos(b) - \cos(a)\sin(b) \tag{2.5.18}$$

In the future you may also find it useful to use the associated cosine formula

$$\cos(a - b) = \cos(a)\cos(b) + \sin(a)\sin(b) \tag{2.5.19}$$

If we apply (2.5.18) to (2.5.17), we'll have

$$a\cos(\omega t) + b\sin(\omega t) = |\bar{x}|\left(\sin(\omega t)\cos(\phi) - \cos(\omega t)\sin(\phi)\right) \tag{2.5.20}$$

Just as we did when we solved the forced, undamped problem, we must equate the coefficients of the cosine and sine terms separately in order to satisfy (2.5.20). This means that

$$a = -|\bar{x}|\sin(\phi) \tag{2.5.21}$$

and

$$b = |\bar{x}|\cos(\phi) \tag{2.5.22}$$

If we square both sides of (2.5.21) and (2.5.22) and add them, we obtain

$$|\bar{x}|^2 = a^2 + b^2 \tag{2.5.23}$$

or

$$|\bar{x}| = \sqrt{a^2 + b^2} \tag{2.5.24}$$

Dividing (2.5.21) by (2.5.22) gives us

$$-\tan(\phi) = \frac{a}{b} \tag{2.5.25}$$

or

$$\phi = -\tan^{-1}\left(\frac{a}{b}\right) \tag{2.5.26}$$

Note that the amplitude of the resultant wave is equal to the square root of the sum of the squares of the sine's and cosine's amplitudes. This is similar to how we find the length of a right triangle's hypotenuse: the length of the hypotenuse is equal to the square root of the sum of the squares of the other sides. This is also the way we find a vector's length. The length of a vector is equal to the square root of the squares of its projection onto the $x$ and $y$ axes. This geometric view is applicable to our problem. People will often think of a sinusoidal response as being a vector that rotates about the origin. We'll get back to this when we start discussing the alternative way of finding the forced oscillator's response.

If we substitute the actual values of $a$ and $b$, our final results are

$$\phi = \tan^{-1}\left(\frac{2\zeta\omega\omega_n}{\omega_n^2 - \omega^2}\right) \tag{2.5.27}$$

and

$$|\bar{x}| = \frac{\bar{f}}{m} \frac{1}{\sqrt{(\omega_n^2 - \omega^2)^2 + (2\zeta\omega_n\omega)^2}} \tag{2.5.28}$$

We can find the magnitude of the system's transfer function by dividing (2.5.28) by $\bar{f}$

$$|g(\omega)| = \frac{1}{m\sqrt{(\omega_n^2 - \omega^2)^2 + (2\zeta\omega_n\omega)^2}} \tag{2.5.29}$$

Finally, we can reexpress the magnitude and phase in terms of the nondimensional frequency $\Omega$, which you'll remember is equal to $\dfrac{\omega}{\omega_n}$:

$$\phi = \tan^{-1}\left(\frac{2\zeta\Omega}{1 - \Omega^2}\right) \tag{2.5.30}$$

and

$$|g(\Omega)| = \frac{1}{k\sqrt{(1 - \Omega^2)^2 + (2\zeta\Omega)^2}} \tag{2.5.31}$$

These results are worth examining. Let's start by looking at what happens to the phase angle as we increase $\omega$ from zero. For very low values of $\omega$ we see that $\phi$ is close to zero. As $\omega$ increases, the denominator of (2.5.30) approaches zero (as $\omega$ approaches $\omega_n$). Thus the overall fraction becomes very large. This tells us that the phase angle is approaching 90 degrees. In fact, when $\omega = \omega_n$, the phase angle is exactly equal to 90 degrees. Since our assumed form for the solution subtracted $\phi$ from the sine output ($x(t) = |\bar{x}| \sin(\omega t - \phi)$), we realize that the solution is *lagging* the input force. This is always going to be the case. When you add damping to a system such as the one we're examining, the output waveform will lag the input. Note also that the particular value of $\zeta$ doesn't affect the phase angle when $\omega = \omega_n$. All directly forced systems will have phase shifts of 90 degrees when forced at their natural frequencies. This fact can be used as an identification tool, and we'll look into this in Section 2.9.

As $\omega$ increases just a bit more, the magnitude of $\dfrac{2\zeta \omega \omega_n}{\omega_n^2 - \omega^2}$ is still huge, but now is negative. Therefore the phase angle continues to increase beyond 90 degrees. As $\omega$ approaches infinity, $\omega_n^2 - \omega^2$, it becomes large, thus driving $\dfrac{2\zeta \omega \omega_n}{\omega_n^2 - \omega^2}$ to 0. Therefore $\phi$ approaches 180 degrees and the ouput lags the input by almost 180 degrees. The speed at which the phase angle goes from 0 to 180 degrees depends upon the damping factor $\zeta$. Figure 2.15*b* shows what the phase behavior looks like as a function of frequency for several $\zeta$ values. Note that for small values of $\zeta$ the phase response resembles that of an undamped oscillator. Below the natural frequency the response is essentially in phase and above it the response is mostly out of phase. As $\zeta$ is increased, the change from in-phase response to out-of-phase response becomes much slower.

Now that we've looked at the phase response, it's time to look at the magnitude response. Figure 2.15*a* shows what $k|g(\Omega)|$ looks like for a variety of $\zeta$ values ( $k|g(\Omega)|$ is more convenient to plot than $|g(\Omega)|$ because it's nondimensional).

Just as in the case of the phase response, here the displacement response resembles that of the undamped oscillator for small values of $\zeta$. The magnitude starts at the static-deflection value, increases slowly at first, grows to a large value (but not infinity), and then goes to zero. As already noted, the fact that the response doesn't go to infinity, even at resonance, is one of the advantages that damping brings us.

You'll note that for low values of $\zeta$, the peak response amplitude is greater than the zero frequency value and that it occurs around the system's natural frequency. As the damping is increased, the peak value decreases and the frequency at which the peak occurs is reduced. Finally, at $\zeta = \frac{1}{\sqrt{2}}$, we lose the peak entirely. As the damping is increased yet further, the response becomes smaller and smaller at all frequencies.

This basic response is the key to understanding a whole range of applications. For instance, those of you interested in hi–fi systems know that amplifiers are often referred to as having a flat response from 0 to 100 kHz (or some other upper limit). The response starts off pretty much flat, and then, when the natural frequency is reached, the response begins to peak noticeably. Of course, an amplifier is a lot more complicated than a simple SDOF system; however the essential features are the same. Amplifiers have their own internal resonances and, when the excitation frequency reaches them, it causes the response to peak. The same thing is seen in loudspeaker specifications; they're fine until you try to play too high a frequency through them.

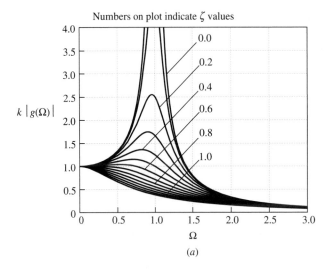

Numbers on plot indicate $\zeta$ values

$k\,|g(\Omega)|$

$(a)$

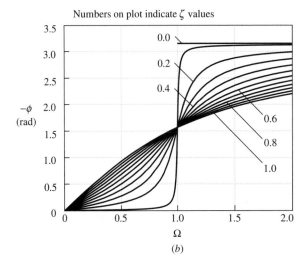

Numbers on plot indicate $\zeta$ values

$-\phi$
(rad)

$(b)$

**Figure 2.15** Amplitude and phase response of direct-force-excited, spring-mass damper (varying $\zeta$)

The degree to which a system peaks at resonance is closely tied to the type of system being examined. For instance, if you were designing an oscillator circuit, you'd want the peak to be very high and you'd want to be operating within the peaked region. However, if you were designing a robotic manipulator, you'd want a much lower peak. As in all our work, engineering trade-offs must be made, and there's no way to get everything you might want. For example, the reason a manipulator should have a low peak response is because it is being used to manipulate something, say a welder in an automobile factory. A peaky response at resonance implies that the arm won't quickly move to position and stay there, but rather it will oscillate a bit—something that's bad for accuracy and speed. Decreasing the peak response (by increasing the damping) will fix this problem, but it means that the speed with which the arm can move is reduced. Of course, this is a transient behavior and can be evaluated using the descriptors of Figure 2.6. Logically, we should be able to examine either the steady-state forced response or the transient response in order to deduce what the relevant natural frequency, damping factor, etc., might be. Indeed, both these approaches are possible and the steady-state approach is the second major avenue through which identification is done. We'll examine this approach in Section 2.9.

**Example 2.4**

**Problem**    Consider the system shown in Figure 2.14. Let the mass be equal to 4 kg, the damping coefficient be 80 N/m/s, and the spring coefficient be 40,000 N/m. How much will the output lag the input if the forcing is at 80 rad/s?

**Solution**    The natural frequency for the system is equal to

$$\omega_n^2 = \tfrac{40,000}{4} = 10,000 \tag{2.5.32}$$

Since $2\zeta\omega_n = \tfrac{c}{m}$ we have

$$\zeta = \frac{c}{m}\frac{1}{2\omega_n} = \frac{80}{4}\frac{1}{200} = .10 \tag{2.5.33}$$

and therefore have (from 2.5.27)

$$\phi = \tan^{-1}\left(\tfrac{(2)(.1)(80)(100)}{(100^2-80^2)}\right) = .418 \text{ rad} \tag{2.5.34}$$

Thus the output lags the input by .418 rad.

## 2.6  COMPLEX REPRESENTATIONS

Now that we've taken an in-depth look at how a directly forced system behaves, it's time to introduce a new way of looking at vibratory responses. Although representing a system's response as

$$x(t) = a\cos(\omega t) + b\sin(\omega t) \tag{2.6.1}$$

is physically attractive, it's a bit awkward mathematically, as you've seen in the previous section. Our solution to this problem will be to represent $x(t)$ in a new way, namely

$$x(t) = \bar{x}e^{i\omega t} \tag{2.6.2}$$

where $\bar{x}$ can be complex.

   With this representation we're using a completely exponential description of the motion, rather than a sinusoidal one. To illustrate why this is a good idea, let's again examine a directly excited, damped system,

$$\ddot{x} + 2\zeta\omega_n\dot{x} + \omega_n^2 x = \frac{f(t)}{m} \tag{2.6.3}$$

Furthermore, let's represent the forcing as $\bar{f}e^{i\omega t}$ and the response as $\bar{x}e^{i\omega t}$. This will give us

$$\left(-\omega^2 \bar{x} + 2i\zeta\omega\omega_n\bar{x} + \omega^2\bar{x}\right)e^{i\omega t} = \frac{\bar{f}}{m}e^{i\omega t} \tag{2.6.4}$$

It's a simple matter to cancel the common factor of $e^{i\omega t}$, leaving us with

$$\left(\omega_n^2 - \omega^2 + 2i\zeta\omega\omega_n\right)\bar{x} = \frac{\bar{f}}{m} \tag{2.6.5}$$

Recalling that the transfer function is just the output divided by the input, we can see from (2.6.5) that it's equal to

$$g(\omega) \equiv \frac{\bar{x}}{\bar{f}} = \frac{1}{m\left(\omega_n^2 - \omega^2 + 2i\zeta\omega\omega_n\right)} \tag{2.6.6}$$

Note that this is a complex quantity. If we express it in the form

$$g(\omega) = |g|e^{-i\phi} \tag{2.6.7}$$

that is, a magnitude and phase component, we'll have

$$|g| = \frac{1}{m\sqrt{(\omega_n^2 - \omega^2)^2 + (2\zeta\omega_n\omega)^2}} \tag{2.6.8}$$

and

$$\phi = \tan^{-1}\left(\frac{2\zeta\omega\omega_n}{\omega_n^2 - \omega^2}\right) \tag{2.6.9}$$

Compare how easily we obtained (2.6.8) and (2.6.9) with the work that was required to obtain (2.5.30) and (2.5.31). Instead of a few pages, it took us only a few lines. That's the beauty of a complex representation; it's compact and quick to use. Also note that the phase shift is always negative. $\phi$ itself is positive in (2.5.17) and (2.6.7), but you'll note that it's multiplied by $-1$ in the final result, $\sin(\omega t - \phi)$ and $e^{-i\phi}$, respectively. This makes good sense physically, as it implies that the system's response will be lagging the input. The damping acts to "delay" the response a bit, and it's always trying to catch up with the input. It would be strange if this wasn't the case, since a positive phase shift would imply that the system could somehow know what the input was going to be *before* it happened.

Although a complex representation is certainly an efficient way of looking at our problem, there is a downside. The difficulty is that we're no longer dealing with real quantities; our variables are now explicitly complex in nature. Let's take a moment to think about what this means. If we say that $x(t) = \bar{x}e^{i\omega t}$, then we can use our trigonometric identities to obtain

$$x(t) = \bar{x}\left(\cos(\omega t) + i\sin(\omega t)\right) \tag{2.6.10}$$

Our first thought might be to say "Hold on a second! $x(t)$ is a *physical* displacement. How can it have an imaginary component?"

The answer is—it can't. If $x(t)$ is a physical displacement, it must always be real. So must $f(t)$. However, what's stopping us from using $f(t) = \bar{f}e^{i\omega t}$ and remembering that we're only going to use the *real* part of it? If we take $\bar{f}$ to be real, then the real part of $\bar{f}e^{i\omega t}$ is just $\bar{f}\cos(\omega t)$.

This actually gives us some flexibility. For instance, if we wanted our input to be $\sin(\omega t)$, then we simply have to use the imaginary part of $\bar{f}e^{i\omega t}$. Of course, once we've decided that we're going to pay attention to a particular part of the input (say the real part), then we also have to look only at the real part of the output. Is this difficult? Not at all. Our response is $\bar{x}e^{i\omega t}$. Since $\bar{x}$ is complex in general, we can write it as

$$\bar{x} = \bar{x}_r + i\bar{x}_i \tag{2.6.11}$$

Substituting (2.6.11) into (2.6.10) gives us

$$x(t) = (\bar{x}_r + i\bar{x}_i)\big(\cos(\omega t) + i\sin(\omega t)\big) \tag{2.6.12}$$

Expanding this out yields

$$x(t) = \bar{x}_r \cos(\omega t) - \bar{x}_i \sin(\omega t) + i\big(\bar{x}_r \sin(\omega t) + \bar{x}_i \cos(\omega t)\big) \tag{2.6.13}$$

Taking the real part leaves us with

$$x(t) = \bar{x}_r \cos(\omega t) - \bar{x}_i \sin(\omega t) \tag{2.6.14}$$

How do we find $x_r$ and $x_i$? Just use our recent results. From (2.6.5) we'll have

$$\bar{x} = \frac{\frac{\bar{f}}{m}}{\omega_n^2 - \omega^2 + 2i\zeta\omega\omega_n} \tag{2.6.15}$$

If $\bar{f}$ is real, then this can be rewritten as

$$\bar{x} = \frac{\frac{\bar{f}}{m}(\omega_n^2 - \omega^2 - 2i\zeta\omega\omega_n)}{(\omega_n^2 - \omega^2)^2 - (2\zeta\omega\omega_n)^2} \tag{2.6.16}$$

We can immediately identify $x_r$ and $x_i$ as

$$\bar{x}_r = \frac{\frac{\bar{f}}{m}(\omega_n^2 - \omega^2)}{(\omega_n^2 - \omega^2)^2 + (2\zeta\omega\omega_n)^2} \tag{2.6.17}$$

and

$$\bar{x}_i = \frac{-\frac{\bar{f}}{m}(2\zeta\omega\omega_n)}{(\omega_n^2 - \omega^2)^2 + (2\zeta\omega\omega_n)^2} \tag{2.6.18}$$

A method that some people find useful to visualize this is to plot the force and displacement vectors on the complex plane. Consider for now just the forcing $f = \bar{f}e^{i\omega t}$, with $\bar{f}$ being real. Expanding this out in terms of the real and imaginary parts gives us

$$f = \bar{f}\big(\cos(\omega t) + i \sin(\omega t)\big) \qquad (2.6.19)$$

Both real and imaginary components are sinusoidal, and they're exactly 90 degrees out of phase with each other. The overall magnitude is $\bar{f}$, and the phase angle on the complex plane is given by

$$\phi = \tan^{-1}\big(\tan(\omega t)\big) = \omega t \qquad (2.6.20)$$

Thus the phase angle is increasing linearly with time. A series of snapshots of the forcing vector is shown in Figure 2.16 for the case of $\omega = \frac{\pi}{4}$. As you can see, the magnitude is a constant and the vector simply rotates in a counterclockwise direction. If we look at the response vector for a few chosen frequencies (fixing the time $t$), we see from Figure 2.17 that it simply lags the force vector (as previously discussed) and has a magnitude that varies with frequency.

In Figure 2.17 the force vector is real. For low frequencies the response is almost in phase with the force. Near resonance ($\omega = 10$ rad/s) the amplitude is larger than its low-frequency magnitude and lags the force by a bit more than 90 degrees. For a forcing frequency well above resonance (30 rad/s) the response is almost completely out of phase with the force and has a small magnitude ($|\bar{x}| \to 0$ as $\omega \to \infty$).

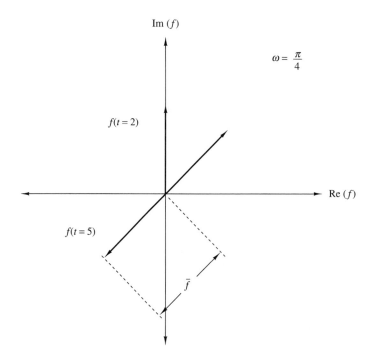

**Figure 2.16**  Complex representation of forcing vector

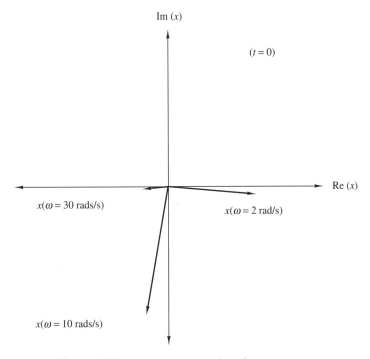

**Figure 2.17**   Complex representation of response vector

Another use of this sort of graphical viewpoint is for visualizing the relationship between displacement, velocity, and acceleration. If we let the displacement for a particular coordinate be

$$x(t) = \bar{x}e^{i\omega t} \qquad (2.6.21)$$

then we can differentiate with respect to $t$ in order to find the velocity and acceleration,

$$\dot{x}(t) = i\omega\bar{x}e^{i\omega t} \qquad (2.6.22)$$

$$\ddot{x}(t) = -\omega^2\bar{x}e^{i\omega t} \qquad (2.6.23)$$

We can easily plot all three on the complex plane, as shown in Figure 2.18. In this figure we've used two values for $\omega$, one less than and one greater than one. What we can see is that the velocity leads the displacement by 90 degrees and, in turn, the acceleration leads the velocity by 90 degrees. If the oscillation frequency is less than one, then the magnitude of the velocity is *less* than that of the displacement whereas the velocity's magnitude is *greater* than that of the displacement when the frequency is greater than one. This means that, even if a part is vibrating with a small displacement, the accelerative loads might be significant if the oscillation frequency is large.

To recap, you may well find it easier to analyze a system by using complex representations and taking the real (or imaginary) part in order to obtain the actual response, rather than explicitly

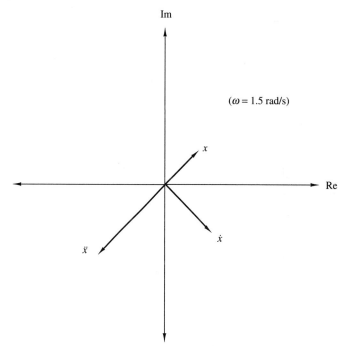

**Figure 2.18** Complex representation of displacement, velocity, and acceleration

using sines and cosines. This efficiency is not limited to SDOF systems, but also applies to the MDOF vibration problems that we'll be encountering.

**Example 2.5**

> **Problem**  Find the response of the sinusoidally excited system governed by
>
> $$\ddot{x} + 2\zeta\omega_n\dot{x} + \omega_n^2 x = 2\sin\omega t$$
>
> where $\zeta = .1$, $\omega_n = 10$ rad/s, and $\omega = 6$ rad/s.
>
> **Solution**   If we let the forcing be given by
>
> $$\frac{f(t)}{m} = 2e^{6it} \qquad \left(\frac{\bar{f}}{m} = 2\right)$$
>
> then we need to take the imaginary part of the response.
>     From (2.6.13) we have
>
> $$x(t) = \bar{x}_r \sin(\omega t) + \bar{x}_i \cos(\omega t)$$
>
> From (2.6.17) we have
>
> $$\bar{x}_r = \frac{2(10^2 - 6^2)}{(10^2 - 6^2)^2 + (2(.1)(6)(10))^2} = .0302$$

and from (2.6.18) we have

$$\bar{x}_i = \frac{-2(2(.1)(6)(10))}{(10^2 - 6^2)^2 + (2(.1)(6)(10))^2} = -.00566$$

Thus $x(t) = .0302 \sin(6t) - .00566 \cos(6t)$

## 2.7 DAMPED SEISMIC MOTION

Recall that there was no qualitative difference in the form of the equations of motion for an undamped seismically excited system and an undamped, direct-force excited system. This fact allowed us to draw conclusions from our seismic system that also held for the force-excited one. Unfortunately, this happy state of affairs ends when damping is added. The equations of motion for a seismically excited system with damping (Figure 2.3) have already been presented in (2.2.2) and are repeated here for convenience.

$$\ddot{x} + 2\zeta\omega_n\dot{x} + \omega_n^2 x = \omega_n^2 y + 2\zeta\omega_n\dot{y} \tag{2.7.1}$$

The part of this equation that qualitatively differs from the direct-force excited equation is in the $2\zeta\omega_n\dot{y}$ term. Since the excitation is transmitted from the ground motion to the mass by means of a damper, as well as a spring, this velocity-dependent forcing term arises. Luckily for us, our new grasp of complex representations means that solving this new problem will not be overly difficult. If we assume that $x = \bar{x}e^{i\omega t}$ and $y = \bar{y}e^{i\omega t}$, then (2.7.1) becomes

$$\bar{x}\left(\omega_n^2 - \omega^2 + 2i\zeta\omega\omega_n\right) = \bar{y}(\omega_n^2 + 2i\zeta\omega\omega_n) \tag{2.7.2}$$

The transfer function is simply

$$g(\omega) \equiv \frac{\bar{x}}{\bar{y}} = \frac{\omega_n^2 + 2i\zeta\omega\omega_n}{\omega_n^2 - \omega^2 + 2i\zeta\omega\omega_n} \tag{2.7.3}$$

If we compare (2.7.3) with (2.6.6), we see that the important difference is in the fact that we've gained an additional phase shift from the numerator. (The fact that we have a difference in the overall magnitude isn't too important, since we had that in the undamped case, also.) Going ahead and expressing the numerator and denominator of (2.7.3) in magnitude-phase form, we'll have

$$g(\omega) = \frac{\sqrt{\omega_n^4 + (2\zeta\omega\omega_n)^2}\,e^{i\phi_1}}{\sqrt{(\omega_n^2 - \omega^2)^2 + (2\zeta\omega\omega_n)^2}\,e^{i\phi_2}} \tag{2.7.4}$$

where $\phi_1 = \tan^{-1}\left(\dfrac{2\zeta\omega}{\omega_n}\right)$ and $\phi_2 = \tan^{-1}\left(\dfrac{2\zeta\omega\omega_n}{\omega_n^2 - \omega^2}\right)$.

Pulling the magnitude and phase angles together finally gives us

$$g(\omega) = \sqrt{\frac{\omega_n^4 + (2\zeta\omega\omega_n)^2}{(\omega_n^2 - \omega^2)^2 + (2\zeta\omega\omega_n)^2}}\,e^{i\phi}, \qquad \phi = \phi_1 - \phi_2 \tag{2.7.5}$$

Of course, we can also express this in terms of the nondimensional frequency $\Omega$:

$$g(\Omega) = \sqrt{\frac{1 + (2\zeta\Omega)^2}{(1 - \Omega^2)^2 + (2\zeta\Omega)^2}}\, e^{i\phi}, \qquad \phi = \tan^{-1}(2\zeta\Omega) - \tan^{-1}\left(\frac{2\zeta\Omega}{1 - \Omega^2}\right) \qquad (2.7.6)$$

What conclusions can we draw? Well, the first thing to note is that the magnitude response is similar to the direct-force excited case. The characteristic polynomial $((1 - \Omega^2)^2 + (2\zeta\Omega)^2)$ is the same for both cases. This is reassuring since the characteristic polynomial depends only upon the system being excited, not the excitation itself. Since the system hasn't changed (it's a mass, spring, and damper for both cases), the characteristic polynomial shouldn't alter. Also, we see that the numerator is almost the same as long as the damping is small. As the damping goes to zero, the two cases become more and more similar. When the damping increases, the numerators differ appreciably. In addition, the differences are magnified by the frequency of the excitation. Since the new term $2\zeta\Omega$ contains $\Omega$ explicitly, it will increase linearly as $\Omega$ increases. For the force-excited case, the response was finite at small-forcing frequencies and went to zero, like $\frac{1}{\Omega^2}$, for very high-frequency excitations. For the seismically damped case, the zero frequency limit also yields a finite response. However, as the frequency goes to infinity, the amplitude scales as $\frac{1}{\Omega}$. It still goes to zero; however, it does so more slowly than the force-excited amplitude does. This effect can be seen in Figure 2.19a. Finally, we see that increasing the damping will *decrease* the response if $\Omega$ is less than about 1.4, but will actually *increase* the amplitude if $\Omega$ is above this point. This behavior is fundamentally different from the force-excited case, for which increasing the damping will decrease the response amplitude for all values of forcing frequency. Thus, if you're trying to reduce the vibration levels for a seismically excited system, the frequency of the forcing becomes important, since increasing the damping will increase the vibration levels if you're in the wrong frequency regime.

As far as phase is concerned, we see that the seismic case actually has a positive phase effect due to the $2\zeta\Omega$ term. This almost cancels the phase due to $\frac{2\zeta\Omega}{1 - \Omega^2}$, thus slowing the initial buildup of phase lag. This can be seen in Figure 2.19a, for which the phase at small $\omega$ is growing very slowly as compared with that of Figure 2.15b. As you can see from Figure 2.19b, the phase doesn't approach $-\pi$ as the frequency approaches infinity. This is because the value of $\frac{2\zeta\Omega}{1 - \Omega^2}$ goes to zero while $2\zeta\Omega$ goes to infinity. Thus the overall seismic response doesn't end up lagging the input by 180 degrees at high frequencies, but rather by 90 degrees. Therefore all of the nonzero damping phase curves in Figure 2.19b approach a phase of $-\frac{\pi}{2}$ radians at high frequencies, as seen more clearly in Figure 2.19c.

### Example 2.6

**Problem**   Consider the system shown in Figure 2.3. What is the difference between the amplitude and phase lag for the system with and without damping? Let $m = 2$ kg, $c = 80$ N/m/s, and $k = 20,000$ N/m. The seismic excitation has an amplitude of 5 mm and a frequency of 80 rad/s.

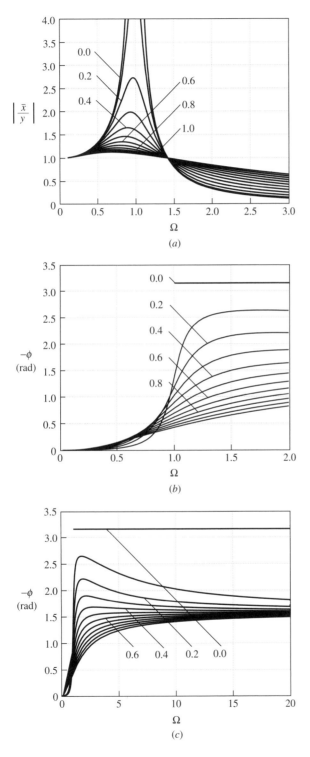

**Figure 2.19**  (*a*) and (*b*) Magnitude and phase response of seismically excited, spring-mass-damper (varying $\zeta$); (*c*) phase response of seismically excited, spring-mass-damper (varying $\zeta$) (Note: Numbers on plot indicate $\zeta$ values)

**Solution** The natural frequency is found from

$$\omega_n = \sqrt{\frac{k}{m}} = 100 \text{ rad/s} \tag{2.7.7}$$

and the damping factor is given by

$$\zeta = \frac{c}{m}\frac{1}{2\omega_n} = .20 \tag{2.7.8}$$

From (2.7.6) we have

$$g(\Omega) = \sqrt{\frac{1 + (2\zeta\Omega)^2}{(1 - \Omega^2)^2 + (2\zeta\Omega)^2}}e^{i\phi}, \qquad \phi = \tan^{-1}(2\zeta\Omega) - \tan^{-1}\left(\frac{2\zeta\Omega}{1 - \Omega^2}\right) \tag{2.7.9}$$

Thus for the damped problem we have

$$|g(\Omega)| = \sqrt{\frac{1 + (2\zeta\Omega)^2}{(1 - \Omega^2)^2 + (2\zeta\Omega)^2}} \tag{2.7.10}$$

which evaluated at $\Omega = \frac{80}{100} = .8$ gives us

$$|g(.8)| = 2.18 \tag{2.7.11}$$

The phase angle is

$$\phi = -.417 \text{ rad} \tag{2.7.12}$$

Since the input amplitude is 5 mm we have an output amplitude of $5(2.18) = 10.9$ mm. If $\zeta$ is set to zero we'll have

$$g(.8) = 1.\bar{6} \tag{2.7.13}$$

and thus an output amplitude of $8.\bar{3}$ mm along with a phase lag of

$$\phi = 0 \text{ rad} \tag{2.7.14}$$

We see that the amplitude of the damped response has increased by 31 percent over the undamped case and the phase lag has increased from 0 to 24 degrees. Thus the damping has had a very significant effect on the system's response.

## 2.8 ROTATING IMBALANCE

The final type of single-degree-of-freedom system we're going to examine, and perhaps the most important in a practical sense, is a system in which the excitation comes from a rotating (or reciprocating) mass. The number of problems that come under this heading are enormous. One common example is that of a four-cylinder engine. Unlike engines with six and eight cylinders, four-cylinder engines cannot be perfectly balanced and will always tend to vibrate due to the motion of the pistons and crankshaft. This vibration couples into the body through the engine mounts and causes a host of vibration problems for the car designer and for the driver. Another example is

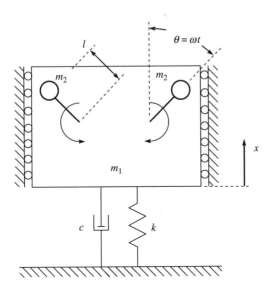

**Figure 2.20** Spring-mass system with rotating imbalance

a clothes washer. Most of us have seen what happens when the washer enters the spin cycle and the wet clothes have all gathered into a compact mass. In extreme cases the washing machine actually rocks its way across the floor because of the resulting vibrations. Even a rotating shaft can experience an inertially driven excitation if the shaft's center of mass isn't precisely centered. If it's off by a bit, then this offset mass will excite vibrations in the shaft just as an independently rotating mass would. You can come up with any number of other examples. The basic example of a rotating excitation is shown in Figure 2.20. For this problem we have a base mass ($m_1$) and two counter-rotating masses ($m_2$). The rotation rate is constant and equal to $\omega$.

In order to find the equations of motion for this system, we need to break it up into its component parts. The appropriate free-body diagrams are shown in Figure 2.21. Note that since the two arms are counter-rotating, we will have no excitations in the $y$ direction, just in the $x$. Because

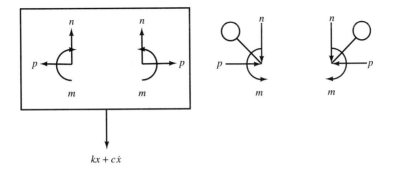

**Figure 2.21** Free-body diagram for a spring-mass system with rotating imbalances.

the problem is symmetric, we know that the forces due to the arm on the left will be the same as those for the right arm, canceling each other in the $y$ direction and adding together in the $x$.

Applying Newton's second law for the main mass gives us

$$m_1\ddot{x} = -kx - c\dot{x} + 2n \tag{2.8.1}$$

For completeness, the forces $p$ and moments $m$ are also shown in Figure 2.21, but, as just mentioned, they're equal and opposed and so will not affect the equation of motion for the main mass.

We need to look at either rotating arm to determine the unknown interaction force $n$. Taking the right arm, we have

$$m_2 a_x = -n \tag{2.8.2}$$

where $a_x$ is the acceleration of the center of mass of the arm in the $x$ direction. Since the arm is rotating at a constant rate with respect to the main mass, the only acceleration term is the centripetal acceleration, which is equal to $l\omega^2$ and is directed toward the arm's attachment point to the body. Thus the total acceleration of the arm's center of mass is equal to $\ddot{x} - l\omega^2\cos(\omega t)$. Therefore

$$m_2\big(\ddot{x} - l\omega^2\cos(\omega t)\big) = -n \tag{2.8.3}$$

Combining both (2.8.1) and (2.8.3) yields the final equation

$$(m_1 + 2m_2)\ddot{x} + c\dot{x} + kx = 2m_2 l\omega^2\cos(\omega t) \tag{2.8.4}$$

Although two counter-rotating masses were used in this example so that only forces in the $x$ direction were present, we could have easily considered just a single rotating force as long as the body was constrained to only allow motion in the $x$ direction. In this case we'd have had constraint forces between the moving body and the rigid walls. Thus, whether we really have two counter-rotating masses or simply a single rotating mass with freedom to move in only one direction, the end result in terms of the moving mass will be the same. If we wish to consider a single rotating mass, then we could use $m$ as the total rotating mass component (replacing our previous $2m$). Many authors also use $e$ for the distance that $m$ is from the center of rotation. Making both these substitutions will give us

$$(m_1 + m)\ddot{x} + c\dot{x} + kx = me\omega^2\cos(\omega t) \tag{2.8.5}$$

The unique aspect of this problem is that the rotational rate appears to the second power in the forcing term. This has a significant effect on the overall response, as we'll now see. If we let the forcing in (2.8.5) be given by $me\omega^2 e^{i\omega t}$ (using a complex representation and remembering that the actual excitation is the real part of $me\omega^2 e^{i\omega t}$) and assume a solution of the form $x(t) = \bar{x}e^{i\omega t}$, then after canceling the common exponential factor and solving for $\bar{x}$ we'll get

$$\bar{x} = \frac{me\omega^2}{-(m_1 + m)\omega^2 + k + i\omega c} \tag{2.8.6}$$

Both the direct-force excited and seismically excited systems had a finite amplitude response at low frequencies and had responses that went to zero as the forcing frequency became very large. But in the present example we see that the amplitude response goes to zero as the forcing frequency becomes small and approaches a finite value $\left(\dfrac{-me}{m_1 + m}\right)$ as the forcing frequency becomes large.

By dividing both numerator and denominator by $m_1 + m$, we obtain our final form, involving the system's natural frequency and damping factor

$$\bar{x} = \frac{e\beta\omega^2}{\omega_n^2 - \omega^2 + 2i\zeta\omega\omega_n} \tag{2.8.7}$$

where $\beta = \dfrac{m}{m_1 + m}$. $\omega_n$ and $\zeta$ are defined with respect to the the total mass, $m_1 + m$. We can make this a bit more general by using a nondimensional frequency, $\Omega = \dfrac{\omega}{\omega_n}$, giving us

$$\bar{x} = \frac{e\beta\Omega^2}{1 - \Omega^2 + 2i\zeta\Omega} \tag{2.8.8}$$

The normalized magnitude response is given by $\dfrac{|\bar{x}|}{e\beta}$ and this is plotted versus $\Omega$ for a variety of damping values in Figure 2.22.

At this point it is interesting to draw a parallel between the response of a rotating imbalance problem and the response of our direct-force excited system. You'll recall that the amplitude response of the two situations differs by a factor of $\omega^2$ (and a constant term). So it seems that no parallel exists. However, this conclusion is premature. What if we're not really interested in the amplitude response of the system, but rather in the acceleration? This would certainly be the case if we were concerned with limiting the accelerations that our system undergoes, something

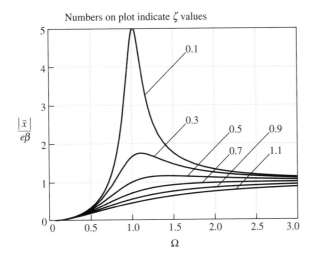

**Figure 2.22**  Normalized magnitude response for damped, rotating imbalance (varying $\zeta$)

that every engineer involved with packaging and shipping worries about. In this case we'll be looking at

$$a(t) = \ddot{x}(t) = \frac{d^2}{dt^2}(\bar{x}e^{i\omega t}) = -\omega^2 \bar{x}e^{i\omega t} \tag{2.8.9}$$

We've already found $\bar{x}$ (2.6.15), and using this result can find the magnitude of the acceleration ($\bar{a}$)

$$\bar{a} = \frac{-\frac{\bar{f}}{m}\omega^2}{\omega_n^2 - \omega^2 + 2i\zeta\omega\omega_n} \tag{2.8.10}$$

As you can see, (2.8.10) and (2.8.7) have the same form. Thus the acceleration response of a direct-force excited system behaves in the same manner as a system with a rotating imbalance. Actually, this shouldn't be all that surprising, since the force due to the rotating mass involves the acceleration of that mass.

**Example 2.7**

**Problem**   One of the most common examples of a rotating imbalance is associated with an automobile's tire/wheel combination. If you own a car you probably know that tire stores will offer to balance your tires for you, whenever you purchase new ones. The way in which the balance is effected is by adding little lead weights to the inside rim of the wheel. If this wasn't done, the system would behave exactly as predicted above, leading to a resonant response at certain rotation rates, which, for a car, means at certain forward velocities. Consider the system illustrated in Figure 2.23. The circle represents the tire/wheel system, $k_2$ represents the stiffness of the spring connecting the wheel to the car, and $k_1$ represents the stiffness of the tire. $c$ represents any damping in the system. We want to determine the response of the tire as a function of frequency.

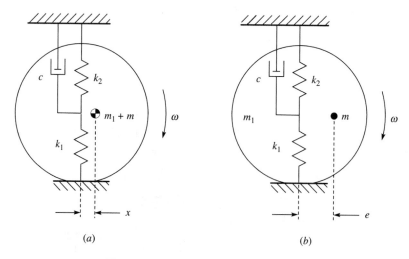

**Figure 2.23**   Rotating tire/wheel assembly

**Solution**   We'll look at the problem in two ways, since this kind of problem often causes confusion. The first viewpoint is shown in Figure 2.23$b$. We'll consider the mass of the tire/wheel to be $m_1$ and let $m$ be the mass of the rotating imbalance. This matches our analyses from above. It may occur to you, however, that tire/wheel systems don't really have rotating imbalances: the problem is simply that, due to manufacturing imperfections, the center of mass isn't exactly at the center of the wheel. Thus the problem is more correctly one in which the mass of the system being excited is zero and the rotating imbalance is actually the mass of the entire system (shown in Figure 2.23$b$). As we'll see, both these viewpoints will lead to the same result.

The key to understanding how both views are related is to ask where the center of mass is for the *system* of a perfectly balanced, tire/wheel combination of mass $m_1$ and a rotating imbalance of mass $m$ located $e$ away from the center. $h$ is the location of the total center of mass. From elementary statics we know that

$$hm_1 = (e - h)m \tag{2.8.11}$$

which gives us

$$x = \frac{em}{m_1 + m} \tag{2.8.12}$$

Thus the location of the center of mass for Figure 2.23$a$ is now known. Applying (2.8.6) for case ($b$) gives us

$$\bar{x} = \frac{me\omega^2}{-(m_1 + m)\omega^2 + k_1 + k_2 + c\omega i} \tag{2.8.13}$$

If we apply (2.8.6) for case ($a$) (remembering what $h$ is equal to), we'll get

$$\bar{x} = \frac{(m_1 + m)\frac{em}{m_1 + m}\omega^2}{-(0 + m_1 + m)\omega^2 + k_1 + k_2 + c\omega i} \tag{2.8.14}$$

which becomes

$$\bar{x} = \frac{me\omega^2}{-(m_1 + m)\omega^2 + k_1 + k_2 + c\omega i} \tag{2.8.15}$$

Thus we see that both views give us the same result. We've either got a small imbalance exciting the overall system with a large moment arm or a large imbalance exciting the system with a small moment arm. The final equations are identical.

**Example 2.8**

**Problem**   Having analyzed an imbalanced tire in Example 2.6, let's look at some particular parametric values. We'll consider the case for which the suspension stiffness $k_2$ is equal to 20,000 N/m, the tire stiffness is equal to 180,000 N/m, the mass of the tire/wheel assembly is 15 kg, the mass of the stationary pieces (brakes, struts, etc.) is 20 kg, and the deviation of the center of mass from the center of the wheel is 2 mm. The distance from the ground to the center of the wheel's hub is .28 m. The damper for the car has been damaged and only gives us 265 N/m/s. We'll determine what the maximum vibration amplitude will be for this system and at what vehicle speed it will occur.

**Solution**   Using (2.8.6) gives us

$$\bar{x} = \frac{(.002)(15)\omega^2}{-(35)\omega^2 + 200,000 + 265\omega i} \tag{2.8.16}$$

Note that the mass in the numerator is 15 kg because this is the total rotating mass, whereas the mass in the denominator is 35 kg (the total mass, including rotating, and nonrotating parts). To find the maximal response, we can differentiate $\bar{x}$ with respect to $\omega$, set the resultant equal to zero, and solve for $\omega$. An easier way to get the same result is to square $\bar{x}$ and differentiate with respect to $\omega^2$. The peak response will occur at the same frequency, but the mathematical manipulations are simpler in the second approach.

Carrying through will finally yield (for small $\zeta$)

$$\omega_{\text{rp}} = \frac{\omega_n}{\sqrt{1 - 2\zeta^2}} \tag{2.8.17}$$

The subscript rp stands for rotating peak.

For this problem $\omega_n^2 = \dfrac{200,000}{35}$ and so $\omega_n = 75.59$ rad/s. From $\frac{c}{m} = 2\zeta\omega_n$ we can find that $\zeta = .05$. Thus $\omega_{\text{rp}} = 75.78$ rad/s. Using this in (2.8.16) gives us a response magnitude of

$$|\bar{x}| = 8.57 \text{ mm} \tag{2.8.18}$$

Since the distance from the ground to the hub of the wheel is .28 m, we can find the car's velocity from

$$v = (75.78 \text{ rad/s})(.28 \text{ m}) = 21.2 \text{ m/s} \tag{2.8.19}$$

or 47 mph, for those of you still accustomed to English units.


## 2.9 IDENTIFICATION OF DAMPING AND NATURAL FREQUENCY

A topic of great practical importance is that of identification. In the real world, nobody is going to tell you what the damping factor or natural frequency is for the system you're examining. Thus it's up to you to figure these quantities out for yourself. The field of system identification is a huge one, and we won't even begin to cover it in this book. However, we will introduce a few simple techniques that can give us good approximations to the quantities we're looking for.

The first response type we'll look at is the unforced system response. If the damping is relatively low, we can get a good estimate of both the natural frequency and the damping factor by examining just how the response decays after an initial excitation. If the damping is zero, we know that the natural frequency is simply the frequency of oscillation of the system. And if the damping is small, then the damped oscillation frequency $\omega_d$ (which is equal to $\omega_n\sqrt{1 - \zeta^2}$) is very close to $\omega_n$. We can, however, do much better than simply approximating our answers. Let's recall our solution to the unforced problem (1.4.15). If we include only the cosine term (implying an initial displacement with zero initial velocity), then at $t = 0$ the response is equal to $b_1$. Next, look at the response one period later, that is at $t = t_p = \dfrac{2\pi}{\omega_d}$. Now the response is equal to

$$x(t_p) = b_1 e^{\frac{-2\pi\zeta}{\sqrt{1-\zeta^2}}} \tag{2.9.1}$$

If we take the ratio of these two values, we obtain

$$\frac{x(0)}{x(t_p)} = e^{\frac{2\pi\zeta}{\sqrt{1-\zeta^2}}} \tag{2.9.2}$$

Taking the natural log of both sides yields

$$\ln\left(\frac{x(0)}{x(t_p)}\right) = \frac{2\pi\zeta}{\sqrt{1-\zeta^2}} \tag{2.9.3}$$

This result is known as the *logarithmic decrement*, and is usually denoted by $\sigma$:

$$\sigma \equiv \ln\left(\frac{x(0)}{x(t_p)}\right) \tag{2.9.4}$$

Although we derived this from a signal that was maximal at $t = 0$, it can be found easily by taking the ratio of any two successive peaks. Once we have $\sigma$, it becomes a simple matter to solve for $\zeta$:

$$\zeta = \frac{\sigma}{\sqrt{4\pi^2 + \sigma^2}} \tag{2.9.5}$$

This is further simplified if the damping factor is small. In this case the denominator is closely approximated by $2\pi$ and our expression for $\zeta$ becomes

$$\zeta = \frac{\sigma}{2\pi} \tag{2.9.6}$$

If we wish to use more oscillations, the formula need be changed only slightly. As long as we're letting an integer number of oscillations elapse between samples, the cosine term is always equal to 1.0. Thus the only change is that the response has decayed due to the exponential envelope. If we had waited for $n$ periods, the response amplitude would have been

$$x(nt_p) = b_1 e^{\frac{-2n\pi\zeta}{\sqrt{1-\zeta^2}}} \tag{2.9.7}$$

In this case, the ratio becomes

$$\frac{x(0)}{x(nt_p)} = e^{\frac{2n\pi\zeta}{\sqrt{1-\zeta^2}}} \tag{2.9.8}$$

Taking the natural log of this ratio leads to

$$\ln\left(\frac{x(0)}{x(nt_p)}\right) = \frac{2n\pi\zeta}{\sqrt{1-\zeta^2}} \tag{2.9.9}$$

This is equal to $n$ times the log decrement as determined from two successive peaks. Thus

$$\sigma = \frac{1}{n}\ln\left(\frac{x(0)}{x(nt_p)}\right) \tag{2.9.10}$$

Once $\sigma$ has been determined, we would find $\zeta$ using either (2.9.5) or (2.9.6). This way of determining $\zeta$ is also a very good means of determining whether the damping for your particular system is linear or not. If it is linear then the damping factor found between peak one and two of the response should be the same as that found between peak two and three. If the results differ, then this implies that the damping isn't linear.

What we've now seen is that a free vibration test will allow us to identify the damping factor in a single-degree-of-freedom system by means of the log decrement. Once the damping factor is known, we can find the natural frequency because we know the damped frequency of oscillation ($\omega_d = \omega_n\sqrt{1 - \zeta^2}$). However, we're not limited to free vibration tests. We can glean the same information from a forced test. What we're looking for in these tests are the transfer functions of the system. We can obtain these transfer functions in a number of ways, and we'll look at some of them in Chapter 8. For now we'll just assume that somehow we've obtained them, and they're available for use in our identification tests.

If you look at Figure 2.15$b$ you'll note that, for the direct-force-excited system, the phase angle goes through $-90$ degrees exactly at the system's natural frequency, regardless of the damping value. Thus, all we need do is look at the phase, determine when it goes through $-90$ degrees, and read off the frequency at which this occurs. This will give us $\omega_n$.

Interestingly, there are even more indicators of $\omega_n$ and $\zeta$. The first one is found by comparing the actual peak amplitude response of the system to the response at zero frequency. From (2.5.29) we know that the transfer function of a direct-force-excited system is given by

$$|g(\omega)| = \frac{1}{m\sqrt{(\omega_n^2 - \omega^2)^2 + (2\zeta\omega_n\omega)^2}} \qquad (2.9.11)$$

Thinking back to your first calculus class, you'll recall that the slope of a curve is zero at a local maximum. Thus, to determine the frequency at which our amplitude-response curve is a maximum (which we'll call $\omega_p$), we need only differentiate with respect to $\omega$ and set the result equal to zero. Actually, although this approach will give us the correct answer, we can simplify matters even further. Rather than differentiating with respect to $\omega$, we can differentiate with respect to $\omega^2$, since only even powers of $\omega$ show up in the equation. This will give us $\omega_p^2$ instead of $\omega_p$. We can easily take the square root to find $\omega_p$. However, we can simplify yet further. Taking the derivative of the reciprocal of the square root of a function is somewhat complicated. So instead, let's square $g(\omega)$. Since the function is positive everywhere anyway, this won't change our results. And finally, rather than looking at $g^2(\omega)$, let's differentiate $\frac{1}{g^2(\omega)}$. Thus, instead of looking for the maximum of $g(\omega)$ (shown in Figure 2.24), we'll be looking for the minimum of $h(\omega^2) \equiv \frac{1}{g^2(\omega)}$ (Figure 2.25). Thus we want

$$\frac{dh(\omega^2)}{d\omega^2} = 0 \qquad (2.9.12)$$

or

$$2(\omega_n^2 - \omega^2)(-1) + (2\zeta\omega_n)^2 = 0 \qquad (2.9.13)$$

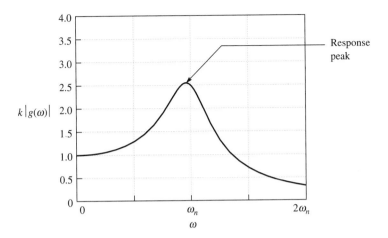

**Figure 2.24**   Typical underdamped response showing a peaked response

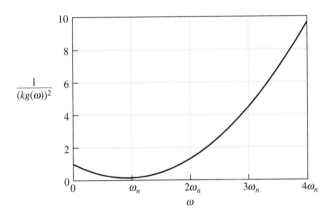

**Figure 2.25**   Graph whose minimum corresponds to the maximum amplitude response

Solving for $\omega_p$, we obtain

$$\omega_p = \omega_n\sqrt{1 - 2\zeta^2} \qquad (2.9.14)$$

For small values of damping, $\omega_p$ is approximately equal to $\omega_n$. So another way of estimating $\omega_n$ for lightly damped systems is to use the frequency at which the peak response occurs.

We can also use this result to verify what we can see from Figure 2.15a. We've already mentioned that the response increases as we approach $\omega_n$ when the damping is small, but decreases as we approach $\omega_n$ when the damping is large. We can also see from Figure 2.15a that the peak moves to the left as the damping increases.

All these observations are reflected in (2.9.14). Clearly, as $\zeta$ is increased, $\omega_p$ drops. Furthermore, when $\zeta = \frac{1}{\sqrt{2}}$, $\omega_p = 0$, that is, the peak occurs at $\omega = 0$. For values of the damping factor that are greater than $\frac{1}{\sqrt{2}}$, no peak occurs, and the response amplitude decays monotonically with increasing frequency.

Next, consider the ratio of the peak response amplitude to that at zero frequency $\frac{|g(\omega_p)|}{|g(0)|}$. Recalling that the response amplitude for a direct-force-excited system (2.5.28) is

$$|\bar{x}| = \frac{\bar{f}}{m} \frac{1}{\sqrt{(\omega_n^2 - \omega^2)^2 + (2\zeta\omega_n\omega)^2}} \qquad (2.9.15)$$

we see that this ratio is equal to

$$\frac{\dfrac{\bar{f}}{m} \dfrac{1}{\sqrt{(\omega_n^2-\omega_p^2)^2+(2\zeta\omega_n\omega_p)^2}}}{\dfrac{\bar{f}}{m\omega_n^2}} \qquad (2.9.16)$$

We've already found that $\omega_p = \omega_n\sqrt{1 - 2\zeta^2}$, and if we substitute this into (2.9.16) and simplify, we'll obtain

$$\frac{|g(\omega_p)|}{|g(0)|} = \frac{1}{2\zeta\sqrt{1 - \zeta^2}} \qquad (2.9.17)$$

Since $\zeta^2 \ll 1$ for systems in which the response is appreciably peaked at $\omega_p$, we can neglect the $\zeta^2$ term and finally obtain

$$\frac{|g(\omega_p)|}{|g(0)|} = \frac{1}{2\zeta} \qquad (2.9.18)$$

Now this is a nifty result. It tells us that if we've got the transfer function for our system, we can determine the damping factor just by finding the peak response and dividing by the zero frequency (or static) response.

The final way of finding the damping factor is to look at the *half-power points* of the response curve. By determining the frequencies at which the amplitude of the response drops down by a factor of $\sqrt{2}$ from its peak value, we can determine the damping factor. It certainly isn't obvious why this should be so, but it is, as we'll now demonstrate.

We've already determined that the peak response of a mass-forced system occurs at $\omega_p = \omega_n\sqrt{1 - 2\zeta^2}$. What we now want is to determine the frequencies, $\omega_1$ and $\omega_2$, at which the response amplitude is equal to $\frac{1}{\sqrt{2}}|g(\omega_p)|$. We'll call our unknown half-power frequencies $\omega_h$. Using our previous results, we see that we'll need to solve

$$\frac{\bar{f}}{m} \frac{1}{\sqrt{(\omega_n^2 - \omega_h^2)^2 + (2\zeta\omega_n\omega_h)^2}} = \frac{\bar{f}}{m} \frac{1}{2\omega_n^2\zeta\sqrt{(2 - 2\zeta^2)}} \qquad (2.9.19)$$

The term on the left is the general expression for the amplitude at a given frequency (2.5.28), while the term on the right is equal to $\frac{1}{\sqrt{2}}$ times the peak response amplitude (from 2.9.17).

To solve this equation we can square both sides, cross multiply (to get everything in the numerator), and end up with the equation

$$\omega_h^4 + \omega_h^2(4\zeta^2\omega_n^2 - 2\omega_n^2) + \omega_n^4 - 8\omega_n^4\zeta^2(1 - \zeta^2) = 0 \qquad (2.9.20)$$

If we solve this quadratic expression, we'll get two values for $\omega_h^2$

$$\omega_{h_1}^2 = \omega_n^2(1 - 2\zeta^2) - 2\omega_n^2\zeta\sqrt{1 - \zeta^2} \qquad (2.9.21)$$

and

$$\omega_{h_2}^2 = \omega_n^2(1 - 2\zeta^2) + 2\omega_n^2\zeta\sqrt{1 - \zeta^2} \qquad (2.9.22)$$

Once we take their square roots, we can subtract $\omega_{h_1}$ from $\omega_{h_2}$. If we also neglect terms that involve $\zeta^2$ or higher powers (since they're negligible compared to $\zeta$ terms), we finally obtain

$$\omega_{h_2} - \omega_{h_1} = 2\omega_n\zeta \qquad (2.9.23)$$

Defining $\delta_h$ as

$$\delta_h \equiv \omega_{h_2} - \omega_{h_1} \qquad (2.9.24)$$

we can write

$$\zeta = \frac{\delta_h}{2\omega_n} \qquad (2.9.25)$$

Here is yet another way to determine $\zeta$. For this identification you must find the frequency of the peak response, find the frequencies at which the response has dropped down by a factor of $\sqrt{2}$, take their difference, and divide by $2\omega_n$ in order to obtain the damping factor. Note that this method is tougher to accomplish in actual practice than the peak-response method because the slope of the response curve is quite steep at the half-power points when the damping is low. Thus it's going to be hard to find the half-power frequencies very accurately.

**Example 2.9**

**Problem**    Consider the response shown in Figure 2.26. This response corresponds to a spring-mass damper that has been struck by an impact hammer, resulting in the vibrations shown. Determine the damping factor by means of the log decrement.

**Solution**    The first peak looks to be about .093 and the second is around .068. Thus $\sigma = \ln\left(\dfrac{.093}{.068}\right)$
$= .313$. Equation (2.9.5) lets us solve for $\zeta$ and leads to an estimate of $\zeta = .05$.

    If we try to estimate $\zeta$ from two successive peaks, we can use the first peak of .093 and the third peak of .05. This will produce a $\sigma$ of .310, which will give a $\zeta$ of .049.

    Both these estimates are excellent, as the equation of motion used to generate the illustrated response was

$$\ddot{x} + \dot{x} + 100x = 0 \qquad (2.9.26)$$

for which $\omega_n = 10$ and $\zeta = .05$.

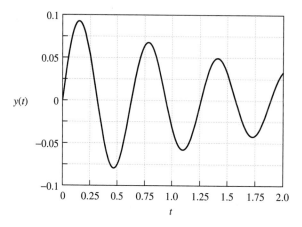

**Figure 2.26** Free response of SDOF system

**Example 2.10**

In this example we'll try to determine the damping factor of a spring-mass damper under direct-force excitation through use of the half-power points. Figure 2.27*b* shows the absolute value of the transfer function as a function of frequency. Calculate the desired parameter values.

**Solution**    We know that the magnitude of the transfer function for a direct-force-excited system is given by

$$|g(\omega)| = \frac{\bar{f}}{m} \frac{1}{\sqrt{(\omega_n^2 - \omega^2)^2 + (2\zeta\omega\omega_n)^2}} \tag{2.9.27}$$

We also know that the damping factor is approximated by $\dfrac{\delta_h}{2\omega_n}$ where $\delta_h$ is the difference in frequency of the two half-power points.

From the graph we can estimate that $|g|_{\max} = 12.5$ and occurs at 20 rad/s. The half-power points correspond to an amplitude of $\dfrac{12.5}{\sqrt{2}} = 8.84$. The frequencies corresponding to this are around

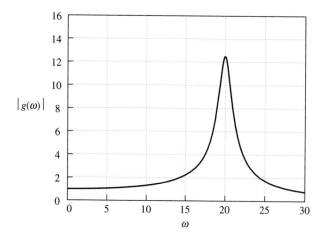

**Figure 2.27** Absolute value of transfer function for underdamped spring-mass-damper

20.8 and 19.1. Thus we have $\delta_h = 20.8 - 19.1 = 1.7$ rad/s. We can then find $\zeta$ from the half-power formula

$$\zeta = \frac{1.7}{(2)(20)} = .0425 \qquad (2.9.28)$$

We can validate this result by using the ratio of $|g_{max}|$ to $|g_0|$:

$$\frac{1}{2\zeta} = \frac{|g_{max}|}{|g_0|} = \frac{12.5}{1} = 12.5 \qquad (2.9.29)$$

Solving for $\zeta$ gives us $\zeta = .04$, which is in close agreement with the damping factor value we found from the half-power analysis.

Just for interest, the values used to generate Figure 2.27 were $\bar{f} = 400$ N, $m = 1$ kg, $\omega_n = 20$ rad/s, and $\zeta = .04$.

## 2.10 OTHER TYPES OF DAMPING

It's already been mentioned that viscous damping, although convenient from a mathematical point of view, doesn't really model the damping found in actual structures very well. What we'll do in this section is look at other damping models and see how their inclusion affects the response of our systems.

The first type of damping we'll look at is dry-friction (also called Coulomb damping). You've undoubtedly already seen this type of damping in your physics and dynamics classes. The simplest dry-friction model assumes that two contacting bodies that are moving relative to each other will experience a constant friction force that always acts to oppose the motion. Figure 2.28 illustrates this behavior. The magnitude of the force is equal to $\mu W$, where $W$ is the weight of the object that is sliding, and $\mu$ is the coefficient of friction between the object and the surface it's sliding on. This force can be represented mathematically as

$$F_d = \mu W \text{sign}(\dot{x}) \qquad (2.10.1)$$

where $\dot{x}$ is the relative velocity and sign($-$) indicates the sign function (equal to 1.0 when the argument is greater than zero and $-1.0$ when the argument is less than zero).

If we use this sort of damping with our spring-mass system, our equation of motion will be

$$m\ddot{x} + d\text{sign}(\dot{x}) + kx = 0 \qquad (2.10.2)$$

In the above equation we've used $d$ to stand for $\mu W$ for simplicity.

The free-vibration behavior of this system is rather interesting. Since the damping force is constant (as long as the velocity's sign doesn't change), we can rewrite (2.10.2) as

$$\begin{aligned} m\ddot{x} + kx = -d, \qquad \dot{x} > 0 \\ m\ddot{x} + kx = d, \qquad \dot{x} < 0 \end{aligned} \qquad (2.10.3)$$

What this tells us is that we can consider the system to be a piecewise, linear, forced system, that is, the system looks like a linear spring-mass system being driven by a constant force. The only

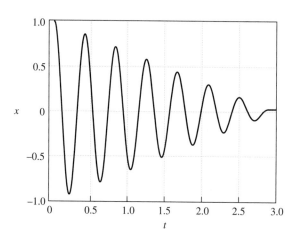

**Figure 2.28**  Force versus velocity relationship for dry friction

difference between this system and a completely linear one is that the forcing changes sign when the velocity changes sign. If we give this system an initial displacement and then release it, the envelope of the resulting oscillations is itself linear. This linear decay of amplitude is quite different from what we saw with linear damping, which produced an exponentially decaying envelope for the oscillations. A typical plot of a dry-friction damped response is shown in Figure 2.29. Note that the oscillations stop in a finite time, whereas in the case of linear damping the oscillations continue

**Figure 2.29**  Time response of a dry-friction damped system forever (although they get very small). Also note that the response stops at a nonzero value. This is because the oscillations eventually become so small that the spring force is unable to overcome the dry-friction force. Thus the mass gets "stuck" at some deflected position. The precise position will depend upon the initial conditions.

A dry-friction model is often used to approximate the damping that occurs when parts rub against each other. Thus if you were trying to estimate the damping that might exist in a large-space structure, something made up of many truss elements, you might try to estimate the dry-friction losses due to relative motions of the truss elements with the joint assemblies that hold them together.

**Example 2.11**

**Problem**   Figure 2.29 shows the response of a spring-mass system with a dry-friction interface. Determine the linear-damping estimates that come from examining the change in the peak response between peak 1 and 2, peak 2 and 3, and peak 1 and 3. Is the linear estimate found from two oscillation cycles (peak 1 to 3) equal to the average of that found for the peak 1 to 2 and peak 2 to 3 estimates?

**Solution**   We know from (2.9.5) that

$$\zeta = \frac{\sigma}{\sqrt{4\pi^2 + \sigma^2}} \tag{2.10.4}$$

and from (2.9.10) that

$$\sigma = \frac{1}{n}\ln\left(\frac{x(0)}{x(nt_p)}\right) \tag{2.10.5}$$

Using the data from Figure 2.29 we see that

$$\sigma_{1-2} = \ln\left(\frac{.87}{.73}\right) = .175 \tag{2.10.6}$$

and therefore

$$\zeta_{1-2} = .028 \tag{2.10.7}$$

Looking between peak 2 and 3 gives us

$$\sigma_{2-3} = \ln\left(\frac{.73}{.59}\right) = .213 \tag{2.10.8}$$

Thus we have

$$\zeta_{2-3} = .034 \tag{2.10.9}$$

Lastly, the log decrement between peak 1 and 3 is

$$\sigma_{1-3} = \ln\left(\frac{.87}{.59}\right) = .388 \tag{2.10.10}$$

and the associated linear damping estimate is

$$\zeta_{1-3} = .062 \tag{2.10.11}$$

We can immediately see that the damping couldn't have been linear in the original system. If the damping had been linear, all the linear damping estimates would have been identical (assuming no experimental error), as would all the log decrements. What can be seen in these results is that the log decrement over two cycles is equal to the sum of the log decrements from peak 1 to 2 and from peak 2 to 3. Furthermore, the linear damping estimate from peak 1 to 3 is equal to the sum of those from peak 1 to 2 and from peak 2 to 3. Thus now we see what trend should tip us off to the presence of dry friction. If the damping estimates increase as time increases (as ours did, going from .028 to .034), then we may have dry friction. If the overall damping estimate between three peaks is equal to the sum of those found from the two adjacent peaks, then we have even more reason to suspect dry friction. Of course, we may have a combination of linear and dry friction, so we must try to identify the amount of each particular component.

The other kind of damping we'll consider is called *structural* or *hysteretic* damping. This sort of damping does a reasonable job of approximating the sort of energy loss that is present in materials that are repeatedly loaded and unloaded. In testing for this type of damping, a specimen is cyclically loaded, and the testing machine records the stress developed during the loading. If the system has internal damping, the response of such a test will look like that shown in Figure 2.30. When the damping test is run, the curve is traversed in a clockwise manner. After one full cycle of testing, the material's state is back where it started. However, the fact that there is a finite area enclosed by the curve implies that energy was lost during the test. If no damping had been present, the system would have unloaded along the same path that it took when loading was applied. When materials unload along a different path (as is illustrated), the effect is called *hysteresis*, which tells us why this damping is also called hysteretic damping.

You can see that the area within the stress-strain curve is proportional to energy loss when you realize that stress is a force quantity and strain is a displacement quantity. Thus the integral represents the product of force and displacement, that is, a work term. Therefore, the area within the curve is proportional to the work done on the specimen and indicates how much energy was dissipated by the specimen during the test.

It turns out that the stress-strain curve doesn't depend upon frequency (at least within a reasonable range of frequencies). You can also see that doubling the amplitude of the strain will double the entire curve (both horizontally and vertically). The result of this is that the energy dissipated over one cycle will be proportional to the square of the amplitude of the motion. (The

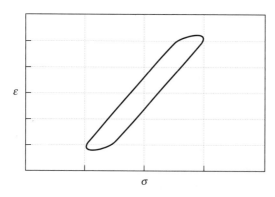

**Figure 2.30** Stress-strain curve for hysteretically damped material

energy is proportional to the area enclosed in the curve and the area is proportional to the square of the amplitude.)

Our next question is how we can deal with this kind of damping. We'd like to come up with the equivalent viscous damping that will match the hysteretic damping in some sense. The most convenient approach to this question is to try to get an equivalent viscous damper that would dissipate the same amount of energy over one cycle as the hysteretically damped material would. Therefore, our immediate task is to find out how much energy is dissipated over a cycle by a viscous damper.

As usual, we'll use complex notation to obtain our solution. We can assume that the forcing is actually equal to $\bar{f} \cos(\omega t)$ (with $\bar{f}$ real), and thus our solution is given by the real part of $x$ that results from solving

$$m\ddot{x} + c\dot{x} + kx = \bar{f}e^{i\omega t} \tag{2.10.12}$$

We've solved this problem earlier in the chapter, and the complete solution is given by

$$x(t) = \bar{f}|g(\omega)|e^{(i\omega t - \phi)} \tag{2.10.13}$$

where $|g|(\omega)$ and $\phi$ are defined in (2.6.8) and (2.6.9). The velocity is found by differentiating the displacement response, yielding

$$\dot{x}(t) = i\omega \bar{f}|g(\omega)|e^{(i\omega t - \phi)} \tag{2.10.14}$$

Now that we have expressions for the applied force and the velocity we can determine the energy dissipated over one cycle. The work done over one cycle is given by

$$\Delta w = \int f \, dx \tag{2.10.15}$$

where the integral is taken over the entire displacement history. We can reexpress this in terms of time by realizing that $dx = \dot{x}dt$. Thus our work equation becomes

$$\Delta w = \int_0^{\frac{2\pi}{\omega}} f\dot{x} \, dt \tag{2.10.16}$$

Combining (2.10.14) and (2.10.16) gives us

$$\Delta w_v = \int_0^{\frac{2\pi}{\omega}} \mathrm{re}(\bar{f}e^{i\omega t}) \cdot \mathrm{re}(i\omega \bar{f}|g|e^{(i\omega t - \phi)}) \, dt \tag{2.10.17}$$

$$= -\bar{f}^2|g|\omega \int_0^{\frac{2\pi}{\omega}} \cos(\omega t) \sin(\omega t - \phi) \, dt \tag{2.10.18}$$

$$= \bar{f}^2|g|\pi \sin(\phi) \tag{2.10.19}$$

where re(-) indicates that the real part of the argument is being considered and $w_v$ indicates that the energy dissipated is due to viscous damping. Recalling (2.6.9), we can find $\sin(\phi)$ to be

$$\sin(\phi) = \frac{2\zeta \omega \omega_n}{\sqrt{(\omega_n^2 - \omega^2)^2 + (2\zeta \omega \omega_n)^2}} \tag{2.10.20}$$

Comparing this with (2.6.8) gives us

$$\sin(\phi) = 2\zeta\omega\omega_n m|g| \tag{2.10.21}$$

Thus (2.10.19) becomes

$$\Delta w_v = (\bar{f}|g|)^2 \pi c\omega \tag{2.10.22}$$

since $2\zeta\omega_n = \frac{c}{m}$. Finally, $\bar{f}|g|$ is simply the magnitude of the resultant oscillations $|\bar{x}|$. Thus we have

$$\Delta w_v = |\bar{x}|^2 \pi c\omega \tag{2.10.23}$$

This result tells us that the energy absorbed due to viscous damping over one cycle of loading is proportional to the square of the amplitude of the response ($|\bar{x}|^2$), and it varies linearly with the damping coefficient $c$ and with the frequency of the test $\omega$. Thus, if we double the frequency, we'll double the energy absorbed. This makes sense since the force of a viscous damper is proportional to velocity, and doubling $\omega$ will double the velocity. It's also logical that the amount of damping will vary with the damping coefficient. You'd expect that doubling the damping coefficient should double the energy absorbed.

Now that we've seen how energy is consumed by a viscous damper, we can go ahead and determine what an equivalent viscous damper might be for hysteretic damping. We've already mentioned that the energy absorbed by a hysteretic damper is proportional to the square of the oscillation amplitude and isn't proportional to the forcing frequency, while we've just seen that the energy absorbed by a viscous damper has a frequency dependence. If the damping force generated by the viscous damper had been proportional to displacement, instead of velocity, then we wouldn't have had a frequency dependence. However, a force that's simply proportional to displacement is a spring force and doesn't dissipate energy. It's necessary for the force to be out of phase with respect to the displacement for energy dissipation to occur. Realizing this, why don't we assume that hysteretic damping is proportional to displacement, but is out of phase with it. A way to do this would be to use complex exponentials as our solutions and to assume that the force developed is proportional to $ix$ (with the spring force proportional to $x$ and a viscous damper's force proportional to $i\omega x$). Of course, this would be tough to justify on physical gounds. However, let's run with this idea anyway and see where it leads.

If we take the damping force to be $ik\gamma x$, then our equation of motion becomes

$$m\ddot{x} + kx + ik\gamma x = f \tag{2.10.24}$$

Since the spring and damper are both proportional to displacement, we can combine them as follows:

$$m\ddot{x} + (1 + i\gamma)kx = f \tag{2.10.25}$$

$k(1 + i\gamma)$ is called the *complex stiffness* and $\gamma$ is the *structural damping factor*.

We can quickly go through the same sort of analysis we've just finished for the viscous case in order to find how much energy is dissipated by hysteretic damping. If we let $f = \bar{f}e^{i\omega t}$ (with

$\bar{f}$ real) and $x = \bar{x}e^{i\omega t}$, we can solve (2.10.25) to find

$$\bar{x} = \frac{\bar{f}}{k - \omega^2 m + i\gamma k} \tag{2.10.26}$$

Multiplying the real part of the force and the real part of the velocity we can form the dissipated energy

$$\Delta w_s = \int_0^{\frac{2\pi}{\omega}} \bar{f}\cos(\omega t)(-\omega\bar{f})\frac{(k - \omega^2 m)\sin(\omega t) - \gamma k\cos(\omega t)}{(k - \omega^2 m)^2 + (\gamma k)^2} \tag{2.10.27}$$

Evaluating this gives us

$$\Delta w_s = \frac{\pi\gamma k\bar{f}^2}{(k - \omega^2 m)^2 + (\gamma k)^2} \tag{2.10.28}$$

Finally, recognizing that $|\bar{x}|^2$ (2.10.26) shows up in our work expression, we can form the final result

$$\Delta w_s = \pi\gamma k|\bar{x}|^2 \tag{2.10.29}$$

This is exactly what we wanted, a dissipation that's proportional to the square of the amplitude and that has no frequency dependence. Thus we've got a conceptual model for structural damping, that is, a damping force that varies with displacement and is 90 degrees out of phase with it. If we compare the energy dissipated by viscous and structural damping

$$\Delta w_v = c\pi\omega|\bar{x}|^2$$

$$\Delta w_s = \gamma k\pi|\bar{x}|^2$$

we see that we can come up with an equivalent viscous damping coefficient $c_{eq}$ by letting

$$c_{eq}\pi\omega|\bar{x}|^2 = \gamma k\pi|\bar{x}|^2 \tag{2.10.30}$$

or

$$c_{eq} = \frac{\gamma k}{\omega} \tag{2.10.31}$$

How would you go about actually determining $c_{eq}$? First you must run a stress-strain test to determine what the energy absorption is for your material. Since you'll know the amplitude of your displacement during the test, you'll be able to determine $\gamma$ using (2.10.29). Then you'd use (2.10.31) to calculate $c_{eq}$. You'll note that you also need $\omega$. What this implies is that the $c_{eq}$ will depend on the frequency at which your system will be oscillating. This is generally okay since there is usually a primary vibration taking place that you'll be concerned about. However, it's important to realize that your equivalent viscous coefficient is only good at this particular frequency. If the frequency changes, you must recalculate $c_{eq}$.

A related observation is that these damping analyses are only applicable to a system that's vibrating in a sinusoidal way. It would be nice if we could also determine the response of the system to a nonsinusoidal input (a step or impulse), as we can with viscous damping. Unfortunately, that's not possible to do without changing our damping model drastically and increasing the complexity of the analysis by a very significant amount.

**Example 2.12**

> **Problem**  A sinusoidal stress-strain test has been run on a system, and you're to determine the structural damping factor. The linear stiffness $k$ was equal to 100,000 N/m, the driving force had an amplitude of 350 N, and the amplitude of the response was .01 m. Furthermore, the output differed from the input by 90 degrees.
>
> **Solution**  An examination of (2.10.26) shows that the output and input differ by 90 degrees when $\omega = \sqrt{\dfrac{k}{m}}$. For this situation we have
>
> $$|\bar{x}| = \frac{\bar{f}}{|\gamma k|} \tag{2.10.32}$$
>
> Using the given data this means that
>
> $$.01 = \frac{350}{100,000\gamma} \tag{2.10.33}$$
>
> which gives us $\gamma = .35$.

**Example 2.13**

> **Problem**  For the system discussed in Example 2.11, determine the energy dissipated during one cycle of the stress-strain test.
>
> **Solution**  The energy dissipated is given by
>
> $$\Delta w_s = \pi \gamma k |\bar{x}|^2 \tag{2.10.34}$$
>
> As we have already determined the value for $\gamma$ and are given $k$ and $|\bar{x}|$, we can immediately solve for the energy dissipated
>
> $$\Delta w_s = \pi(.35)(100,000)(.01)^2 = 11 \text{ Nm} \tag{2.10.35}$$

## 2.11 ACCELEROMETERS AND SEISMOMETERS

*Transducer* is the word given to describe those devices that allow us to measure the relevant variables in a vibrations experiment. One of the early decisions that must be made in the course of a vibrations test is determining which transducer is most appropriate for the particular problem at hand. In this section we'll discuss two kinds of transducers and give some insight into their uses and limitations.

Perhaps the most widely used vibrational transducer is the accelerometer. Although there is more than one way to build accelerometers, the most common is the piezoelectric, shown for

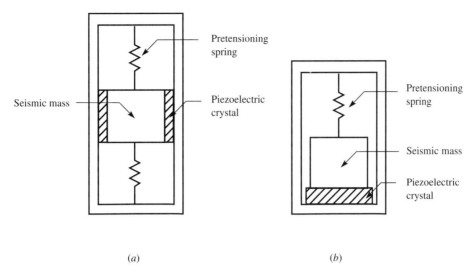

(a)                                                (b)

**Figure 2.31**   Accelerometer configurations

two configurations in Figure 2.31. The basic mechanism at work for both is a mass that acts on a piezoelectric crystal. Because piezomaterials generate an electric charge that is proportional to their deformation, we can use them to determine the acceleration of the object on which the accelerometer is attached. In Figure 2.31*a* the piezoelectric material is sheared as the seismic mass moves relative to the surrounding casing, thus generating a charge. The accelerometer shown in Figure 2.31*b* induces a charge in the piezoelectric material by compressing it.

Both of the configurations of Figure 2.31 can be viewed as simple single-degree-of-freedom systems (a mass attached to the accelerometer casing through a spring). Referring to Figure 2.32, we see that the internal movable mass within the accelerometer body is denoted by $m$, the spring that connects this mass to the accelerometer is denoted by $k$, the actual motion of the accelerometer body is given by $y$, and the relative motion of the mass $m$ with respect to the accelerometer body is given by $x$. We've also included a damper $c$ to account for frictional dissipation. It is very

**Figure 2.32**   Simple accelerometer model

important to understand that $x$ is the *relative* motion of the internal mass, not its actual motion with respect to an outside observer.

Referring to Figure 2.32, and remembering that the absolute motion of the seismic mass is given by $x + y$, we can perform a force balance to find

$$m(\ddot{x} + \ddot{y}) = -kx - c\dot{x} \tag{2.11.1}$$

Rearranging this yields

$$m\ddot{x} + c\dot{x} + kx = -m\ddot{y} \tag{2.11.2}$$

and dividing by $m$ brings us to

$$\ddot{x} + 2\zeta\omega_n\dot{x} + \omega_n^2 x = -\ddot{y} \tag{2.11.3}$$

If we assume that the accelerometer body is oscillating sinusoidally, then using $y(t) = \bar{y}e^{i\omega t}$ and $x(t) = \bar{x}e^{i\omega t}$ will lead to

$$\left(\omega_n^2 - \omega^2 + 2i\zeta\omega_n\omega\right)\bar{x} = \omega^2\bar{y} \tag{2.11.4}$$

Therefore we have

$$x(t) = \frac{\omega^2\bar{y}e^{i\omega t}}{\omega_n^2 - \omega^2 + 2i\zeta\omega_n\omega} \tag{2.11.5}$$

This is the governing equation for our accelerometer. Although this doesn't look particularly simple, this situation changes when we look at a limiting case for which the excitation frequency is much smaller than the natural frequency of the accelerometer, that is, $\omega \ll \omega_n$. In this case the dominant term in the denominator of (2.11.5) is $\omega_n^2$, and (2.11.5) becomes

$$x(t) = \frac{1}{\omega_n^2}(\omega^2\bar{y}e^{i\omega t}) \tag{2.11.6}$$

What we now must recall is that the accelerometer is mounted on a surface that is experiencing an oscillation equal to $\bar{y}e^{i\omega t}$. Therefore the surface's acceleration is given by $-\omega^2\bar{y}e^{i\omega t}$. Thus we can write the displacement of the mass as

$$x(t) = -\frac{1}{\omega_n^2}\ddot{y}(t) \tag{2.11.7}$$

Thus we have a direct measure of the object's acceleration: we need simply multiply $x(t)$ by $-\omega_n^2$ to obtain it.

This simple formula holds only as long as $\omega \ll \omega_n$, thus telling us what the limitation on the usable frequency range is equal to.

In addition to being used as an accelerometer, this same setup can be used to obtain a different kind of measurement, namely the displacement of the mounting surface. Consider the case in which the natural frequency of our device is much lower than the exciting frequency. This is the situation that corresponds to point 5 on our response graph of a seismically excited mass (Figure 2.7). You'll recall that what occurred in this case was that your hand moved up and down while the mass stayed fixed. Thus the relative motion between the mass and your hand was exactly the same as the actual motion of your hand, just 180 degrees out of phase. This is the result we should expect the math to give us.

If $\omega \gg \omega_n$ in (2.11.5) we obtain

$$x(t) = -\bar{y}e^{i\omega t} \tag{2.11.8}$$

Thus, as expected, our transducer gives us a reading that's directly proportional to the mounting surface's motion. When used in this manner, the transducer is called a seismometer.

Physically, what the constraint $\omega \gg \omega_n$ means is that the natural frequency of the seismometer must be small, implying either a large mass or a weak spring. Consequently seismometers are usually substantially larger than accelerometers.

**Example 2.14**

**Problem**    Assume a given accelerometer has a moving mass of 10 g and a spring constant of $3.95 \times 10^7$ N/m. We'll assume that there is no damping present. What is the upper usable frequency range for this accelerometer if the maximal error that we'll allow is 5 percent?

**Solution**    The general solution for an accelerometer was given by (2.11.5),

$$x(t) = \frac{\omega^2 \bar{y}e^{i\omega t}}{\omega_n^2 - \omega^2 + 2i\zeta\omega_n\omega} \tag{2.11.9}$$

We can recognize the numerator as simply $-1 \times$ the acceleration of the base. Thus the ratio of $-x$ response to acceleration is given by

$$-\frac{x(t)}{\ddot{y}} = \frac{1}{\omega_n^2 \left(1 - \frac{\omega^2}{\omega_n^2}\right)} \tag{2.11.10}$$

As the frequency of excitation increases, the ratio of $x$ to acceleration grows. Therefore a 5 percent error means we'll have

$$1.05 = \frac{1}{1 - \frac{\omega^2}{\omega_n^2}} \tag{2.11.11}$$

or

$$\frac{\omega}{\omega_n} = .2182 \tag{2.11.12}$$

Since $\omega_n = 62,850$ rad/s (using the given values for $k$ and $m$), this means $\omega = 13,700$ rad/s or 2180 Hz. Thus the maximum usable frequency range is about 20 percent of the system's natural frequency.

## 2.12 HOMEWORK PROBLEMS

### Section 2.2

1. We're going to analyze the motions of a unicyclist. The actual system is shown below, as is our approximate model. If the unicyclist is traveling at 3 m/s and hits the bump illustrated below, what will the cyclist's maximum deflection from $x_{eq}$ be?

2. Consider the unicyclist of Problem 1. The road she's traveling on has a sinusoidal profile, as shown below. What is the worst speed for her to travel at?

3. A machine on its suspension deflects .8 cm under static conditions. What is the ratio of the mass's oscillation amplitude to the base oscillation amplitude if the base is oscillating at 30 Hz?

4. An 18 kg mass has to be protected from vibration and the illustrated system has been designed to do this. The mass is to be placed within a protective container and isolated from the walls by two springs. The force characteristics of the springs are

$$f_1 = 5x \qquad \text{(spring1)}$$
$$f_2 = x + x^3 \quad \text{(spring2)} \qquad\qquad (2.12.13)$$
$$f_3 = 31x \qquad \text{(spring3)}$$

In all three cases the displacement units are assumed to be meters and the forces are newtons. Initially the mass-spring unit is 12 m long. The available space inside the protective box is 8 m. Two possible configurations are being considered, shown below. Which configuration will be most effective in isolating the mass from external disturbances, assuming that these disturbances occur around $\omega = 1.2$ and are of small amplitude $(y = \epsilon \sin(3.0t))$? (Note: Best isolation here means minimum movement.)

5. Consider a simple model of a car traveling along a road. (The same model as in Problem 1.) We'll look only at translational motions (vertical oscillations). The road profile is sinusoidal with amplitude .01 m and wavelength 8 m. The speed limit is 35 km/hr. Would you experience a smaller oscillation amplitude if you reduced your speed to 25 km/hr? $m = 1000$ kg and $k = 70,224$ N/m.

6. Consider the illustrated system. A rigid ring with mass $m$ is connected via four massless, inextensible wires to a central hub (radius $r$). Each wire has a tension $T$. If the motion of the hub is given by $\theta(t) = \theta_0 \cos \omega t$ ($\theta_0$ small), determine the angular response of the ring.

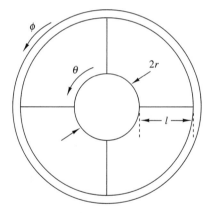

7. Shown below is the steady-state time response of an undamped SDOF system under seismic excitation (transients are ignored). Determine $\omega_n$ for this system. (Note: You *don't* need to try to back out the answer by plugging amplitudes and frequencies into the relevant dynamic equations.)

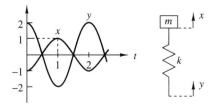

**8.** Shown below is an enclosure that contains a tensioned spring that presses on a mass. The initial compression of the spring is equal to $h$. The enclosure is subjected to a sinusoidally varying displacement $x(t) = a\cos(\omega t)$. Determine the frequency $\omega$ for which the mass will lose contact with the enclosure (consider $m, k, h$, and $a$ fixed, and ignore gravity).

**9.** Consider a seismically excited spring-mass system (see Figure 2.1). Find the range of oscillation frequencies $\omega$ for which the force transmitted to the mass $m$ is less than or equal to 200 N. $m = 1$ kg, $k = 20,000$ N/m, and $y(t) = \cos(\omega t)$ ($y$ measured in centimeters).

**10.** Consider the system shown in text Figure 2.10. Let $f$ be zero and let $x = .003\sin(\omega t)$ (measured in meters). What is the force transmitted to the ground? $m = 10$ kg and $k = 5000$ N/m.

**11.** As shown below, the spring $k$ pushes the mass $m$ against a moving surface. The motion of the surface is given by $x(t) = a(\cos\omega t)$. When $x(t) = 0$, the mass is being pressed into the surface with a force $kh$ (i.e., the spring is precompressed on amount $h$). Determine the critical speed $\omega$ for which the mass will first lose contact with the surface. (Note: Identify the critical $\omega$ for $a > h$ and $a < h$.)

**12.** Given the displacement input $y(t) = \bar{y}\cos(\omega t)$, determine the equation of motion for the illustrated system (neglect gravity). What is the natural frequency for the system?

## Section 2.3

**13.** Given the data shown below, determine the system's natural frequency.

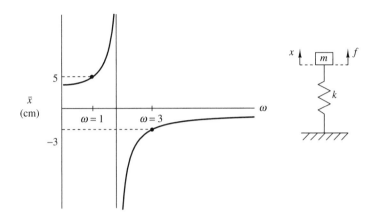

**14.** Determine the actual response of a spring-mass system when forced at its natural frequency. The relevant equation is

$$\ddot{x} + \omega_n^2 x = \bar{f} \cos(\omega_n t)$$

Assume a solution $x(t) = a_1 \cos(\omega_n t) + a_2 \sin \omega_n t + a_3 t \sin(\omega_n t)$.

**15.** Find the equations of motion and natural frequency for the illustrated system (neglect gravity). What would $f$ need to be for there to be no overall excitation of the system?

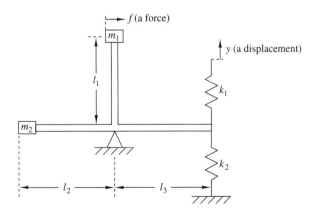

**16.** Derive the system equations of motion if the displacement input is given by $y(t) = \bar{y} \cos(\omega t)$ (neglect gravity).

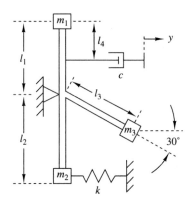

**17.** The figure below is a model of the valves in an engine. $m$ is the mass of a valve, $k$ is the restraining spring, and the rotating body is the cam that drives the valves. The profile of the cam gives an $x$ deflection of the valve equal to $x_0 + a \cos(\omega t)$. Determine at what rotational velocity the inertial forces of the piston become so great that the piston loses contact with the cam. When the cam is in the indicated position, the total spring deflection is equal to $x_c$.

$$x_0 = \frac{h_1 + h_2}{2}$$

$$a = \frac{h_1 + h_2}{2}$$

**18.** Does the force transmitted to the floor reach a maximum as $\omega$ is increased or does it increase indefinitely? At high $\omega$ what is more important to the transmitted force, $k$ or $c$?

**19.** Given the force input $f = \bar{f} \sin(\omega t)$, determine the equation of motion for the illustrated system. What is the system's natural frequency? (Neglect gravity.)

## Section 2.4

**20.** Determine the transfer function for the illustrated system that relates input force at the mass to output force felt at the floor.

**21.** Find the transfer function between the displacement input $y$ and the displacement output $x$.

**22.** Find the transfer function between the input displacement $y$ and the output force against the wall.

## Section 2.5

**23.** Consider the spring-mass damper system shown below. You must determine the actual values for $k$ and $c$, but only have a limited amount of information. Given the data below, determine $k$ and $c$.

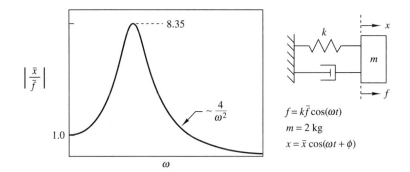

$f = k\bar{f}\cos(\omega t)$

$m = 2$ kg

$x = \bar{x}\cos(\omega t + \phi)$

**24.** Is the plot shown below (actual displacement versus frequency for a direct-force-excited damped oscillator) a reasonable one?

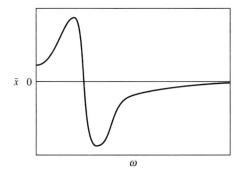

**25.** Shown below is the steady-state response (transients are ignored) of a direct-force-excited SDOF. The mass is equal to 2 kg. What other system parameters can you determine?

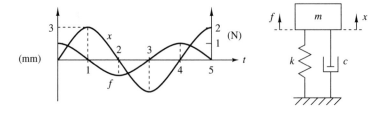

**26.** Let $l_1 = 1$ m, $l_2 = .5$ m, $l_3 = 2$ m, $m_1 = 1.1$ kg, $k = 100$ N/m, and $c = 3$ N/m/s. Determine the equation of motion for the illustrated system in terms of the parameters $\omega_n$ and $\zeta$ (neglect gravity).

## Section 2.6

**27.** Put $x(t) = 2\cos(3t) + 3\sin(3t)$ into complex form so that the real part of the complex number gives us $x(t)$.

**28.** Can $x(t) = 4\cos(4t) + 2\sin(2t)$ be put into complex form so that $x(t)$ is expressed as the real part of a single complex term?

**29.** Express $x(t) = \cos(t) - \sin(t)$ in purely complex form, that is, so that there is no need to take the real or imaginary part of the complex variables.

**30.** Find the complex transfer function between the input displacement $y$ and the output force against the ground for the system shown in Figure 2.3.

**31.** Find the complex transfer function between the velocity input $\dot{y}$ and the displacement output $x$ for the following system.

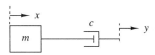

## Section 2.7

**32.** We now add a damper to the unicycle of Problem 2. Let $c = 88$ N/m/s. Under steady-state conditions, what will the maximum amplitude response be and at what velocity will it occur?

**33.** Assume that a seismically excited system is accurately modeled by Figure 2.3. You must determine whether or not the weld connecting the spring to the mass will hold. The maximum stress that the weld can withstand is $1.0 \times 10^6$ N/m². The mass is 10 kg and the spring constant is equal to 4000 N/m. The damping constant is equal to 40 N/m/s. Is it reasonable to allow the excitation frequency to reach 10 rad/s? The attachment area of the spring is 1 mm $\times$ 1 mm and the base amplitude is 1 mm.

**34.** Consider the system shown in Figure 2.3. What is the magnitude of $m$'s response to a ground excitation of $y(t) = .01 \cos(30t)$ for $k = 9000$ N/m, $m = 10$ kg, and $c = .6$ N/m/s? This should give you a general feel for how very lightly damped structures would respond in an earthquake.

**35.** Repeat Problem 34 but use $c = 6$ N/m/s. Does this improve the situation?

**36.** When the damping is small, increasing the damping by a factor of $\alpha$ will reduce the peak response by the same factor. Does this hold true at large values of damping? Why?

## Section 2.8

**37.** The system illustrated below is a centrifuge. The plot shows its amplitude response as a function of frequency with a specimen, $m_2$, in the machine. The mass of the centrifuge is $m_1$ and it is restrained by a spring of stiffness $k$. Given that the radius of the spinning chamber is .1 m and that the mass $m_1$ is equal to 100 kg, find the mass of the specimen as well as the spring constant $k$.

**38.** Find the response amplitude of the illustrated system of terms $\omega$, $l_1$, $l_2$, $m_1, m_2$, and $k$.

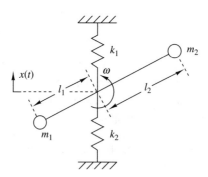

**39.** An overhead fan (shown below) isn't dynamically balanced. Because of this, the center of mass of the rotating fan-blade assembly lies off the center of rotation by a distance $\ell$. The rotating mass has a magnitude $m_R$ while the nonrotating motor housing has a mass $m_N$. Treat the support rod as a massless. The radius of gyration of the blade assembly about the point of rotation is equal to $k_R$. Determine the response of the fan when the blades are rotating at a fixed speed $\omega$. (Hint: The spring stiffness in the problem comes from gravity. Look at the problem as a spring mass being excited by a rotating mass. Since the angle is small ($\phi \ll 1$) you can just look at $x, y$ motions.)

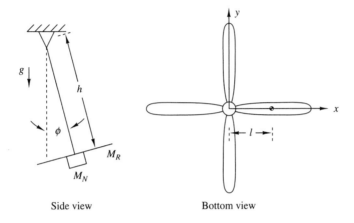

Side view                          Bottom view

**40.** The figure below illustrates a top view of a top-loading washing machine. The rotational speed of the drum is $\omega = 30$ rad/s and the radius of the drum is $r = .3$ m. Assume that the wet clothes have all gathered into a ball of mass 6 kg (approximated by a point mass). Determine the lateral forces generated by the washing machine. Assume that the mass of the drum is equal to 3 kg.

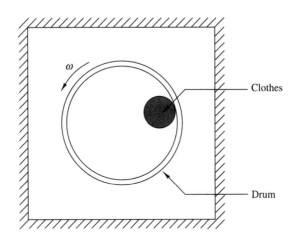

**41.** Consider the system of Problem 40, but refine the approximation. The drum rotates within a nonrotating sleeve and the total assembly has a mass equal to 3 kg. The outer assembly is held in place by a spring, as shown in the figure below. Note that motion is only allowed in the $x$ direction. The spring constant $k$ is equal to 500 N/m. The ball of clothes is still 6 kg and $\omega = 30$ rad/s. Determine the force applied to the ground.

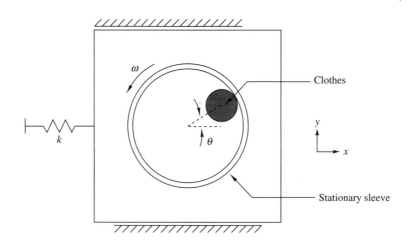

**42.** A simplified model of a rotating shaft within its bearings is shown in the accompanying figure. If the shaft is imbalanced it will experience a time-varying force due to the rotating mass. The operating frequency is equal to 60 rad/s, the support stiffness $k_1$ is equal to 170,000 N/m, and the total rotating mass is equal to 100 kg. Determine the amplitude of vibration, given the fact that at high frequencies ($\omega \to \infty$) the oscillation amplitude is equal to .001 m.

## Section 2.9

**43.** All you have left from a vibration study is a scrap of paper, illustrated below. Can you use this data to reconstruct what the peak-response amplitude was for the system? An answer that's within 5 percent is good enough. The system was a direct-force-excited spring-mass system.

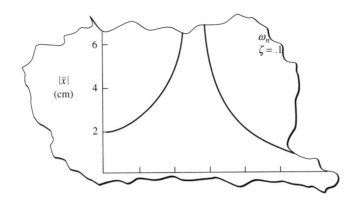

**44.** Say you have a shock absorber on your car that has a different value of viscous damping on the extension stroke than on the compression stroke. If the value of the damping factor is $\zeta = .1$ during compression, what must it be on the extension stroke in order for the overall damping factor to be .2? (Note: Ignore changes to $\omega_d$ due to varying damping rates; they'll be small for the values of $\zeta$ you're considering.)

**45.** Can you deduce the natural frequency and damping factor of the given system from the following data? The damping factor is small ($\zeta \ll 1$) and the governing equation is

$$\ddot{x} + 2\zeta \omega_n \dot{x} + \omega_n^2 x = \omega_n^2 \bar{f} \sin(\omega t) = f(t)$$

**46.** Can you determine the value of $\zeta$ for the illustrated system and responses? The responses shown are two samples from a large number of tests. The governing equation is given by

$$m\ddot{x} + c\dot{x} + kx = f(t) = \bar{f}\sin(\omega t)$$

The second test produced the maximal response amplitudes of all the tests.

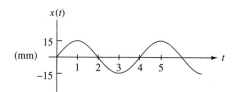

**47.** My father recently purchased a new car. An annoying flaw with the vehicle is a vibration that becomes noticeable when traveling about 52 mph. The technicians took the car out for a test and gathered acceleration data from an accelerometer. Back at the shop they looked at the data and saw that a strong peak in the vibrational energy was present at 12 Hz when the car had been traveling at 52 mph, just as my father had claimed. I suggested to my father that the vibration must be due to a resonance in the car that was excited due to the rotation of the tires. Why did I think this? (Note: The distance from the center of the wheel to ground for my father's car is equal to 1 ft.)

## Section 2.10

**48.** In this exercise you'll determine the amount of energy dissipated over one cycle of forcing for Coulomb friction. Assume the response is given by $x(t) = a\sin(\omega t)$, and thus $\dot{x}(t) = a\omega\cos(\omega t)$. The equation of motion is

$$m\ddot{x} + \mu mg\,\mathrm{sign}(\dot{x}) + kx = f_s\sin(\omega t) + f_c\cos(\omega t)$$

The forcing has been broken into distinct sine and cosine components so that our assumption of a purely sinusoidal response for $x$ will hold. Calculate the response and only retain those terms corresponding to sinusoidal oscillations at the frequency $\omega$. This information, along with (2.10.16), will allow you to compute the energy dissipated over one cycle of motion.

**49.** Using the same approximation to the sign function as in Problem 48, show that the steady-state amplitude of oscillation for a direct-forced mass with Coulomb friction is given by

$$a = \frac{\sqrt{f^2 - \left(\dfrac{4\mu mg}{\pi}\right)^2}}{k - m\omega^2}$$

where $f$ is the magnitude of the forcing and $a$ is the response magnitude.

**50.** Using the response magnitude relation of Problem 49, determine what the magnitude of forcing must be for finite oscillations to exist. What if the forcing is below this limit?

**51.** Consider a direct-force-excited spring-mass system with Coulomb friction. $k = 10,000$ N/m, $m = 5$ kg, $f = 100$ N, $\mu = .1$, and $\omega = 10$ rad/s. Determine the magnitude of the steady-state response using the information in Problem 49.

**52.** How much will the magnitude of oscillation be reduced if the system of Problem 51 has its coefficient of friction increased from .1 to .6?

**53.** How much will the magnitude of oscillation be altered if the system of Problem 51 has the forcing frequency increased to 20 rad/s?

**54.** Consider the static problem of a spring-mass system with Coulomb friction. If $k = 10,000$ N/m, $m = 10$ kg, and $\mu = .35$, what is the maximal initial deflection of the mass that will not result in subsequent motion of the mass?

**55.** Consider a spring-mass system with Coulomb damping and direct-force excitation. Let $f = 50$ N, $m = 1$ kg, $k = 1000$ N/m, and $\omega = 10$ rad/s. You've run two tests. For the first you conclude that the oscillation magnitude is 55.5 mm and the second you obtain 55.6 mm. Both readings would seem to be within a reasonable experimental error. Calculate the value of $\mu$ that would correspond to these measurements and comment on whether determining $\mu$ from such a test is a reliable approach.

**56.** Consider a spring-mass system with Coulomb damping and direct force excitation. For $f = 20$ N, $\mu = .4$, and $m = 5$ kg the system remains in static equilibrium. What should $\mu$ be reduced to in order to permit oscillations in the system?

**57.** Consider a spring-mass system under direct-force excitation with Coulomb damping. As in Problem 55, you've run two tests and recorded the oscillation magnitude. For the first run you obtain a magnitude of 41.5 mm and for the second obtain 42 mm. Use this information to calculate $\mu$ and comment on whether, for these parameters, this is a reliable means of determining $\mu$. $k = 1000$ N/m, $m = 2$ kg, $\omega = 23$ rad/s, and $f = 5.1$ N.

## Section 2.11

**58.** Consider the accelerometer described in Example 2.14. Assume that we've included a damper in the accelerometer, with $\zeta = .3$. With all other parameters the same, what will the percentage error be for the accelerometer's prediction of the magnitude of the acceleration. What phase error will this damping induce?

**59.** Consider the accelerometer described in Example 2.14. Assume that a viscous damper ($\zeta = .15$) is included in parallel with the spring. At what frequency will the phase difference between the actual acceleration and the accelerometer's measurement differ by 15 degrees?

**60.** If an accelerometer is being approximated as an undamped spring-mass system, what percentage of $\omega_n$ is the maximum frequency for which the sensor error is less than or equal to 1 percent.

**61.** Consider a seismometer for which $m = .2$ kg and $k = 200$ N/m. What is the lowest frequency of vibration at which the seismometer can be used if we require accuracy of at least 15 percent? Neglect any damping for this problem.

**62.** Consider Problem 61. If the mass of the seismometer is doubled, how much lower will the lowest allowable frequency be? Again, neglect damping and allow a maximum of 15 percent error.

# 3

# Nonsinusoidal
# Excitations

## 3.1 INTRODUCTION

This chapter considers the problem of how to find a system's response when the input excitation is not a simple sinusoid. If you think about it, it seems pretty obvious that almost no real system will have a purely sinusoidal input. Many, many of the inputs will be periodic, that is, the input is made up of an endlessly repeating waveform. But the chances that the waveform is a pure sine wave is pretty low. For instance, the repeated firings inside the cylinder of an internal combustion engine will produce a periodic, but not sinusoidal, loading on a piston. Thus we're motivated to figure out how we can solve for a system's steady-state response for these cases.

Furthermore, the input to our system might not even be periodic. Oftentimes the loading will be very transient in nature. An earthquake is a good example of a transient loading. The ground will shake for a few seconds and then stop. Our problem then consists of determining a structure's response to a loading that exists for only a finite amount of time. If the earthquake lasts for a long time, then we might actually view the problem in a steady-state way. But if the earthquake is of short duration, then a transient viewpoint is more appropriate. A similar problem will occur when an airplane encounters a momentary updraft. The updraft will load the wings only for a short time, and the wings will then bend (and the airplane displace) in response to the input.

The loading might even be **very** transient, that is, be a shock load such as the case shown above. An explosion will create a high level of force that will last only for a very short time. Similarly, if a system is accidentally dropped to the ground, it will experience a very short, very severe loading. We'll certainly want to know how to handle such cases to ensure that our systems are able to withstand normal usage. This type of analysis takes place in industry when packaging concerns are discussed. We want a product to withstand the normal operating environment and also be able to withstand the loads that may occur prior to installation. For instance, some of the most severe loadings are applied to automobiles, not during normal driving, but while the cars are being shipped to the dealers. Automobile manufacturers spend a good deal of effort to determine the nonsinusoidal loads that a vehicle will experience during shipment and work to ensure that they survive them.

The final category of loading, random loadings, falls between the first two. Periodic excitations have a repeating pattern and don't die away. Transient loadings have no periodicity and last only a finite time. Random loadings are not periodic (like transient loads), but don't die away. These loadings are very common in real life. The force of wind on a building will be random. There's always some wind, but the direction and force vary randomly. The vibration of a floor in a manufacturing facility will be random, the inputs due to people walking about, parts being dropped or moved occasionally, doors being opened and closed, etc. If an earthquake goes on for a very long time, the loading can be viewed as random.

We'll start the chapter by looking at the most immediately accessible extension to what we already know, the response of a vibratory system to a periodic, but not sinusoidal, input. Following this, we'll consider what needs to be done to deal with transient inputs. Finally, we'll look at the problem of random loading.

## 3.2 FOURIER SERIES ANALYSIS

This section will cover how to analyze systems for which the excitation is periodic, but isn't made up of a single sinusoid. We've already mentioned that superposition lets us easily determine the output of a system to an input $y_1 + y_2$ if we already know the corresponding outputs $x_1$ and $x_2$ for the individual inputs (it's simply equal to $x_1 + x_2$). Just to clarify this idea a bit more, let's work out a particular example. Say we want to know the response of a seismically excited system for the input

$$y(t) = \frac{8}{\pi^2} \sin(\omega t) - \frac{32}{36\pi^2} \sin(3\omega t) \tag{3.2.1}$$

A plot of $y(t)$ versus time is shown in Figure 3.1. Note that the addition of the $\sin(3\omega t)$ term has served to sharpen up the base sinusoid, making it more triangular. This isn't a coincidence and is due to the particular choice of amplitudes used. We'll look at this later in this chapter.

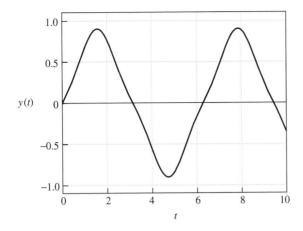

**Figure 3.1**  Sum of two sinusoids

The governing equation of motion (2.2.11) is repeated here for convenience,

$$\ddot{x} + \omega_n^2 x = \omega_n^2 y \tag{3.2.2}$$

Since the input consists of two linearly independent excitations and the dynamical system has no damping, we can assume a response

$$x(t) = b_1 \sin(\omega t) + b_2 \sin(3\omega t) \tag{3.2.3}$$

which, when differentiated twice and substituted into (3.2.2), leads to

$$b_1(\omega_n^2 - \omega^2)\sin(\omega t) + b_2(\omega_n^2 - 9\omega^2)\sin(3\omega t) = \frac{8\omega_n^2}{\pi^2}\sin(\omega t) - \frac{32\omega_n^2}{36\pi^2}\sin(3\omega t) \tag{3.2.4}$$

Since $\sin(3\omega t)$ and $\sin(\omega t)$ are linearly independent, the only way that (3.2.4) can be satisfied is if the coefficients of each sinusoid on the right and left side of the equation are equal to each other, or

$$b_1(\omega_n^2 - \omega^2) = \frac{8\omega_n^2}{\pi^2} \tag{3.2.5}$$

and

$$b_2(\omega_n^2 - 9\omega^2) = -\frac{32\omega_n^2}{36\pi^2} \tag{3.2.6}$$

Thus we can easily solve for $b_1$ and $b_2$:

$$b_1 = \frac{8}{\pi^2}\frac{\omega_n^2}{(\omega_n^2 - \omega^2)} \tag{3.2.7}$$

$$b_2 = -\frac{32}{36\pi^2}\frac{\omega_n^2}{(\omega_n^2 - 9\omega^2)} \tag{3.2.8}$$

Therefore

$$x(t) = \frac{8}{\pi^2}\frac{\omega_n^2}{(\omega_n^2 - \omega^2)}\sin(\omega t) - \frac{32}{36\pi^2}\frac{\omega_n^2}{(\omega_n^2 - 9\omega^2)}\sin(3\omega t) \tag{3.2.9}$$

As advertised, this is exactly the same result as if we had solved

$$\ddot{x} + \omega_n^2 x = \frac{8\omega_n^2}{\pi^2} \sin(\omega t) \tag{3.2.10}$$

to obtain

$$x(t) = \frac{8}{\pi^2} \frac{\omega_n^2}{(\omega_n^2 - \omega^2)} \sin(\omega t) \tag{3.2.11}$$

and had solved

$$\ddot{x} + \omega_n^2 x = \frac{32\omega_n^2}{36\pi^2} \sin(3\omega t) \tag{3.2.12}$$

to obtain

$$x(t) = -\frac{32}{36\pi^2} \frac{\omega_n^2}{(\omega_n^2 - 9\omega^2)} \sin(3\omega t) \tag{3.2.13}$$

and added the two solutions together. Note that the solutions have been written in the form

$$\text{(amplitude of forcing)(transfer function)(time variation of forcing)} \tag{3.2.14}$$

This is to make it doubly clear that the response of a linear system is only the transfer function multiplied by the excitation.

Now we can begin to see why superposition is such a powerful concept. It implies that we can easily find the response of dynamical systems as long as we can consider the inputs to be made up of individual sinusoids. Thus the obvious question is: When can we view an input to a system as being made up of individual sinusoidal components? Clearly $y(t) = e^{-2t}$ isn't a likely candidate since it is patently nonsinusoidal. But what about a triangular input, such as that illustrated in Figure 3.2? Is this made up of sinusoids? We might be tempted to say yes, since this figure looks a lot like Figure 3.1, which we've just seen is composed of $\sin(\omega t)$ and $\sin(3\omega t)$. In fact, Jean Baptiste Joseph Fourier, in the early 1800s, showed that we can approximate any well-behaved,

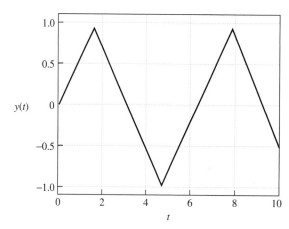

**Figure 3.2**  Triangle wave

periodic signal as the infinite sum of a series of pure sinusoids. Note the *periodic* in the previous sentence. At this point we will consider only inputs that are periodic, that is, if $f(t)$ is the function under consideration, then $f(t + T) = f(t)$, where $T$ is the period of the function. Later in the book we'll consider what can be done in the cases where the functions of interest are not periodic.

According to Fourier, we should be able to represent a periodic signal, $f(t)$, as

$$f(t) = \sum_{n=0}^{\infty} a_n \cos(n\omega_0 t) + \sum_{n=1}^{\infty} b_n \sin(n\omega_0 t) \qquad (3.2.15)$$

The $\omega_0$ component is the lowest frequency (longest wavelength) present in the signal $f(t)$ and is related to the signal's period $T$ by

$$\omega_0 = \frac{2\pi}{T} \qquad (3.2.16)$$

You'll note that the summations in the preceding equation go to infinity. This means that if we really want to precisely model an arbitrary periodic signal, we might need an infinite number of sinusoids to do so. Luckily, we will never need an infinite number of terms for our analyses, for reasons that will become clear momentarily.

Assuming that we can indeed approximate $f(t)$ by a series of sinusoids, the problem still remains how to determine $a_n$ and $b_n$, the coefficients that multiply the sinuosoids. In order to make the general procedure clear, let's first work out a specific example. The function that we are going to analyze is

$$f_1(t) = 2\sin(\omega_0 t) - 3\cos(2\omega_0 t) \qquad (3.2.17)$$

Thus we have

$$\sum_{n=0}^{\infty} a_n \cos(n\omega_0 t) + \sum_{n=1}^{\infty} b_n \sin(n\omega_0 t) = f_1(t) = 2\sin(\omega_0 t) - 3\cos(2\omega_0 t) \qquad (3.2.18)$$

Of course, it's obvious that we can just read off the correct answers ($a_2 = -3$, $b_1 = 2$, and all other $a_n$ and $b_n = 0$), but for the purpose of this discussion we'll assume that we don't know the answers in this way, that we'll have to calculate them from $f(t)$. This procedure is what we'd actually have to do in a real Fourier series analysis, one in which we wouldn't know the sinusoidal contributions beforehand.

Here's how we'll do it. First, multiply both sides of (3.2.18) by $\sin(m\omega_0 t)$. This is the sinusoid associated with $b_m$. For our particular example we'll expect answers only for $m = 1$ since that's the only sine component in $f(t)$. This gives us

$$\sum_{n=0}^{\infty} a_n \cos(n\omega_0 t) \sin(m\omega_0 t) + \sum_{n=1}^{\infty} b_n \sin(n\omega_0 t) \sin(m\omega_0 t)$$

$$= 2\sin(\omega_0 t)\sin(m\omega_0 t) - 3\cos(2\omega_0 t)\sin(m\omega_0 t) \qquad (3.2.19)$$

We can simplify matters a bit by reexpressing these products of sines and cosines. Recall that

$$\sin(m\omega_0 t) = \tfrac{1}{2i}(e^{im\omega_0 t} - e^{-im\omega_0 t}) \qquad (3.2.20)$$

and

$$\cos(m\omega_0 t) = \tfrac{1}{2}(e^{im\omega_0 t} + e^{-im\omega_0 t}) \qquad (3.2.21)$$

Using (3.2.20) and (3.2.21) we can expand (3.2.19) out into its exponential parts and then combine these exponentials back into sines and cosines:

$$\sum_{n=0}^{\infty} a_n \left(\frac{1}{2}\right) [\sin((n+m)\omega_0 t) - \sin((n-m)\omega_0 t)]$$

$$+ \sum_{n=1}^{\infty} b_n \left(\frac{1}{2}\right) [-\cos((n+m)\omega_0 t) + \cos((n-m)\omega_0 t)]$$

$$= 2\left(\frac{1}{2}\right) [-\cos((1+m)\omega_0 t) + \cos((1-m)\omega_0 t)]$$

$$- 3\left(\frac{1}{2}\right) [\sin((2+m)\omega_0 t) - \sin((2-m)\omega_0 t)] \qquad (3.2.22)$$

Next we're going to exploit a property called orthogonality, something we haven't talked about yet, but that will be easy to understand once you see it at work. To make use of orthogonality, we will integrate both sides of (3.2.22) from $t = 0$ to $t = T$, that is, over one period. So we'll have:

$$\sum_{n=0}^{\infty} a_n \left(\frac{1}{2}\right) \left[ \int_0^{\frac{2\pi}{\omega_0}} \sin((n+m)\omega_0 t)\, dt - \int_0^{\frac{2\pi}{\omega_0}} \sin((n-m)\omega_0 t)\, dt \right]$$

$$+ \sum_{n=1}^{\infty} b_n \left(\frac{1}{2}\right) \left[ -\int_0^{\frac{2\pi}{\omega_0}} \cos((n+m)\omega_0 t)\, dt + \int_0^{\frac{2\pi}{\omega_0}} \cos((n-m)\omega_0 t)\, dt \right]$$

$$= 2\left(\frac{1}{2}\right) \left[ -\int_0^{\frac{2\pi}{\omega_0}} \cos((1+m)\omega_0 t)\, dt + \int_0^{\frac{2\pi}{\omega_0}} \cos((1-m)\omega_0 t)\, dt \right]$$

$$- 3\left(\frac{1}{2}\right) \left[ \int_0^{\frac{2\pi}{\omega_0}} \sin((2+m)\omega_0 t)\, dt + \int_0^{\frac{2\pi}{\omega_0}} \sin((2-m)\omega_0 t)\, dt \right] \qquad (3.2.23)$$

Taken all together, this looks **really** messy. However, if we look at each term individually we notice a remarkable thing. It seems that almost every term is equal to zero since, for each term, all we have is an integration over one or more periods of a pure sinusoid. And we know that integrating a cosine or sine over one or more periods yields zero as a result. For instance, if $n = 1$ and $m = 5$ then the first integral of the equation is

$$\int_0^{\frac{2\pi}{\omega_0}} \sin(6\omega_0 t)\, dt = -\frac{\cos(6\omega_0 t)}{6\omega_0}\bigg|_0^{\frac{2\pi}{\omega_0}} = 0 \qquad (3.2.24)$$

This will be true for any $n$ and $m$ combination, except those for which $n = m$. If $n = m$, then $\sin((n - m)\omega_0 t) = 0$. Thus it isn't a finite sinusoid anymore. However, it's equal to zero and thus gives us no trouble. However, $\cos((n - m)\omega_0 t) \to 1$ when $n = m$. Therefore the fourth integral on the left side of (3.2.23) becomes

$$\int_0^{\frac{2\pi}{\omega_0}} \cos((n - m)\omega_0 t)\, dt = \int_0^{\frac{2\pi}{\omega_0}} dt = \frac{2\pi}{\omega_0} \tag{3.2.25}$$

Let's take a second to recall what we've found. We already know that only two harmonics show up in our signal, namely $2\sin(\omega_0 t)$ and $-3\cos(2\omega_0 t)$. We formed a general Fourier series, set it equal to our signal, multiplied by $\sin(m\omega_0 t)$, and integrated over a cycle. We then found that almost every term went to zero. If $m$ isn't equal to 1, then all the terms on the right side of (3.2.23) are equal to zero. Only if $m = 1$ do we obtain a finite result.

If we choose $m = 1$, then (3.2.23) becomes

$$b_1 \frac{\pi}{\omega_0} = 2 \left( \frac{1}{2} \right) \left( \frac{2\pi}{\omega_0} \right) = \frac{2\pi}{\omega_0} \tag{3.2.26}$$

or

$$b_1 = 2 \tag{3.2.27}$$

This is the correct result! Of course, we knew when we started that the coefficient of the Fourier series for $\sin(\omega_0 t)$ must be 2 since that's how much of that particular harmonic was in the signal originally. The point is that we usually don't know ahead of time how much of a particular harmonic there will be, and the procedure we've just gone over will give us the answer. We could go ahead and find the cosine harmonics in the same way, simply multiply by $\cos(m\omega_0 t)$, integrate over a period, and vary $m$ until we get a finite answer. The general procedure for determining all the sine and cosine coefficients can be written as

$$a_n = \frac{\omega_0}{\pi} \int_0^{\frac{2\pi}{\omega_0}} f(t) \cos(n\omega_0 t)\, dt \tag{3.2.28}$$

and

$$b_n = \frac{\omega_0}{\pi} \int_0^{\frac{2\pi}{\omega_0}} f(t) \sin(n\omega_0 t)\, dt \tag{3.2.29}$$

The only other twist you'll need to remember is that if the signal you're analyzing doesn't average out to zero, we consider it to have a bias or a constant offset. Since constants are even functions, we can view this as a cosine with zero frequency, and thus we'll have

$$a_0 = \frac{\omega_0}{2\pi} \int_0^{\frac{2\pi}{\omega_0}} f(t)\, dt \tag{3.2.30}$$

These coefficients (3.2.28, 3.2.29, and 3.2.30) are then used in

$$f(t) = \sum_{n=0}^{\infty} a_n \cos(n\omega_0 t) + \sum_{n=1}^{\infty} b_n \sin(n\omega_0 t) \qquad (3.2.31)$$

in order to define our Fourier series.

You're probably thinking that this looks a bit tedious to do by hand. And it is. But it's a snap to do on a computer. You just need a short subroutine that takes in a signal and then decomposes it into its individual harmonics, using pretty much the same approach that we've just discussed.

Please don't dismiss all this lightly. The ideas behind the Fourier series will pop up again and again in the future. What we're doing is using (3.2.28–3.2.30) as *filters*. The indicated integration operations really do sift through $f(t)$ and filter out everything except the harmonic we're looking for. If that harmonic is in the signal, then these calculations will find it. This is a very powerful idea and finds uses not only in linear vibrations, but in nonlinear vibrations as well.

By determining the Fourier series representation of the signal, we have done a very significant thing: We have moved from a time-domain description into the frequency domain. We now can determine how much of a certain frequency is present in a given signal. This viewpoint is very prevalent in a wide range of applications. For instance, a great deal of control work is done almost entirely in the frequency domain. The control designer specifically looks for control strategies that will reduce the amount of motion in a system at specific frequencies and is willing to let the motion increase at other frequencies in order to do so. Another example of where a frequency viewpoint is useful can be seen in the Tacoma-Narrows Bridge disaster. Most of you have already seen film clips of the Tacoma-Narrows Bridge collapse in high school. It involved the spectacular destruction of the Tacoma-Narrows Bridge, several decades ago, due to excessive motions of the bridge. (Really excessive!) The bridge failed because the designers didn't consider what sort of dynamic responses the bridge might have under different wind conditions. If they had known what to look for, they would have realized that a particular frequency of motion could be excited by the correct wind conditions and that the wind would pump energy into the bridge's motion until the bridge was destroyed (which is what occurred). The fundamental behavior is one that is most easily viewed from a frequency standpoint.

Finally, almost all studies of structural systems are concerned with the frequency content of the system. The reason for this is that periodic excitations close to the system's natural response frequencies can cause extremely large deflections of the system, deflections that can threaten the structural integrity of the system.

We'll examine spectral representations more fully later in the chapter. For now we must be content with knowing that any periodic forcing of our system can be broken down into several sinusoidal inputs and that we can, therefore, determine the response of our system by first determining the individual responses to all the separate harmonic components of the forcing, then adding them together, a process that is graphically shown in Figure 3.3. On the top of the figure we see a complex periodic input broken up into its constituent harmonic parts by means of a Fourier series analysis. Next, each harmonic is fed through the system being analyzed. As a result, the

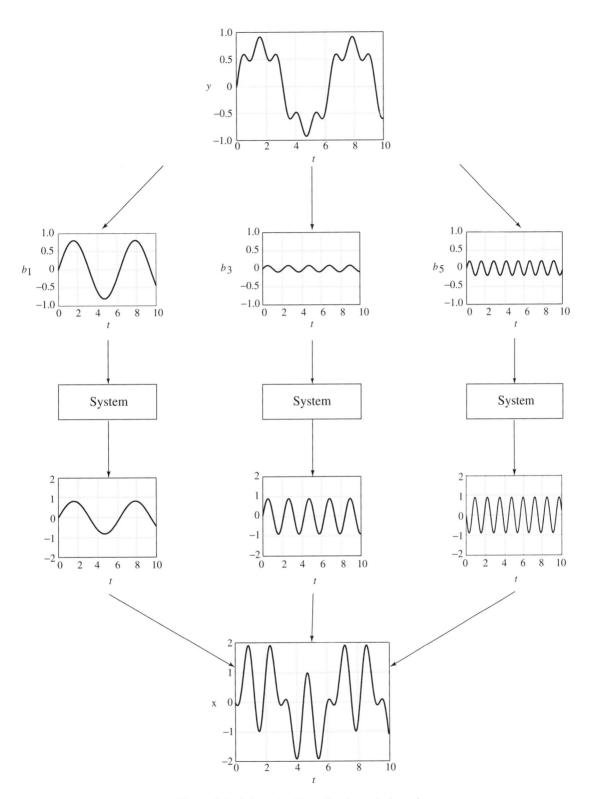

**Figure 3.3** Schematic of how Fourier analysis works

amplitude and phase of each component are altered. Finally, the resultant outputs are summed into the final response. Note that the input looks like a single harmonic with higher harmonic "wiggles," while the output is much more irregular. This change occurred because the harmonic near the system's natural frequency was greatly amplified due to resonance (due to the system's low damping). Thus we see that a low-magnitude component of the input signal can become significant or even dominant in the output.

It's worth reemphasizing that this Fourier series analysis technique is one of the powerful results that comes of having a linear system. Only when the governing differential equation is linear (all dependent variables and their derivatives appear only to the first power) can we break up a signal, calculate the individual responses, add them up, and end up with the actual solution. In a nonlinear system this would not work.

**Example 3.1**

> **Problem**   Determine the response of the system shown in Figure 3.4 when the base motion is in the form of the illustrated triangle wave.
>
> **Solution**   Although we could go ahead and just "plug and chug" using (3.2.28–3.2.31), we'll spend some time thinking about the problem and see how doing so can sometimes reduce our work level. The first thing to notice about the input forcing is that it is an odd function. Therefore we know immediately that all the $a_n$ coefficients in (3.2.31) are going to equal zero. (If the function is odd, then it can't contain any even components.) Next, we can see that only the odd harmonics ($n = 1, 3, 5, \ldots$) will appear because of the function's symmetry. Figure 3.5 shows why this is so. If an even harmonic ($n = 2, 4, \ldots$) was present, it would skew the function, making it look lopsided. Odd harmonics won't disturb the function's symmetry. Thus we need only concern ourselves with the odd harmonics. Furthermore, when we multiply the function by $\sin(n\omega_0 t)$ and integrate, we'll get an equal contribution from the section from 0 to $\frac{\pi}{\omega_0}$, as we do from the section from $\frac{\pi}{\omega_0}$ to $\frac{2\pi}{\omega_0}$. Thus we can calculate the integral over just one interval and double the answer. And finally,

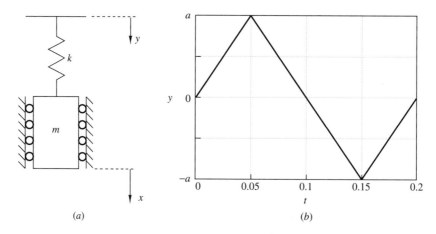

(a)                                                   (b)

**Figure 3.4**   Seismically excited system, triangle wave input

**Figure 3.5**    Adding odd and even-order harmonics destroys symmetry

we can see that the same argument holds for our remaining interval. The integral from 0 to $\frac{\pi}{2\omega_0}$ is the same as that from $\frac{\pi}{2\omega_0}$ to $\frac{\pi}{\omega_0}$. Therefore we need only calculate the integral from 0 to $\frac{\pi}{2\omega_0}$ and multiply the answer by 4. Figure 3.6 shows this. The sectors labeled A all contain the same area.

Based on this analysis, our problem has simplified to solving

$$b_n = 4\frac{\omega_0}{\pi} \int_0^{\frac{\pi}{2\omega_0}} \sin(n\omega_0 t)\left(\frac{2a\omega_0 t}{\pi}\right) dt, \qquad n = 1,\ 3,\ 5,\ \ldots \tag{3.2.32}$$

This isn't actually all that difficult to solve. If we integrate by parts, we'll have

$$b_n = \frac{8a\omega_0^2}{\pi^2}\left(-\frac{\cos(n\omega_0 t)}{n\omega_0}\Big|_0^{\frac{\pi}{2\omega_0}} + \int_0^{\frac{\pi}{2\omega_0}}\frac{\cos(n\omega_0 t)}{n\omega_0}dt\right), \qquad n = 1,\ 3,\ 5,\ \ldots \tag{3.2.33}$$

The first term is always equal to zero for odd values of $n$ and the second term integrates out to

$$\frac{8a\omega_0^2}{\pi^2}\left(\frac{\sin(n\omega_0 t)}{(n\omega_0)^2}\Big|_0^{\frac{\pi}{2\omega_0}}\right), \qquad n = 1,\ 3,\ 5,\ \ldots \tag{3.2.34}$$

The sine can only yield 1.0 or $-1.0$, depending on the value of $n$, and so the final solution becomes

$$b_{2n-1} = \frac{8a}{\pi^2}\frac{(-1)^{n+1}}{(2n-1)^2}, \qquad n = 1,\ 2,\ \ldots \tag{3.2.35}$$

Now that we've got the $b_n$ coefficients, we can put them together to obtain our Fourier series representation of the triangle wave

$$y(t) = \frac{8a}{\pi^2}\sum_{n=1}^{\infty}\frac{(-1)^{n+1}}{(2n-1)^2}\sin\left((2n-1)\omega_0 t\right), \qquad n = 1,\ 2,\ \ldots \tag{3.2.36}$$

The rest is reasonably straightforward. We know that the transfer function for the illustrated system is

$$g(\omega) = \frac{\omega_n^2}{\omega_n^2 - \omega^2} \tag{3.2.37}$$

Therefore we can immediately write the complete response by combining (3.2.36) and (3.2.37) to obtain

$$x(t) = \frac{8a}{\pi^2}\sum_{n=1}^{\infty}\left(\frac{(-1)^{n+1}}{(2n-1)^2}\right)\left(\frac{\omega_n^2}{\omega_n^2 - (2n-1)^2\omega_0^2}\right)\sin\left((2n-1)\omega_0 t\right), \qquad n = 1,\ 2,\ \ldots \tag{3.2.38}$$

a result that's graphically illustrated in Figure 3.7.

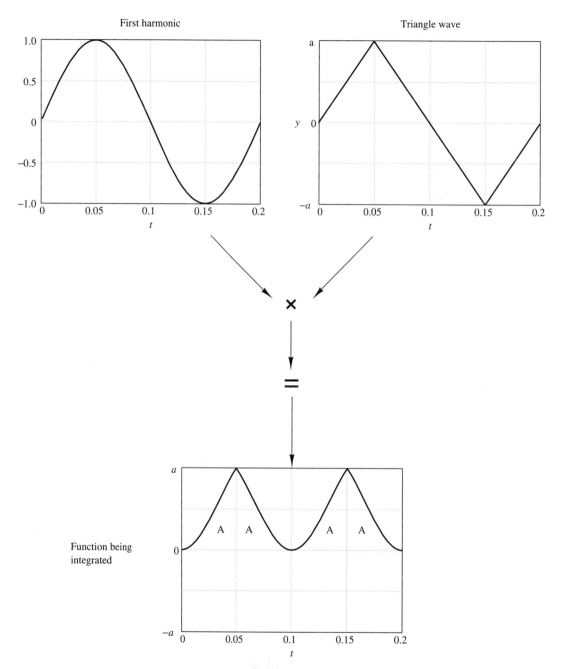

**Figure 3.6**  Symmetry in the integrals of a particular Fourier decomposition

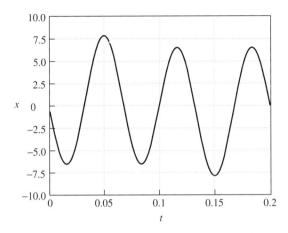

**Figure 3.7**  Steady-state response of spring-mass system to triangle wave

We've seen that the response harmonics in an input can occur near a system's natural frequency, causing a strong amplification of that harmonic. To make matters more complicated, multiple-degree-of-freedom systems and continuous systems have more than one natural frequency, any of which could potentially be excited by a harmonic of the input. Thus structural analysts must be aware of all the harmonics running around the system and how they might interact with the system's natural frequencies. If an excessive response due to resonance is found, then either the input or the system itself must be modified in order to move the excitation harmonic away from the natural frequency that's causing the problem.

## 3.3 FORCED RESPONSE VIA THE CONVOLUTION INTEGRAL

We've been concentrating on periodic excitations up until now because these are among the most common kind of system inputs you'll be dealing with. There are, however, many situations in which the excitation isn't periodic, and thus the approaches we've looked at aren't applicable. You know that the general solution to a forced differential equation involves both a transient and steady-state solution. So far, we've been consistently neglecting the transient motions, the claim being that they'll die out due to dissipation, leaving us with the purely steady-state response. Thus we were able to assume the existence of a steady-state response (having an unknown amplitude and phase) and directly solve for this response. This section takes a different point of view. We're not going to presuppose some steady-state response that will qualitatively match the excitation. Instead, we'll consider what the precise response is for *any* input and determine what the solution is equal to.

Just so you know where we're aiming, we're first going to find out how our systems respond to an impulsive force. Roughly speaking, an impulsive force is one that occurs over a very short span of time, such as a quick tap on the mass of the system. Once we have this response, we're going to claim that *any* forcing we might run into can be approximated as a string of impulses. Thus we'll just figure out the response due to all the individual impulses and sum them together. It's similar, in a way, to how we calculated the response of the system to a periodic, but nonsinusoidal input, using a Fourier series approach. We figured out what particular sinusoids made up the periodic signal, determined the system's response to each individually, and then summed up all the responses.

Our first question should be "What precisely is meant by an impulse?" Simply (and confusingly) stated, an impulse is an input that takes place over zero time and has infinite magnitude. Clearly, such an object can't really exist. Infinite magnitudes over zero time intervals aren't the sort of thing you can produce in the real world. This is an important point—impulses are mathematical abstractions. All we can do in the real world is approximate them. There are many ways of approximating them, and the one we'll use is probably the most easily grasped. We'll start by looking at a forcing function that's equal to $\frac{1}{\epsilon}$ and that lasts only from $t = 0$ to $t = \epsilon$ (Figure 3.8):

$$\begin{aligned} f_\epsilon &= \frac{1}{\epsilon} \qquad 0 < t \le \epsilon \\ f_\epsilon &= 0 \qquad t > \epsilon \end{aligned} \qquad (3.3.1)$$

$\epsilon$ is small to start and we'll make it smaller as we continue. You can see that as $\epsilon$ gets smaller, the magnitude of $f_\epsilon$ grows and the length of time it is nonzero shrinks. Thus, as $\epsilon$ approaches zero, the magnitude goes to infinity and the input takes place over zero time. We call such a function a Dirac delta function [1], $\delta(t)$, and represent it graphically by a vertical arrow, as shown in Figure 3.9. Although we can't actually attain such an input in real life, we can come close. The easiest way to approximate such an input is to take a hammer and sharply strike the mass in our system. The hammer will hit the mass and almost immediately bounce back, imparting a large force to the mass that lasts only for a short time. As we'll see later, this approach is a very common tool in vibrational analysis.

You'll note that the area under the $f_\epsilon$ function is always equal to 1.0, regardless of how small $\epsilon$ is. Do we have any way of knowing what such a forcing will do when applied to a dynamical system? Actually, you probably already have learned the answer in a dynamics course. Any course in two-dimensional dynamics will cover the topic of momentum for a system and how it changes. What's taught is that the change in momentum of a mass $m$ is equal to $\int f(t)dt$, where $f(t)$ is an

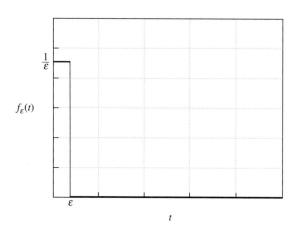

**Figure 3.8** First approximation to an impulse function

**Figure 3.9**  Dirac delta function

externally applied force. Since this is simply the area under the $f_\epsilon$ curve, we see that the linear momentum must be altered as a result of the force. Thus, applying the force shown in Figure 3.8 will mean that

$$m(\dot{x}(\epsilon) - \dot{x}(0)) = \int_0^\epsilon \frac{1}{\epsilon} dt = 1 \tag{3.3.2}$$

If the initial velocity is zero, then this implies that the velocity at $t = \epsilon$ is equal to

$$\dot{x}(\epsilon) = \tfrac{1}{m} \tag{3.3.3}$$

As we take $\epsilon$ smaller and smaller (going to a limit of zero), we approach a simple change of initial conditions. As $\epsilon$ approaches zero, $\dot{x}(\epsilon)$ approaches $\dot{x}(0)$. What about the displacement $x$, though? Is it possible that this might change also? To answer this, let's assume that it does change due to the impulse and goes to a value $x_0$. We know that the velocity at time $t$ is found from $\dfrac{x(t+\epsilon) - x(t)}{\delta}$. Thus, in the limit, as $\epsilon$ goes to zero, we'd have

$$\dot{x}(0) = \lim_{\epsilon \to 0} \frac{x_0}{\epsilon} \to \infty \tag{3.3.4}$$

Since we know the velocity can't go to infinity, we conclude that the displacement must actually still be zero after the impulse has been applied.

You can see these same results in a slightly different way by explicitly solving

$$m\ddot{x} = f_\epsilon(t) \tag{3.3.5}$$

where $f_\epsilon(t)$ is shown in Figure 3.8. Assume initial conditions $x(0) = 0$ and $\dot{x}(0) = 0$. We can integrate both sides to obtain

$$\dot{x}(t) = \frac{1}{\epsilon} t \tag{3.3.6}$$

and

$$x(t) = \frac{1}{\epsilon} \frac{t^2}{2} \tag{3.3.7}$$

for $t$ between 0 and $\epsilon$. (Note that the initial conditions were used in the solution to eliminate any constants of integration.) Let's assume that $\epsilon$ is small, say .01. Then $\dot{x}(.01) = 1.0$ and $x(.01) = .005$. It's pretty clear that the velocity is finite and the change in position is essentially zero. As $\epsilon$ goes to zero, this effect becomes even more pronounced, finally leading to a finite change in initial velocity and no change in displacement as $\epsilon$ hits zero.

What about using a Dirac delta function as the input for a more complicated system? To see, let's determine what will happen to a general spring-mass damper when an impulse is applied. In this case the equation of motion would be given by

$$m\ddot{x} + c\dot{x} + kx = \delta(t) \tag{3.3.8}$$

To aid in our physical understanding, let's go ahead and use the $f_\epsilon$ approximation for the delta. Integrating once, from $t = 0$ to $t = \epsilon$, gives us

$$m\left(\dot{x}(\epsilon) - \dot{x}(0)\right) + c\left(x(\epsilon) - x(0)\right) + k\left(\int_0^\epsilon x\,dt\right) = 1 \tag{3.3.9}$$

Can $\int_0^\epsilon x\,dt$ be finite (and thus part of the left-hand-side terms that add to 1.0)? If so, then $x(\epsilon)$ must go to infinity like $\frac{a}{\epsilon}$, since otherwise the integral would vanish. As we already know the mass isn't going to go off to infinity in zero time, we know that the integral must therefore equal zero. Next we can ask if the displacement can be finite, letting the $c\left(x(\epsilon) - x(0)\right)$ term contribute toward making up the 1.0 on the right-hand side. We've already discussed the fact that this would imply an infinite velocity and is, therefore, not possible. Thus we're left with the same result as before, namely that

$$m\left(\dot{x}(\epsilon) - \dot{x}(0)\right) = 1 \tag{3.3.10}$$

Therefore the velocity undergoes a step change. In the limit of $\epsilon \rightarrow 0$, this again produces the result that the effect of an applied impulsive force is to alter the initial velocity and leave the displacement undisturbed.

Now you can see why using a hammer to imply an approximate delta function of force to a test object is a nice idea. If a Dirac delta is applied, then the only result is to change the initial conditions of the system. Thus we'll end up with an *unforced* problem, one for which the displacement is zero and the velocity is finite. We can then observe the free response and, from it, deduce the damping, natural frequency, etc. Many techniques of modal analysis rely on just such an approach.

The other way in which the Dirac delta function is useful is in providing us with a method for determining the forced response of our system to any input. Figure 3.10 illustrates the first step in this approach. Figure 3.10$a$ shows the original continuous forcing function. The next step is to discretize the time line into segments $\Delta$ seconds long, shown in Figure 3.10$b$. If $\Delta$ is small, the area within each four-sided box is given by the value of the function multiplied by the time interval, that is, the first box has area $\Delta f(0)$, the second has area $\Delta f(\Delta)$, etc. The final step is to replace the continuous function by a train of Dirac delta functions, as shown in Figure 3.10$c$.

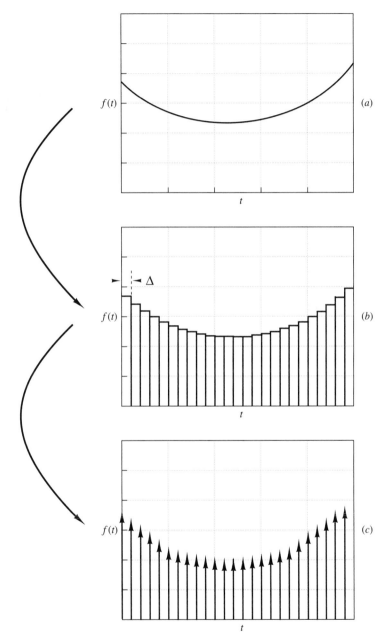

**Figure 3.10**   Approximation of a continuous function by a train of delta functions

Each unit delta function is multiplied by the appropriate factor to make it match the actual function values. Thus the first delta is multiplied by $\Delta f(0)$, the second by $\Delta f(\Delta)$, etc.

The idea here is to hit our system with this train of delta functions, rather than using the original, continuous, function. We do this because we've already figured out how the system will respond to a delta function: it will have a step change in velocity and then undergo free vibration. So what we'll have is a succession of free-vibration responses. The system will undergo free vibrations for $\Delta$ seconds, have a change in velocity due to the next impulse, and then freely vibrate for another $\Delta$ seconds. We can, therefore, sum up all these free-vibration responses to find out how the overall system responds.

Before doing this, let's pin down exactly what the free-vibration response of a spring-mass damper is. Say our system is acted on by a Dirac delta function, with magnitude $\bar{f}$. Our equation of motion is

$$m\ddot{x} + c\dot{x} + kx = \bar{f}\delta(t) \tag{3.3.11}$$

We've seen that this is equivalent to

$$m\ddot{x} + c\dot{x} + kx = 0 \tag{3.3.12}$$

with initial conditions $x(0) = 0$ and $\dot{x}(0) = \frac{\bar{f}}{m}$.

The general solution to an unforced spring-mass damper system is given by

$$x(t) = e^{-\zeta\omega t}(a_1 \cos(\omega_d t) + a_2 \sin(\omega_d t)) \tag{3.3.13}$$

where $\omega_n^2 = \frac{k}{m}$ and $\omega_d = \omega_n\sqrt{1 - \zeta^2}$. Applying the $x(0) = 0$ initial condition sets $a_1$ to zero, and applying the $\dot{x}(0) = \frac{1}{m}$ initial condition gives us $a_2 = \frac{1}{m\omega_d}$ Thus the total solution is given by

$$x(t) = \frac{1}{m\omega_d}e^{-\zeta\omega t}\sin(\omega_d t) \tag{3.3.14}$$

We'll use the symbol $h(t)$ to represent this response and remember that $h(t)$ is the *impulse response* of any spring-mass damper

$$h(t) \equiv \frac{e^{-\zeta\omega_n t}}{m\omega_d}\sin(\omega_d t) \tag{3.3.15}$$

Now let's look at our original problem again. We've already approximated our input function by a string of delta functions. Therefore we know that the response from $t = 0$ to $t = \Delta$ is given by

$$x(t) = \frac{\Delta f(0)e^{-\zeta\omega_n t}}{m\omega_d}\sin(\omega_d t) = \Delta f(0)h(t) \tag{3.3.16}$$

From $t = \Delta$ to $t = 2\Delta$, the response is given by

$$x(t) = \Delta f(0)h(t) + \frac{\Delta f(\Delta)e^{-\zeta\omega_n(t-\Delta)}}{m\omega_d}\sin(\omega_d(t - \Delta))$$

$$= \Delta f(0)h(t) + \Delta f(\Delta)h(t - \Delta) \tag{3.3.17}$$

You can see the pattern shaping up here. We're going to get a longer and longer string of responses as $t$ increases. Jumping to the response at the $n$th interval, we've got

$$x(t) = \sum_{i=0}^{n-1} f(i\Delta)h(t - i\Delta)\Delta \qquad (3.3.18)$$

which holds from $t = (n - 1)\Delta$ to $t = n\Delta$. Look what happens if we now let the interval $\Delta$ get smaller and smaller. The summation becomes an integral, the $\Delta$ becomes a differential, and we're left with

$$x(t) = \int_0^t f(\tau)h(t - \tau)d\tau \qquad (3.3.19)$$

For our problem this becomes

$$x(t) = \int_0^t f(\tau)\frac{e^{-\zeta\omega_n(t-\tau)}}{m\omega_d} \sin(\omega_d(t - \tau))\,d\tau \qquad (3.3.20)$$

Equation (3.3.19) is called a *convolution integral* and, not surprisingly, this approach is called the *convolution method*. This may look pretty heavy-duty, but if you remember where it came from, it should seem a bit more manageable. All we're doing is asking what the *impulse* response is at any particular point in time and then letting this response remain. It's even easier to see what's going on if we use a particular time $t_1$ rather than $t$. In this case he response at $t_1$ is given by

$$x(t_1) = \int_0^{t_1} f(\tau)h(t_1 - \tau)d\tau \qquad (3.3.21)$$

If we think about it in this way we see that the response at $t_1$ depends upon $f(\tau)$, with $f(\tau)$ going from $f(0)$ to $f(t_1)$ and upon $h(t_1 - \tau)$, with $h(t_1 - \tau)$ going from $h(t_1)$ to $h(0)$. For instance, take the first part of the integral, that is, where $\tau = 0$. This is approximately equal to $f(0)h(t_1)\Delta$, where we're taking only a $\Delta$-wide slice of time. This is merely the response due to an initial impulse. Since we're looking at the response at time $t_1$, we let the system continue to vibrate for that long (i.e., we use $h(t_1)$).

Now look at the end of the integral. In this case $\tau$ is almost equal to $t_1$, say $t_1 - \Delta$, and the chunk of the integral near the end is equal to $f(t_1 - \Delta)h(\Delta)\Delta$. This indicates that some of the response is due to the latest excitation. This excitation occured at $t_1 - \Delta$ and is given by $f(t_1 - \Delta)\Delta$. The free vibration only lasts for $\Delta$ seconds in this case, thus we use $h(\Delta)$.

If you look at the discrete way we first approached the integral, and then think about what we've just gone over, you'll see they indicate the same result. Taken in total, this integral operation says to take the response due to the early forcing and remember it. Then add to this the response due to the force that occurred a little bit later. Then add the part due to even later forcings. Continue doing this until you're at the time you want.

Although (3.3.19) is totally valid, you may find it easier to express it in a slightly different way. To derive the new form, we'll define a new variable $\gamma$, where we'll say that $\gamma = t - \tau$. Remember that $t$ is fixed in the convolution. Therefore, if we take a differential of $\gamma$, we'll get

$$d\gamma = -d\tau \qquad (3.3.22)$$

Eliminating $\tau$ in favor of $\gamma$ in (3.3.19) gives us

$$x(t) = -\int_t^0 f(t-\gamma)h(\gamma)d\gamma \qquad (3.3.23)$$

We found the new limits of integration by substituting what $\tau$ is equal to at these limits (0 and $t$) into our expression for $\gamma$. We can flip the limits of integration (which simply generates a $-1$ factor), cancel this $-1$ with the existing $-1$ in front of the integral, and get

$$x(t) = \int_0^t f(t-\gamma)h(\gamma)d\gamma \qquad (3.3.24)$$

Since $\gamma$ is just a dummy variable (it doesn't show up in the final result), we can rename it $\tau$ so that the equation looks more like (3.3.19):

$$x(t) = \int_0^t f(t-\tau)h(\tau)d\tau \qquad (3.3.25)$$

This is identical to (3.3.19) except for the fact that we're putting the time shift into the forcing function $f$, rather than into the impulse response $h$. You should use whichever form is most convenient for your particular problem.

What we've now derived is a general way of determining the response of a system to any input *if* you've already found out the response of the system to a Dirac delta input. Let's work out an example to pin it down a little more.

**Example 3.2**

> **Problem**   Consider the damped system and excitation shown in Figure 3.11. The excitation is a relatively simple one, namely a step input. This is the sort of situation that would occur if you could have a mass suspended from a spring in the absence of gravity and, at $t = t_0$, suddenly switch gravity on. For clarity, we'll let $t_0 = 0$. Letting it be any other value doesn't affect the solution at all, except to shift the time that the mass starts to move from $t = 0$ to $t = t_0$. Let's find the response to this input.
>
> **Solution**   To solve this problem by the method outlined in the previous section, we'll need the response of the system to a unit delta function, that is, the solution to
>
> $$m\ddot{x} + c\dot{x} + kx = \delta(t) \qquad (3.3.26)$$
>
> We already know from (3.3.15) that the impulse response for this problem is given by
>
> $$h(t) = \frac{e^{-\zeta\omega_n t}}{m\omega_d}\sin(\omega_d t) \qquad (3.3.27)$$
>
> Knowing $h(t)$, we can go ahead and use the convolution integral (3.3.25) to find the response to the illustrated step input
>
> $$x(t) = \int_0^t h(\tau)f(t-\tau)d\tau = \frac{\bar{f}}{m\omega_d}\int_0^t e^{-\zeta\omega_n \tau}\sin\left(\omega_d(\tau)\right)d\tau \qquad (3.3.28)$$
>
> At this point we can simply use a table of integrals to evaluate (3.3.28). What we'll do instead is show how to calculate the answer. That way you'll be able to carry on even if you don't have an integral table handy.

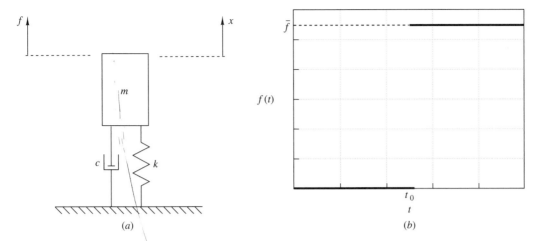

**Figure 3.11**   Damped system subjected to a step input force

If you try integration by parts you'll still end up with an integral involving the product of a sine (or cosine) and an exponential. The trick to getting the answer is to integrate by parts *twice*. In this way we'll end up with an equation having the integral we want to evaluate on the left and also on the right. By combining the two, we'll be able to solve the problem. It goes like this. For the first integration by parts ($\int u\,dv = uv - \int v\,du$) we identify $dv$ as $\sin(\omega_d \tau)d\tau$ and $u$ as $e^{\zeta \omega_n \tau}$. This gives us

$$\frac{\bar{f}}{m\omega_d} \int_0^t e^{-\zeta \omega_n \tau} \sin(\omega_d(\tau))\,d\tau$$

$$= \frac{\bar{f}}{m\omega_d} \left( 1 - e^{-\zeta \omega_n t} \cos(\omega_d t) - \zeta \omega_n \int_0^t e^{-\zeta \omega_n \tau} \cos(\omega_d \tau)d\tau \right) \qquad (3.3.29)$$

When integrating the $e^{-\zeta \omega_n \tau} \cos(\omega_d \tau)d\tau$ term, we'll use $u = e^{-\zeta \omega_n \tau}$ and $dv = \cos(\omega_d \tau)$. This will give us

$$\frac{\bar{f}}{m\omega_d} \int_0^t e^{-\zeta \omega_n \tau} \sin(\omega_d(\tau))\,d\tau = \frac{\bar{f}}{m\omega_d^2} \left( 1 - e^{-\zeta \omega_n t} \cos(\omega_d t) \right)$$

$$- \frac{\zeta \omega_n \bar{f}}{m\omega_d^3} e^{-\zeta \omega_n \tau} \sin(\omega_d \tau)d\tau - \frac{\bar{f}(\zeta \omega_n)^2}{m\omega_d^3} \int_0^t e^{-\zeta \omega_n \tau} \sin(\omega_d \tau)d\tau \qquad (3.3.30)$$

You can see that we've got the same integral on the left and right side. Combining them on the left side of the equation gives us

$$\left( \frac{\bar{f}}{m\omega_d(1 - \zeta^2)} \right) \int_0^t e^{-\zeta \omega \tau} \sin(\omega_d(\tau))\,d\tau$$

$$= \frac{\bar{f}}{k(1 - \zeta^2)} \left( 1 - e^{-\zeta \omega_n t} \cos(\omega_d t) \right) - \frac{\zeta \bar{f}}{k(1 - \zeta^2)^{\frac{3}{2}}} e^{-\zeta \omega_n t} \sin(\omega_d t) \qquad (3.3.31)$$

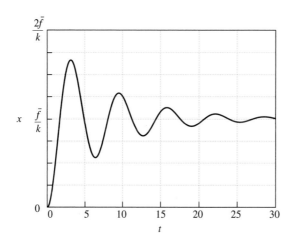

**Figure 3.12** Displacement response for a damped system subjected to a step force input

Some of the $\omega_d$'s were reexpressed in terms of $\omega_n$ and $\zeta$, and $m\omega_n^2$ was reexpressed as $k$ in the above. Multiplying by $(1 - \zeta^2)$ gives us our final expression for $x(t)$

$$x(t) = \frac{\bar{f}}{k}\left(1 - e^{-\zeta\omega_n t}\cos(\omega_d t)\right) - \frac{\zeta\bar{f}}{k\sqrt{1 - \zeta^2}}e^{-\zeta\omega_n t}\sin(\omega_d t) \tag{3.3.32}$$

This solution is illustrated in Figure 3.12. Note that the response starts at zero, grows to a maximum, and then dies down as it oscillates about $\frac{\bar{f}}{k}$. Of course, this steady-state response was predictable from the original differential equation by setting all the time-varying parts ($\ddot{x}$ and $\dot{x}$) to zero and solving the resulting equation. Varying the damping will serve to reduce or increase the rate of decay of the oscillations.

**Example 3.3**

The second example we'll look at will be interesting for two reasons. The first is that we'll see what the response looks like for a finite ramp input force. All right, that really isn't all that interesting, is it? But the second reason is better. After we've found out what the response is to a finite ramp, we'll ensure that the area within the ramp is always equal to 1.0, and we'll reduce the duration of the ramp to zero. This means that the ramp will look a lot like the delta function, with an infinite value, zero duration, and unit area. The interesting question will be whether or not the system responds to this input as it did to the force profile of Figure 3.8.

The force profile is shown in Figure 3.13. As you can see, the force ramps up linearly from zero to a maximum of $\bar{f}$ (at $t = t_0$). Our equation of motion is given by

$$m\ddot{x} + kx = f(t) \tag{3.3.33}$$

or

$$\ddot{x} + \omega_n^2 x = \frac{f(t)}{m} \tag{3.3.34}$$

We can find $h(t)$ either by solving the free-vibration problem subjected to a unit velocity or, more simply, just use (3.3.15) and set $\omega_d = \omega_n$ and $\zeta = 0$. This results in

$$h(t) = \frac{1}{m\omega_n}\sin(\omega_n t) \tag{3.3.35}$$

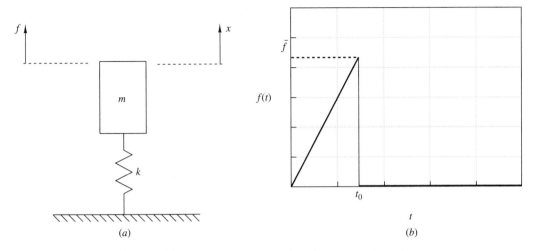

**Figure 3.13**   Undamped system subjected to a a ramp input force

Using $h(t)$ in our convolution integral gives us

$$x(t) = \int_0^t f(t - \tau) \sin(\omega_n \tau) d\tau \tag{3.3.36}$$

If $t$ is less than $t_0$, then we can simply use (3.3.36). Since the forcing function is zero for $t > t_0$, we'd have to solve for $x(t_0)$ and $\dot{x}(t_0)$ using (3.3.36) and then solve for the system's free vibrations for all times $t > t_0$, using $x(t_0)$ and $\dot{x}(t_0)$ as initial conditions.

To solve (3.3.36), we'll need to reexpress $f(t)$ more explicitly. Since $f(t)$ is linearly increasing from 0 to $\bar{f}$, it is given by

$$f(t) = \frac{\bar{f}}{t_0} t, \qquad 0 < t \le t_0 \tag{3.3.37}$$

Using this form for $f(t)$, (3.3.36) becomes

$$x(t) = \frac{\bar{f}}{m \omega_n t_0} \int_0^t \sin(\omega_n \tau)(t - \tau) d\tau \tag{3.3.38}$$

Expanding this yields

$$x(t) = \frac{\bar{f}}{m \omega_n t_0} \left( t \int_0^t \sin(\omega_n \tau) d\tau - \int_0^t \tau \sin(\omega_n \tau)(t - \tau) d\tau \right) \tag{3.3.39}$$

We were able to pull the $t$ out of the first integral because it is considered a constant during the integration with respect to $\tau$. Integrating these (using integration by parts for the second integral) yields

$$x(t) = \frac{\bar{f}}{m \omega_n^2 t_0} \left( t - \frac{\sin(\omega_n t)}{\omega_n} \right) \tag{3.3.40}$$

If we cared to, we could now use this to find $x(t_0)$ and $\dot{x}(t_0)$ and then to apply these as initial conditions to determine the free vibrations of the subsequent motion. Instead, what we'll do is let

$\bar{f} = \frac{2}{t_0}$. For this value of $\bar{f}$ you can see that the area under the $f(t)$ curve is equal to 1.0. The solution for $x(t)$ is now given by

$$x(t) = \frac{2}{m\omega_n^2 t_0^2}\left(t - \frac{\sin(\omega_n t)}{\omega_n}\right) \tag{3.3.41}$$

The next step is to ask what $x(t)$ and $\dot{x}(t)$ are as $t_0$ goes to zero. For $t_0$ small, (3.3.41) becomes

$$x(t) \approx \frac{2}{m\omega_n^2 t_0} - \frac{2}{m\omega_n^3}\frac{\omega_n t_0}{t_0^2} \to 0 \tag{3.3.42}$$

Differentiating (3.3.41) gives us

$$\dot{x}(t) = \frac{2}{m\omega_n^2 t_0^2}(1 - \cos(\omega_n t)) \tag{3.3.43}$$

Letting $t = t_0$ and expanding out $\cos(\omega_n t)$ to include the first two terms gives us

$$\dot{x}(t_0) = \frac{2}{m\omega_n^2 t_0^2}\left(1 - \left(1 - \frac{(\omega_n t_0)^2}{2}\right)\right) \tag{3.3.44}$$

which leads to

$$\dot{x}(t_0) = \frac{2}{m\omega_n^2 t_0^2}\frac{\omega_n^2 t_0^2}{2} = \frac{1}{m} \tag{3.3.45}$$

There we have it! As $t_0$ goes to zero, what we've found is that the application of the ramp input has served to alter the system initial conditions from $x(0) = 0$ and $\dot{x}(0) = 0$ to $x(0) = 0$ and $\dot{x}(0) = \frac{1}{m}$, exactly what we found when we applied the delta that came from $f_\epsilon$. Thus a Dirac delta that's found from taking a step input and letting the duration go to zero, while holding the area fixed at 1.0, has the same effect as a Dirac delta formed from a ramp input in which the duration of the ramp goes to zero and the integral is held at 1.0. As you now might guess, just about any function you can come up with that has vanishingly small duration and area equal to 1.0 will act in the same way. This fact is the reason for much of the success in impact testing in modal analyses. For these tests the test object must be struck with an instrumented hammer. The hope is that the force input of the hammer acts like a Dirac delta function. Sophisticated software can then capture the output of the sensors attached to the system and determine the system properties. Obviously the hammer hit can't really input a Dirac delta. Our results here show that this doesn't matter. As long as the hammer strike is quick, it is well approximated by a Dirac delta function.

Before leaving this section, we'll discuss a couple of additional properties of the Dirac delta function, since they'll be useful to know later on. We've already seen that the Dirac delta function is defined as a function that is zero everywhere except at zero (where it is equal to infinity). The integral over the delta is always equal to 1.0. Thus

$$\int_{-\infty}^{\infty} \delta(t)dt = 1 \tag{3.3.46}$$

We can be a little more general and say that the delta occurs not at $t = 0$, but at any time $t_1$. Such a function is shown in Figure 3.14. Since all we've done is displace the nonzero part of the delta function, the integral over $t$ will still be equal to 1.0.

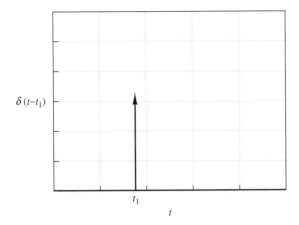

$\delta(t-t_1)$

$t_1$

$t$

**Figure 3.14**  Delta function displaced in time

A helpful property associated with delta functions is the fact that

$$\int_{-\infty}^{\infty} f(t)\delta(t - t_1)dt = f(t_1) \tag{3.3.47}$$

If we recall that the delta function can be viewed as the limiting case of the $f_\epsilon$ function shown in Figure 3.8 (and letting the nonzero part occur at $t = t_1$), this property is pretty easy to see. Since the delta function is zero almost everywhere, most of the integral in (3.3.47) is just equal to zero. The only nonzero contribution occurs right where the delta becomes nonzero. If $\epsilon$ is small, $f(t)$ is approximately equal to $f(t_1)$ over the entire interval. Thus the integral is approximated by

$$\int_{-\infty}^{\infty} f(t)\delta(t - t_1)dt \approx f(t_1)\frac{1}{\epsilon}\epsilon \tag{3.3.48}$$

In the limit of $\epsilon \to 0$, this becomes an equality. Thus we see that the $\epsilon$'s cancel out, leaving us with

$$\int_{-\infty}^{\infty} f(t)\delta(t - t_1)dt = f(t_1) \tag{3.3.49}$$

## 3.4 SHOCK RESPONSE

A special class of transient loadings are those involving shock. As previously mentioned, shock loadings can come about in a variety of ways. Explosions always generate shocks. When an object is suddenly brought to rest after falling, the result is a shock. When a car hits a pothole, a shock is produced that then affects the car and anything in it. Obviously, with sharp loading of this sort, concern over an object's structural integrity becomes acute. The acceleration levels produced by shock loading are high and can easily lead to breakage within the system.

Figure 3.15*a, b,* and *c* shows some common shock profiles. In Figure 3.15*a,* the force input rises quickly to some maximum value and then stays constant at that value. This might be the case

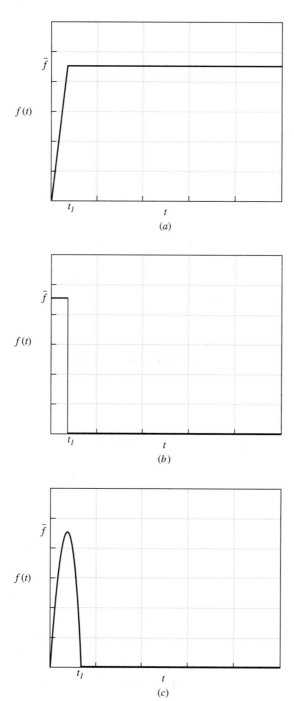

**Figure 3.15** (*a*) Abrupt change in force level; (*b*) impulsive force; and (*c*) half-sine input force

when looking at the loading within a hydraulic system. Once the system is pressurized, the force on an interior surface will be higher than it was before pressurization. If the rise in pressure is rapid, the result is a shock load. In this case the shock is caused by the rapid increase in force, and not by the steady-state force that exists after $t_0$.

Figure 3.15$b$ shows a shock load that consists of a constant force with a short duration $t_0$. This sort of shock can occur as a result of a blast wave, for which the object in the path of the blast will feel a high level of force for a short time.

Finally, Figure 3.15$c$ shows a common approximation to shock loadings, called a half-sine loading. Just as you'd expect from the name, the load has the profile of a half sine wave and then remains at zero for later times. If $t_0$ is small and the amplitude of the half sine is large, the result is a shock load. Engineering experience and insight will often be needed in order to decide whether the loading shown in Figure 3.15$b$ or 3.15$c$ is the most appropriate for the actual shock being considered. If the time duration of the shock is short enough, it might be enough to simply view the input as a Dirac delta function.

Once we know the profile of the shock load, it is easy to apply the convolution approach to determine the response. Alternatively, considering the speed of modern computers, we might simply numerically integrate the equations of motion, using the shock profile as the input excitation. In either event, once we have determined the response, we can then characterize the effect of the shock. Commonly, we are concerned with the *peak* acceleration levels that the shock will induce in the system. Thus we can plot peak acceleration versus some shock parameter (say shock duration for example) and get a feel for how the system will be affected by different shocks.

Let's consider the case for which a test object has been dropped and contacts the ground. We'll determine what the acceleration levels are for the mass and see how this varies as a function of contact velocity. Finally, we'll relate contact velocity to the drop height. This will allow us to set safe limits as to how high the structure should be carried during movement, since the height at which the object is dropped is directly related to the impact velocity. The system to be examined is shown in Figure 3.16, and consists of a spring and a mass. When dropped, it will accelerate due to gravity and eventually contact the floor with a velocity $v$. At the instant it strikes the floor,

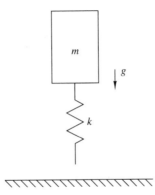

**Figure 3.16**  Spring-mass subjected to shock due to impact

the spring will be uncompressed. We'll measure $x$ from the initial position of the mass when it contacts the floor. We'll also orient $x$ downward so that the initial velocity upon contact is positive. The relevant equation of motion is given by

$$m\ddot{x} + kx = mg \tag{3.4.1}$$

with initial conditions $x(0) = 0$ and $\dot{x}(0) = v$.

      Dividing by $m$ gives us

$$\ddot{x} + \omega_n^2 x = g \tag{3.4.2}$$

The general solution for this problem is given by

$$x(t) = a_1 \cos(\omega_n t) + a_2 \sin(\omega_n t) + \frac{g}{\omega_n^2} \tag{3.4.3}$$

where the sine and cosine are the homogeneous solution and $\frac{g}{\omega_n^2}$ is the particular solution, that is, the one due to the gravity loading. This problem, therefore, falls into the category of Figure 3.14$a$, one in which a constant loading is suddenly applied to the system.

      Applying the initial condition $x(0) = 0$ tells us that $a_1 = -\frac{g}{\omega_n^2}$ and a subsequent application of $\dot{x}(0) = v$ gives us $a_2 = \frac{v}{\omega_n}$. Thus our complete solution is given by

$$x(t) = \frac{g}{\omega_n^2}(1 - \cos(\omega_n t)) + \frac{v}{\omega_n}\sin(\omega_n t) \tag{3.4.4}$$

To find the accelerative response, we need to differentiate twice, yielding

$$\ddot{x}(t) = g\cos(\omega_n t) - v\omega_n \sin(\omega_n)t \tag{3.4.5}$$

We know from our previous work with harmonic vibrations that this response can be expressed as $a\cos(\omega_n t + \phi)$, where $a$ is given by

$$a = \sqrt{g^2 + (v\omega_n)^2} \tag{3.4.6}$$

Therefore the maximum acceleration magnitude is given by $a$.

      It's always nice to normalize our plots, and a convenient normalization to choose for the acceleration level is dividing by gravity. Thus our normalized maximum acceleration is given by $\frac{a}{g}$ or, expressed slightly differently,

$$\left|\frac{\ddot{x}}{g}\right|_{max} = \sqrt{1 + \left(\frac{v\omega_n}{g}\right)^2} \tag{3.4.7}$$

An equation you've undoubtedly seen in your dynamics course is

$$\ddot{x}\,dx = \dot{x}\,d(\dot{x}) \tag{3.4.8}$$

a result that allows us to immediately determine the relationship between the drop height $x_h$ and the impact velocity $v$. Integrating this equation, using a constant acceleration (equal to $g$), gives us

$$gx_h = \frac{v^2}{2}\qquad(3.4.9)$$

Solving for $v$ gives us

$$v = \sqrt{2gx_h}\qquad(3.4.10)$$

Thus our normalized maximum acceleration versus drop height becomes

$$\left|\frac{\ddot{x}}{g}\right|_{max} = \sqrt{1 + \frac{2x_h\omega_n^2}{g}}\qquad(3.4.11)$$

Finally, look again at (3.4.2). As we've discussed previously, the steady-state deflection of the mass due to gravity is found by setting the $\ddot{x}$ term to zero, giving us

$$\omega_n^2 x_{eq} = g\qquad(3.4.12)$$

or

$$x_{eq} = \frac{g}{\omega_n^2}\qquad(3.4.13)$$

Interestingly, you'll see that this term pops up in (3.4.11). Substituting $x_{eq}$ for $\frac{g}{\omega_n^2}$ in (3.4.11) gives us our final expression:

$$\left|\frac{\ddot{x}}{g}\right|_{max} = \sqrt{1 + \frac{2x_h}{x_{eq}}}\qquad(3.4.14)$$

Figure 3.17 shows a plot of this maximal acceleration versus the normalized drop height $\dfrac{x_h}{x_{eq}}$.

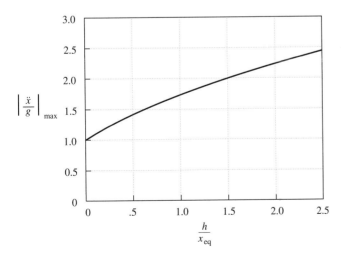

**Figure 3.17**  Normalized maximum acceleration versus normalized distance fallen

You can see that the maximal acceleration increases as we move along the horizontal axis. This certainly makes sense; as the item is dropped from greater heights, the maximal acceleration levels should increase. We can get some more insight by recalling that $\frac{x_h}{x_{eq}}$ is the same as $\frac{1}{2}\left(\frac{v\omega_n}{g}\right)^2$. Since increases in $v$ move us along the horizontal axis, we see that increasing impact velocities increase our maximal accelerations. This fact is directly related to the previous observation. A new observation, however, is the fact that increasing $\omega_n$ also moves us to the right on the horizontal axis, thus increasing maximal acceleration levels. This gives us some useful design insight. We now know that increasing the natural frequency of a system is bad, at least for the impact loads if the system is dropped. We might want a higher natural frequency for other reasons (such as strength), but this tells us that a trade-off we must live with is an increased susceptibility to certain shock loadings. Often we will find that structures that deal with shock well are not particularly good with regard to steady-state excitation and vice versa. Once again, this is a classic case of how we must carefully consider the different factors of a design, weigh their importance, and then come up with the best overall design that produces acceptable performance in all the relevant categories.

**Example 3.4**

> **Problem**   Determine how a mass of 1 kg can be dropped without the maximal acceleration levels exceeding $10g$. Assume the floor is massless and has an effective stiffness coefficient of 10,000 N/m. $g = 9.81$ m/s$^2$.

> **Solution**   From (3.4.11) we have

$$|\ddot{x}|_{max} = g\sqrt{1 + \frac{2x_h\omega_n^2}{g}} \tag{3.4.15}$$

> $\omega_n$ for this example is given by $\sqrt{\frac{10,000}{10}} = 31.6$ rad/s$^2$. We must solve

$$10g = g\sqrt{1 + \frac{2x_h\omega_n^2}{g}} \tag{3.4.16}$$

> for $x_h$. Rearranging this gives us

$$x_h = \frac{99g}{2\omega_n^2} = .486 \text{ m} \tag{3.4.17}$$

As a further illustration of how we can examine shock responses, we'll again look at a falling vibratory system, but this time we'll include damping. Figure 3.18 shows the system configuration. In this case, the analysis is a bit more complex since we must account for the exponential decay of the vibrations.

Thus we won't be able to simply find the overall amplitude of the sinusoidally varying acceleration and say it is equal to the maximum acceleration, as we did in the previous example. For this system we must actually find the time at which the maximal response occurs and then evaluate

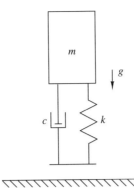

**Figure 3.18**  Spring-mass damper subjected to shock due to impact

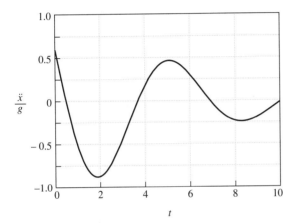

**Figure 3.19**  Normalized acceleration for a damped system impacting a rigid floor

its magnitude. Figure 3.19 shows a typical amplitude and acceleration response ($\zeta = .2$, $\omega_n = 1$ rad/s, and $v = 10$ m/s) for such a system, while Figure 3.20 shows how the acceleration profile changes if the natural frequency of the system is increased to 100 rad/s. Clearly, increasing the natural frequency has increased the overall magnitudes of the accelerations.

In order to determine what that maximal acceleration response is, we must go to more trouble than we did in the undamped case. For both cases shown, the maximal acceleration is associated with the negative acceleration peak (occuring as around $t = 2$ s for $\omega_n = 1$ rad/s and .01 seconds for $\omega_n = 100$ rad/s). We can either determine this peak value analytically or by simply using a program that tracks the acceleration value and stops when the absolute value is a maximum. Note that if the impacts are very gentle, it's possible for the initial acceleration to be the largest one (which didn't happen for these two cases). Thus you'd need to check which value was highest, the initial one or the one occuring some way into the vibration cycle.

For either approach, you first must solve for the acceleration as a function of time. This is done as follows. The equation of motion is given by

$$m\ddot{x} + c\dot{x} + kx = mg \tag{3.4.18}$$

**Figure 3.20** Normalized acceleration for a damped system impacting a rigid floor ($\omega_n = 100$ rad/s)

or

$$\ddot{x} + 2\zeta\omega_n\dot{x} + \omega_n^2 x = g \tag{3.4.19}$$

with $x(0) = 0$ and $\dot{x}(0) = v$. The general solution for this problem is given by

$$x(t) = \frac{g}{\omega_n^2} + e^{-\zeta\omega_n t}(a_1\cos(\omega_d t) + a_2\sin(\omega_d t)) \tag{3.4.20}$$

where $\omega_d$ is the damped frequency of oscillation. Applying the initial conditions gives us

$$a_1 = -\frac{g}{\omega_n^2} \tag{3.4.21}$$

and

$$a_2 = \frac{v - \dfrac{\zeta g}{\omega_n}}{\omega_d} \tag{3.4.22}$$

Differentiating twice give us the acceleration

$$\ddot{x}(t) = e^{-\zeta\omega_n t}\left((g - 2\zeta\omega_n v)\cos(\omega_d t) + \left(\frac{v\omega_n(2\zeta^2 - 1) - g\zeta}{\sqrt{1 - \zeta^2}}\right)\sin(\omega_d t)\right) \tag{3.4.23}$$

As you can see, this is quite a bit more complicated than the undamped case was. If we were just going to numerically check for when the absolute value of the acceleration was a maximum, we wouldn't need to go any further. If we want to determine the peak response analytically, however, we'd need to differentiate once again. Doing so, setting the result to zero (to find the maximum) and rearranging slightly will give us

$$\tan(\omega_d t_{\max}) = \frac{\omega_d b_2 - \omega_n\zeta b_1}{\omega_d b_1 + \omega_n\zeta b_2} \tag{3.4.24}$$

where $b_1 = g - 2\zeta\omega_n v$ and $b_2 = \frac{v\omega_n(2\zeta^2-1)-g\zeta}{\sqrt{1-\zeta^2}}$.

To solve for $t_{max}$, we must take the inverse tangent

$$\omega_d t_{max} = \tan^{-1}\left(\frac{\omega_d b_2 - \omega_n \zeta b_1}{\omega_d b_1 + \omega_n \zeta b_2}\right) \tag{3.4.25}$$

This will then allow us to solve for the time at which the maximum response occurs. Substituting this value into (3.4.23) will then give us the maximal acceleration.

We can also see that the nice way we plotted normalized acceleration versus normalized drop height won't really work here. The variables are too complex to see a simple normalization such as we used before. It's more convenient in this case to plot the acceleration to gravity ratio versus the impact velocity. We'd then need to make a secondary calculation to relate impact velocity to drop height.

As an example of what we'd see by varying different parameters, we see in Figure 3.21 what the absolute value of the maximal acceleration to gravity ration is as a function of impact velocity for a variety of damping values. The natural frequency was taken to be 1.0 rad/s for these responses. There are a few observations that can be made from this plot. First, you'll note that the maximal accleration ratio is sometimes less than 1. This might seem odd since the object was in a 1-$g$ free fall before impact, but what we're looking at is the acceleration ratio just after impact. Once the object has struck the ground, the acceleration is simply a function of the initial velocity and the damping and stiffness characteristics. As you might guess, impacts typically result in much higher accelerative loadings than g, and thus this observation is mainly of academic interest.

The second observation is that the acceleration is growing linearly for increasing values of impact velocity. You can see why this happens from looking as (3.4.23). As $v$ increases it will come to dominate the coefficients of both the cosine and sine term. Since $v$ appears linearly, we'd expect a linear increase in our plot. This fact simply tells us that for large values of $v$, as the impact velocity increases, so does the maximal acceleration.

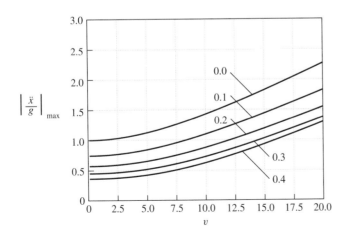

**Figure 3.21**  Normalized maximum acceleration versus impact velocity for damped system (damping factor indicated on graph)

Finally, it is clear for this range of parameters that increasing the damping serves to reduce the maximal acceleration levels. Thus damping is another useful design tool to use in deciding how to build systems to withstand shock loading.

## 3.5 HOMEWORK PROBLEMS

### Section 3.2

**1.** Plot the response of the illustrated system to the illustrated seismic input.

$m = 10$ kg
$k = 9000$ N/m

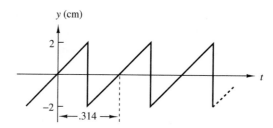

**2.** Given the Fourier series solution for $f(t)$, construct the solution for $g(t)$ in terms of the $a_i$ and $b_i$.

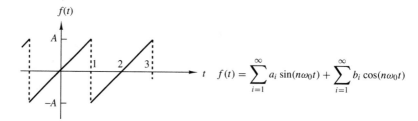

$$f(t) = \sum_{i=1}^{\infty} a_i \sin(n\omega_0 t) + \sum_{i=1}^{\infty} b_i \cos(n\omega_0 t)$$

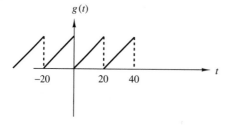

**3.** You are given the spring-mass system below, with mass $m = 10$ kg and spring constant $k = 360$ N/m. The base of the system is given a vertical displacement $y(t)$, in the form of a sawtooth function with amplitude $a = 2$ cm and period $T = 2$ s. Compute the response of the system and identify the dominant response harmonic.

**4.** For the following figure, determine whether the coefficients $b_m$ and $a_m$ are zero or nonzero if the function is approximated by

$$f(t) = \sum_{m=0}^{n} a_m \cos(m\omega_0 t) + \sum_{m=1}^{n} b_m \sin(m\omega_0 t)$$

What is the actual value of $\omega_0$ and $a_0$?

**5.** Give two reasons why we don't retain an infinite number of terms in our Fourier series analyses. What is a reasonable criterion for deciding when we should truncate our series?

**6.** Plot out approximations to the function shown below using a Fourier series analysis. Show how the approximation changes as you go from one included harmonic to several.

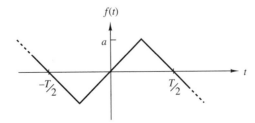

**7.** Consider the forcing shown in Problem 6. Let $a = 5$. Determine the response of a direct-force-excited, spring-mass system to this input. Two cases need to be examined: $T = 3$ sec. and $T = 9.24$ sec. Look at the harmonic content of the input and the harmonic content of the output and use this information to explain why you obtained qualitatively different responses for the two cases. ($m = 5$ kg and $k = 20$ N/m.)

8. Express the following function in terms of a Fourier series representation. It might model the force input of a pair of woodpeckers located on opposite sides of a tree—sharp, regularly spaced impulses. What difficulties will the Fourier series analysis have with such a function?

$$f(t) = \sum_{n=1}^{\infty} (-1)^{n+1} \delta(t - (2n-1)T)$$

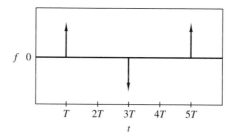

## Section 3.3

9. Determine $x(1)$ and $\dot{x}(1)$ for a direct-force-excited, spring-mass system for the given input via a convolution approach. ($x(0) = \dot{x}(0) = 0$, $m = 1$ kg and $k = 4$ N/m.)

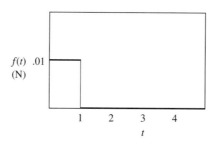

10. Consider a direct-force excited, spring-mass system. Show via a convolution approach that the displacement at $t = .5T$ will be equal to twice the static equilibrium displacement if the force input is constant from $t = 0$ to $t = .5T$.

11. Determine the response of a seismically excited spring-mass system to a step seismic input by using a convolution approach.

**12.** Will the illustrated force input lead to a stable oscillation (no growth in amplitude) for a direct-force-excited, spring-mass system? Use a convolution approach.

## Section 3.4

**13.** Determine the maximal acceleration that the mass shown will experience if dropped onto the illustrated spring. ($m = 10$ kg, $k = 20,000$ N/m, $h = 1$ m, and $g = 9.81$ m/s$^2$.)

**14.** By how much will the maximal acceleration of Problem 13 change if the gravitational acceleration is reduced by a factor of 6 (as would happen if we ran the experiment on Earth and then on the Moon)?

**15.** In an egg drop contest we try to drop an egg in a container from the highest point possible without having the egg break. Assume that the egg doesn't break if dropped from a 2-m height in a container that has a stiffness of 400 N/m. The mass of the egg is equal to .05 kg. How must the supporting stiffness be changed in order to double the height from which the egg can successfully be dropped.

**16.** What is the maximum acceleration felt by a lumped mass that falls onto an arbitrarily stiff surface?

**17.** Will reducing an item's mass decrease the maximal acceleration it will experience if dropped onto an elastic surface?

# 4

## Vibrations Involving More Than One Degree of Freedom

## 4.1 INTRODUCTION

Although some systems can be analyzed from a SDOF standpoint, there are a great many situations for which this viewpoint is simply insufficient. One example, to be studied in some detail in this chapter, is when designers purposely add an additional degree of freedom in order to modify the vibrational response of the original system. The systems that do this are known as vibration absorbers. Automobile designers, when trying to obtain good handling characteristics for their cars, routinely must include more than just one degree of freedom in order to capture the various motions the car exhibits, such as yaw, heave, pitch, and roll. Even the simplest model of a steam turbine should include at least two degrees of freedom, while reasonable models for large space structures routinely involve hundreds of degrees of freedom.

The fundamental difference between a MDOF system and a SDOF system one, obviously, is the fact that we have additional masses to worry about. The important consequence of this fact is that a sinusoidal response of the MDOF system will automatically imply a relationship between the motions of the various masses. For instance, the motion of one mass in a 2-DOF system might be described by a cosine and the motion of the other by a cosine having twice the amplitude of the first. Thus, to describe the motion, we should concern ourselves with the *relative* motion between the two masses. In the case just mentioned, this would mean noting that one mass moves twice as much as the other. This idea of relative motion is a crucial one and we will study it in some detail, since it underlies most of the remaining chapters.

## 4.2 FREE RESPONSE–UNDAMPED SYSTEM

We'll start by analyzing the system shown in Figure 4.1. External forcings and damping have been excluded for clarity and will be examined later in the chapter.

Since 2-DOF problems are easily visualized, can be analytically solved, and can be experimentally demonstrated using the spring included in this book, we'll focus on them when introducing and discussing multiple-degree-of-freedom characteristics. The findings, however, will be equally applicable to problems with any number of degrees of freedom that are desired.

As we can see from the free-body diagram of Figure 4.1b, the equations of motion for this system can be written as

$$m_1\ddot{x}_1 + (k_1 + k_2)x_1 - k_2 x_2 = 0 \qquad (4.2.1)$$

and

$$m_2\ddot{x}_2 + (k_2 + k_3)x_2 - k_2 x_1 = 0 \qquad (4.2.2)$$

At this point we'll make the most important transition of the book. Until now, everything we've

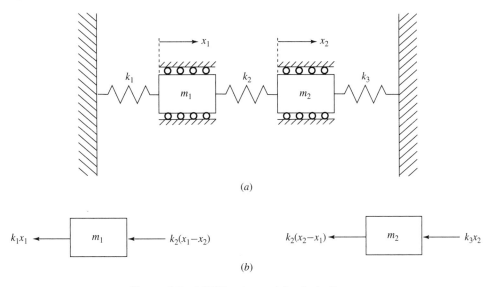

(a)

(b)

**Figure 4.1**   2-DOF system and free-body diagram

been dealing with has involved scalar quantities. But we now have the opportunity to move beyond this stage and into the vector/matrix world, because the equations above can just as easily be reexpressed as

$$\begin{bmatrix} m_1 & 0 \\ 0 & m_2 \end{bmatrix} \begin{Bmatrix} \ddot{x}_1 \\ \ddot{x}_2 \end{Bmatrix} + \begin{bmatrix} k_1 + k_2 & -k_2 \\ -k_2 & k_2 + k_3 \end{bmatrix} \begin{Bmatrix} x_1 \\ x_2 \end{Bmatrix} = \begin{Bmatrix} 0 \\ 0 \end{Bmatrix} \tag{4.2.3}$$

If we define

$$[M] \equiv \begin{bmatrix} m_1 & 0 \\ 0 & m_2 \end{bmatrix}, \; [K] \equiv \begin{bmatrix} k_1 + k_2 & -k_2 \\ -k_2 & k_2 + k_3 \end{bmatrix}, \; X \equiv \begin{Bmatrix} x_1 \\ x_2 \end{Bmatrix}, \; O \equiv \begin{Bmatrix} 0 \\ 0 \end{Bmatrix} \tag{4.2.4}$$

then we can express (4.2.3) as

$$[M]\ddot{X} + [K]X = O \tag{4.2.5}$$

Thus we have a form that is, on the surface, identical to that of our SDOF oscillator. The difference is that now our masses and springs are matrices and our dependent variable is a vector, not a scalar.

The definitions given above follow the convention that will be used throughout the rest of the book. When talking about a scalar quantity, a lowercase letter will be used. Thus, when you see something like $b$, you'll know it is a scalar. Vectors are denoted by an uppercase letter, such as $X$ or $Y$. Finally, matrices are indicated by an uppercase letter surrounded by square brackets, such as $[A]$. When referring to the entries in a vector or matrix, we'll use the letter of the vector/matrix in lowercase form and add a subscript. Thus $a_{1,2}$ is the entry in the first row and second column of $[A]$. As a further example, the first-row, first-column entry in $[K]$ (defined in (4.2.4)) is $k_{1,1}$ and

$$k_{1,1} = k_1 + k_2 \tag{4.2.6}$$

Even though the actual entries are denoted by $k$, you can tell the difference between them and the symbol indicating a particular matrix entry because the matrix entry will have two subscripts separated by a comma (which indicates the row and column number). The only exceptions to this convention are certain parameters that are traditionally written in uppercase letters (such as Young's modulus, $E$). It will be clear from the context that these are parameters and not vectors.

Now that we've entered the matrix/vector world, we're not going to leave it. For the rest of the book, we'll use matrix equations in one form or another. More specifically, we'll use (4.2.5) quite a bit, and thus all the analyses and conclusions we draw in this chapter will be applicable to the problems we'll encounter in the future.

Before analyzing (4.2.5) any further, you should see what you can recall from your class in linear algebra. Almost certainly, part of what you learned included an analysis of the matrix equation

$$[A(\lambda)]X = O \tag{4.2.7}$$

It's also possible that you dealt with a different form of this equation, namely

$$[A]X = \lambda X \tag{4.2.8}$$

In either case, you'll recall that these were called eigenvalue problems and that the solution involved finding both the eigenvalues and eigenvectors for the system. What's important is that (4.2.5) will lead to just such an eigenvalue problem.

To see this, let's presume a solution to (4.2.5) in the form

$$X(t) = \bar{X}e^{i\omega t} = \begin{Bmatrix} \bar{x}_1 \\ \bar{x}_2 \end{Bmatrix} e^{i\omega t} \tag{4.2.9}$$

We know that this is an appropriate form because, as we've stated before, we should always expect linear, constant coefficient, ordinary differential equations to support exponential solutions. We've got three unknowns to deal with, $\bar{x}_1$, $\bar{x}_2$, and $\omega$. We chose $e^{i\omega t}$ instead of $e^{\lambda t}$ because of our previous observation that a spring-mass system supports sinusoidal responses rather than real exponential solutions. However, you should realize that we could use $e^{\lambda t}$ just as easily; it won't affect the final results. You'll notice that we have three unknowns and only two equations. Thus we know that we won't get unique numerical solutions for $\bar{x}_1$, $\bar{x}_2$, and $\omega$. This is fine, as we'll see shortly.

To solve (4.2.5), we'll use our assumed solution (4.2.9), differentiate it twice (for the $\ddot{X}$ term), substitute it in, and find (factoring out the $e^{i\omega t}$ term)

$$- \omega^2[M]\bar{X} + [K]\bar{X} = O \tag{4.2.10}$$

We can now factor out the $\bar{X}$ term to obtain

$$([K] - \omega^2[M])\bar{X} = O \tag{4.2.11}$$

This looks exactly like (4.2.7), (in which $[A]$ is equal to $([K]-\omega^2[M])$. Thus we have an eigenvalue problem. Although $\bar{X} = O$ solves this equation, we recognize this as the trivial solution—trivial in that $\bar{X} = O$ implies no motion at all. Consequently we can ignore this solution as being of no interest. We're therefore left with asking when an equation of the form

$$[A]\bar{X} = O \tag{4.2.12}$$

has a solution for finite $\bar{X}$. The answer is, when $[A]$ is not invertible. And why is this? Well, if $[A]$ was invertible, then we could compute its inverse and, multiplying both sides of (4.2.12) by $[A]^{-1}$, obtain

$$[A]^{-1}[A]\bar{X} = [A]^{-1}O \tag{4.2.13}$$

or

$$\bar{X} = O \tag{4.2.14}$$

Thus, if $[A]$ were invertible, we'd be left with the trivial solution. Therefore, if we want a more interesting solution, $[A]$ must not be invertible. For a matrix not to be invertible, its determinant must be equal to zero. So our requirement for a solution is that

$$\det\left([K] - \omega^2[M]\right) = 0 \tag{4.2.15}$$

Thus

$$\det\left(\begin{bmatrix} k_1 + k_2 - \omega^2 m_1 & -k_2 \\ -k_2 & k_2 + k_3 - \omega^2 m_2 \end{bmatrix}\right) = 0 \qquad (4.2.16)$$

If we evaluate this determinant we'll obtain

$$(k_1 + k_2 - \omega^2 m_1)(k_2 + k_3 - \omega^2 m_2) - k_2^2 = 0 \qquad (4.2.17)$$

Expanding this and grouping the powers of $\omega$ leads to

$$m_1 m_2 \omega^4 - \omega^2((k_2 + k_3)m_1 + (k_1 + k_2)m_2) + k_1 k_2 + k_1 k_3 + k_2 k_3 = 0 \qquad (4.2.18)$$

In order to provide further insight, let's assign specific values to the springs and masses. We'll let both masses have a mass of 1 kg and let the spring constants all be equal to 4 N/m. Under these conditions, (4.2.18) becomes

$$\omega^4 - 16\omega^2 + 48 = 0 \qquad (4.2.19)$$

The illustrated polynomial is called the *characteristic polynomial*, and, when set equal to zero, gives us the *characteristic equation* for the system. Although we'll refer to the roots of this polynomial as our natural frequencies, we'd call them the "poles" of the system if we were looking at it from a controls perspective. So for those of you who have studied (or plan to study) control theory, remember that the poles of a system and an undamped system's natural frequencies are the same thing.

Recognizing that this equation isn't really a full-fledged, fourth-order equation, but is rather quadratic in $\omega^2$, we can solve for $\omega^2$ straightforwardly:

$$\omega_{1,2}^2 = \tfrac{1}{2}(16 \pm \sqrt{16^2 - (4)(48)}) \qquad (4.2.20)$$

or

$$\omega_1^2 = 4 \ (\text{rad/s})^2 \qquad (4.2.21)$$

and

$$\omega_2^2 = 12 \ (\text{rad/s})^2 \qquad (4.2.22)$$

Labeling the natural frequencies in ascending order ($\omega_1 \le \omega_2 \le \omega_3 \ldots$) is the typical convention and the one we'll follow. Just as in the single-degree-of-freedom case, we don't need to worry much about the fact that we'll obtain four values for $\omega$ when we take the square roots of $\omega_1^2$ and $\omega_2^2$, since these will combine to give us sine and cosine solutions in just the same way as they did before. These two values for $\omega^2$ are eigenvalues for our problem and their square roots are equal to the two possible frequencies of oscillation for the system. Note that for the single-degree-of-freedom case the only possible free-vibration frequency was $\omega_n$. Now we have two frequencies.

Once we've found the two solutions for $\omega^2$, we're left with determining $\bar{x}_1$ and $\bar{x}_2$. To go further with our solution, we'll revisit (4.2.11). $\omega^2$ is now known; we simply need specify which of the two solutions we want to use. To start, we'll look at the lower-frequency solution, that is, $\omega_1^2 =$

4 (rad/s)$^2$. Substituting this value for $\omega^2$ and inserting the appropriate values for $m_i$ and $k_i$ gives us

$$\left[ -4 \begin{bmatrix} 1 & 0 \\ 0 & 1 \end{bmatrix} + \begin{bmatrix} 8 & -4 \\ -4 & 8 \end{bmatrix} \right] \left\{ \begin{array}{c} \bar{x}_1 \\ \bar{x}_2 \end{array} \right\} = \left\{ \begin{array}{c} 0 \\ 0 \end{array} \right\} \tag{4.2.23}$$

or more simply

$$\begin{bmatrix} 4 & -4 \\ -4 & 4 \end{bmatrix} \left\{ \begin{array}{c} \bar{x}_1 \\ \bar{x}_2 \end{array} \right\} = \left\{ \begin{array}{c} 0 \\ 0 \end{array} \right\} \tag{4.2.24}$$

Both the equations in (4.2.24) indicate the same thing, namely that $\bar{x}_1$ must be equal to $\bar{x}_2$. The fact that we get the same equation makes sense: it implies that the two rows of the matrix are not independent, something that must be true if the matrix is to be noninvertible.

We are now in a position to write down one of the solutions to our problem. Based on the results so far, we know that one possible solution is

$$X(t) = a_1 \left\{ \begin{array}{c} 1 \\ 1 \end{array} \right\} e^{2it} + a_2 \left\{ \begin{array}{c} 1 \\ 1 \end{array} \right\} e^{-2it} \tag{4.2.25}$$

where $a_1$ and $a_2$ are as yet undetermined constants that we could solve for if we were given specific initial conditions.

A more accessible (and equivalent) form is to use a trigonometric representation instead of the exponential form shown in (4.2.25):

$$X(t) = b_1 \left\{ \begin{array}{c} 1 \\ 1 \end{array} \right\} \cos(2t - \phi_1) \tag{4.2.26}$$

Since we've included a phase shift, this representation can include pure sines and cosines as well as a mix of the two, as discussed earlier.

Equation (4.2.26) represents one of the fundamental responses of the system. Physically, both masses are locked together in their motions; when one moves to the right, so does the other, and to the same degree. You'll also note that the frequency of this oscillation, 2 rad/s, is the same as the natural frequency we'd find from analyzing one of the springs attached to one of the masses. The reason for this is not too difficult to see. Imagine that the middle spring has been removed and that you move both masses the same distance to the left and then release them. Since they're now simply two, independent, single-degree-of-freedom systems, they'll oscillate at their natural frequencies. And since the masses and springs are identical, the natural frequency of oscillation will be the same for each. Thus the masses will move in synchronism with each other and the distance between them will stay fixed. Therefore it doesn't matter if there is a spring between them or not; since the relative distance between them stays constant, the middle spring isn't stretched or compressed anyway. This is a consequence of both the left and right spring-mass systems having the same natural frequencies. Whenever you have this situation, the first natural frequency of the 2-DOF system will be the same as the individual SDOF systems.

We call the vector $\{b_1 \quad b_1\}^T$ an *eigenvector* of our system. You'll also see it called modal vector, eigenmode, natural mode, and sometimes simply a mode. It indicates the relative displacements (and later on will include velocities) of the different masses in the system. Since the system has no damping, the oscillations go on forever, and, if the system is initially displaced into the first eigenmode, it remains there. Note that the amplitude $b_1$ is arbitrary, just as the amplitude was arbitrary in the SDOF case. Only if we consider matching specific initial conditions would we be able to pin down $b_1$ precisely. So the really important fact about this eigenmode is not the particular value of $b_1$, but the fact that both entries are the same. Since the eigenmode is the same no matter what we multiply it by (both entries will remain the same), we'll divide by $b_1$ and define our first eigenvector to be $\{1 \quad 1\}^T$.

We can now solve for the second eigenvector of the system exactly as we solved for the first. In this case $\omega^2$ is equal to 12 (rad/s)$^2$ and our equation governing the second eigenvector is given by

$$\begin{bmatrix} -4 & -4 \\ -4 & -4 \end{bmatrix} \begin{Bmatrix} \bar{x}_1 \\ \bar{x}_2 \end{Bmatrix} = \begin{Bmatrix} 0 \\ 0 \end{Bmatrix} \tag{4.2.27}$$

and our solution is

$$X(t) = a_3 \begin{Bmatrix} 1 \\ -1 \end{Bmatrix} e^{2\sqrt{3}it} + a_4 \begin{Bmatrix} 1 \\ -1 \end{Bmatrix} e^{-2\sqrt{3}it} \tag{4.2.28}$$

or, more familiarly

$$X(t) = b_2 \begin{Bmatrix} 1 \\ -1 \end{Bmatrix} \cos(2\sqrt{3}t - \phi_2) \tag{4.2.29}$$

Using $X_i$ to identify our eigenvectors,

$$X_1 \equiv \begin{Bmatrix} 1 \\ 1 \end{Bmatrix}, \quad X_2 \equiv \begin{Bmatrix} 1 \\ -1 \end{Bmatrix} \tag{4.2.30}$$

we can write our total solution for this unforced, 2-DOF oscillator as

$$X(t) = b_1 X_1 \cos(\omega_1 t - \phi_1) + b_2 X_2 \cos(\omega_2 t - \phi_2) \tag{4.2.31}$$

where $\omega_1$ is the first natural frequency of the system (and is equal to 2 rad/s) and $\omega_2$ is the second natural frequency (equaling $2\sqrt{3}$ rad/s).

As a concrete example of how we would match specific initial conditions, consider the response of the system to the initial conditions

$$x_1(0) = 1 \text{ mm}, \quad \dot{x}_1(0) = 0, \quad x_2(0) = 0, \quad \dot{x}_2(0) = 0 \tag{4.2.32}$$

Note that this initial deflection does not match either the first or second eigenvector's shape. Thus we would predict that both eigenvectors must be involved in the subsequent motion. And we'd be right, because the solution for this case is

$$\phi_1 = 0, \quad \phi_2 = 0, \quad b_1 = .5 \text{ mm}, \quad b_2 = .5 \text{ mm/s} \tag{4.2.33}$$

The total response can be written as

$$X(t) = X_1 \cos(2t) + X_2 \cos(2\sqrt{3}t) \tag{4.2.34}$$

It's important to realize that, even if the initial deflections of the two masses match only one particular eigenvector, both eigenvectors might be involved in the response because of the initial velocities. For instance, the initial conditions

$$x_1(0) = 1, \quad x_2(0) = 1, \quad \dot{x}_1(0) = 2\sqrt{3}, \quad \dot{x}_2(0) = \sqrt{3} \tag{4.2.35}$$

lead to the solution

$$X(t) = X_1 \cos(2t) + X_2 \cos\left(2\sqrt{3}t - \frac{\pi}{2}\right) \tag{4.2.36}$$

even though no trace of the second eigenvector showed up in the initial displacements. Thus both velocities and displacements must correspond to those supported by a particular eigenvector for that to be the sole element in the response.

**Example 4.1**

Now that you've seen what to expect in 2-DOF systems, you can go ahead and use your spring to run your own 2-DOF experiment. First, you'll need to cut the spring into two pieces. Next, you'll need two equal masses. The more symmetric they are, the better, since you'll want the motions to stay purely translational; rocking due to imbalances will just make the experiments harder.

You'll be doing two things in this experiment—experimentally verifying the existence of two eigenvectors (modes) and analytically predicting their natural frequencies from a couple of SDOF tests. Even more amazing, we won't even need to know the actual mass or spring constants. The only additional item you'll need is a watch with a stopwatch function.

To start, attach one of the masses to one spring and the other to the other spring. To minimize confusion, we'll call the mass and spring of the first set $m_1$ and $k_1$ and the second set $m_2$ and $k_2$. Next, hold the free end of the spring against a fixed surface (like a door frame), pull down on the mass, and let go. The mass will then bob up and down. If you start your stopwatch at a high point of the motion and time the motion for 5 or 10 cycles, you can calculate a pretty accurate estimate of the oscillation period by dividing the total time by the number of cycles that occurred. If you call this period $t_1$, then $\sqrt{\frac{k_1}{m_1}}$ is equal to $\frac{2\pi}{t_1}$. You should then repeat this procedure with $m_2$ and $k_2$, leading to an estimate of $\sqrt{\frac{k_2}{m_2}}$. Finally, hook $m_1$ and $k_2$ together and determine $\sqrt{\frac{k_2}{m_1}}$. Now attach $m_1$ to the free end of $k_2$, as shown in Figure 4.2. If you hold the free end of $k_1$, you'll have a 2-DOF system (for vertical motions).

Getting this system to exhibit its first and second eigenvectors isn't difficult. Simply start moving your hand up and down as you did in the SDOF experiments and try to get a resonant response. If you start at a low frequency and slowly increase it, you'll reach a frequency for which very small motions of your hand produce large, in-phase motions of the masses (they both move in the same direction at all times). If you then move your hand faster, you'll reach a point at which the masses move in opposition to each other (one moves up while the other moves down). At this point your hand is oscillating at the system's second natural frequency. Because the damping is so low, the individual mode shapes should be quite apparent. Once the oscillations are established, you can hold your hand against a rigid support and watch the modal vibrations. Figure 4.3 shows what you should be seeing.

All right, now that we've actually seen the eigenmodes in action, it's time to analytically predict what their frequencies should be (and verify the predictions, of course). The equations of

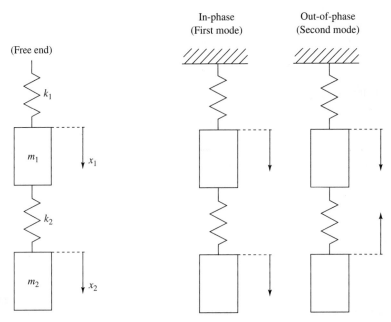

**Figure 4.2** Experimental 2-DOF system

**Figure 4.3** Modal behavior of a physical 2-DOF system

motion for our system are given by

$$
\begin{bmatrix} m_1 & 0 \\ 0 & m_2 \end{bmatrix} \begin{Bmatrix} \ddot{x}_1 \\ \ddot{x}_2 \end{Bmatrix} + \begin{bmatrix} k_1 + k_2 & -k_2 \\ -k_2 & k_2 \end{bmatrix} \begin{Bmatrix} x_1 \\ x_2 \end{Bmatrix} = \begin{Bmatrix} 0 \\ 0 \end{Bmatrix}
$$
(4.2.37)

Now how can we proceed any further without knowing the particular values of the masses and stiffnesses? By dividing through by the masses! If we divide the first equation by $m_1$ and the second by $m_2$, we'll have terms such as $\frac{k_1}{m_1}$ and $\frac{k_2}{m_2}$. These are simply the square of the natural frequencies associated with the simpler SDOF problems we've already analyzed.

Going ahead and dividing by the masses gives us

$$
\begin{bmatrix} 1 & 0 \\ 0 & 1 \end{bmatrix} \begin{Bmatrix} \ddot{x}_1 \\ \ddot{x}_2 \end{Bmatrix} + \begin{bmatrix} \frac{k_1}{m_1} + \frac{k_2}{m_1} & -\frac{k_2}{m_1} \\ -\frac{k_2}{m_2} & \frac{k_2}{m_2} \end{bmatrix} \begin{Bmatrix} x_1 \\ x_2 \end{Bmatrix} = \begin{Bmatrix} 0 \\ 0 \end{Bmatrix}
$$
(4.2.38)

At this point, we must part company, since I have no idea what natural frequencies you have. But just for the record, I'll go through what I get when I run this test. For the masses, I used heavy doughnut-shaped pieces of steel (which came out of the weights of a grandfather clock). Since the two masses I used were identical I had to calculate only the periods associated with $\sqrt{\frac{k_1}{m}}$ and $\sqrt{\frac{k_2}{m}}$. The periods of oscillation for the two, SDOF oscillators were

$$
t_1 = .864 \text{ s}
$$
(4.2.39)

and

$$t_2 = .861 \text{ s} \tag{4.2.40}$$

which correspond to frequencies of

$$\omega_1 = 7.27 \text{ rad/s} \tag{4.2.41}$$

and

$$\omega_2 = 7.30 \text{ rad/s} \tag{4.2.42}$$

A 10-cycle average was used when obtaining these values.

If these results are substituted into (4.2.38), the equations become

$$\begin{bmatrix} 1 & 0 \\ 0 & 1 \end{bmatrix} \begin{Bmatrix} \ddot{x}_1 \\ \ddot{x}_2 \end{Bmatrix} + \begin{bmatrix} 106.2 & -53.3 \\ -53.3 & 53.3 \end{bmatrix} \begin{Bmatrix} x_1 \\ x_2 \end{Bmatrix} = \begin{Bmatrix} 0 \\ 0 \end{Bmatrix} \tag{4.2.43}$$

The characteristic equation is, therefore,

$$\omega^4 - 159.5\omega^2 + 2820 = 0 \tag{4.2.44}$$

which has solutions

$$\omega_1 = 4.50 \text{ rad/s} \tag{4.2.45}$$

and

$$\omega_2 = 11.80 \text{ rad/s} \tag{4.2.46}$$

By exciting the experimental system at resonance, immediately placing the free end of the spring against the top of a door frame, timing 10 oscillations, and then averaging, the following time periods were obtained:

$$t_1 = 1.40 \text{ s} \tag{4.2.47}$$

and

$$t_2 = .538 \text{ s} \tag{4.2.48}$$

These correspond to experimentally derived natural frequencies of

$$\omega_{1_{\text{exp}}} = 4.49 \text{ rad/s} \tag{4.2.49}$$

and

$$\omega_{2_{\text{exp}}} = 11.68 \text{ rad/s} \tag{4.2.50}$$

Comparing the two sets of results, we see that the predicted first natural frequency was within .2 percent of the experimental value, while the second was off by only 1 percent.

These excellent results weren't a fluke; you should have obtained equally accurate answers during your own test. What this experiment shows is that the modeling we've been talking about is really pretty useful and reflects reality very well. Of course, you'll rarely be dealing with masses attached to springs in your professional life. But the basic conclusion still holds. The techniques we've learned (and those still to come) can do an amazing job of predicting real-world behavior, so long as the assumptions that underlie them aren't violated.

Now that we've investigated the eigenvectors, which are a free-response characteristic, we'll move on to consider some forced responses and point out how the natural frequencies and eigenvectors affect the solutions.

## 4.3 FORCED RESPONSE

If we include the applied forces shown in Figure 4.4, our equations become

$$\begin{bmatrix} m_1 & 0 \\ 0 & m_2 \end{bmatrix} \begin{Bmatrix} \ddot{x}_1 \\ \ddot{x}_2 \end{Bmatrix} + \begin{bmatrix} k_1 + k_2 & -k_2 \\ -k_2 & k_2 + k_3 \end{bmatrix} \begin{Bmatrix} x_1 \\ x_2 \end{Bmatrix} = \begin{Bmatrix} f_1 \\ f_2 \end{Bmatrix} \tag{4.3.1}$$

For simplicity, assume that the forcing is in the form

$$\begin{aligned} f_1(t) &= \bar{f}_1 \cos(\omega t) \\ f_2(t) &= \bar{f}_2 \cos(\omega t) \end{aligned} \tag{4.3.2}$$

that is, the magnitudes of the two forces can vary, but the frequency is the same for each. As in the SDOF problem, when we have a sinusoidal forcing we should expect a sinusoidal response. And since the problem has no damping, the output should be in phase or 180 degrees out-of-phase with the input. Of course, we could use complex exponentials to represent the input and output (and we'll do so for the damped case). But for now we'll stick with an explicit cosine forcing, since the mathematics are simpler without any damping and the cosine solution has a more straightforward physical interpretation.

If we let the response be

$$\begin{aligned} x_1(t) &= \bar{x}_1 \cos(\omega t) \\ x_2(t) &= \bar{x}_2 \cos(\omega t) \end{aligned} \tag{4.3.3}$$

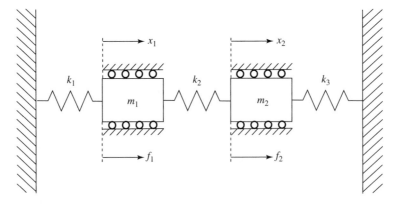

**Figure 4.4**   Forced 2-DOF spring-mass system

and substitute this (along with the forcing) into (4.3.1), then we'll obtain

$$- \omega^2 \begin{bmatrix} m_1 & 0 \\ 0 & m_2 \end{bmatrix} \begin{Bmatrix} \bar{x}_1 \\ \bar{x}_2 \end{Bmatrix} + \begin{bmatrix} k_1 + k_2 & -k_2 \\ -k_2 & k_2 + k_3 \end{bmatrix} \begin{Bmatrix} \bar{x}_1 \\ \bar{x}_2 \end{Bmatrix} = \begin{Bmatrix} \bar{f}_1 \\ \bar{f}_2 \end{Bmatrix} \tag{4.3.4}$$

or

$$[[K] - \omega^2 [M]]\bar{X} = \bar{F} \tag{4.3.5}$$

where $F \equiv \{\bar{f}_1 \quad \bar{f}_2\}^T$.

Unlike (4.2.11), we now have a nonhomogeneous problem, that is, we're not dealing with $[A]\bar{X} = 0$, but with $[A]\bar{X} = \bar{F}$. And unlike the homogeneous case, $[A]$ must be invertible for a solution to exist. If $[A]^{-1}$ exists, then we'll premultiply by it to find $\bar{X} = [A]^{-1}\bar{F}$. For this 2-DOF example, we can work out the solution explicitly, but for more complex problems we'll need to rely on numerical solutions.

You'll recall from your linear algebra course (or from Appendix D) that the inverse of a two-by-two matrix is given by

$$\begin{bmatrix} a & b \\ c & d \end{bmatrix}^{-1} = \begin{bmatrix} d & -b \\ -c & a \end{bmatrix} \frac{1}{(ad - bc)} \tag{4.3.6}$$

Thus our solution $\bar{X}$ is given by

$$\bar{X} = \frac{1}{\Delta(\omega)} \begin{bmatrix} k_2 + k_3 - m_2 \omega^2 & k_2 \\ k_2 & k_1 + k_2 - m_1 \omega^2 \end{bmatrix} \begin{Bmatrix} \bar{f}_1 \\ \bar{f}_2 \end{Bmatrix} \tag{4.3.7}$$

where $\Delta(\omega) \equiv (k_1 + k_2 - m_1 \omega^2)(k_2 + k_3 - m_2 \omega^2) - k_2^2$. The individual components of $\bar{X}$ are given by

$$\bar{x}_1 = \frac{(k_2 + k_3 - m_2 \omega^2)\bar{f}_1 + k_2 \bar{f}_2}{\Delta(\omega)} \tag{4.3.8}$$

and

$$\bar{x}_2 = \frac{k_2 \bar{f}_1 + (k_1 + k_2 - m_1 \omega^2)\bar{f}_2}{\Delta(\omega)} \tag{4.3.9}$$

Notice that the denominator for both terms, $\Delta(\omega)$, is the same, and if it is set to zero we have exactly the same equation that we used to solve for the system natural frequencies (4.2.17). Is this a bizarre coincidence? Of course not! This will invariably be the case when finding the forced response of a linear system such as the one we're dealing with. The denominator is always the characteristic polynomial for our system. If you'll recall what was so special about this polynomial, namely that its roots were the natural frequencies of the system, you'll realize that they play an important role for the forced problem also. Since the characteristic polynomial is equal to zero at the system's natural frequencies, (4.3.8) and (4.3.9) tell us that the denominators go to zero at these frequencies. Thus, if the numerator isn't equal to zero (and it usually isn't), the response will go to infinity. What does this mean? It means that we now have two resonance conditions, one

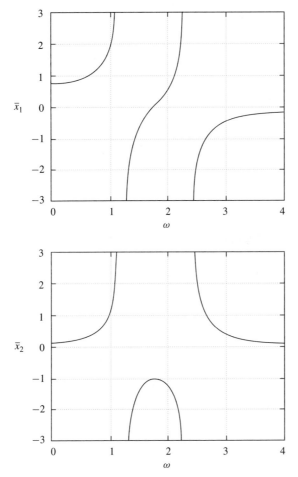

**Figure 4.5**   $\bar{x}_1$ and $\bar{x}_2$ versus $\omega$

for each natural frequency. For the SDOF system we had only one resonant frequency. Now, not surprisingly, we have two. Figure 4.5 shows the response for $\bar{x}_1$ and $\bar{x}_2$ for a representative case (if you're really interested, the parameters $m_1 = 1$, $m_2 = 1$, $k_1 = 1$, $k_2 = 2$, $k_3 = 2$, $\bar{f}_1 = 2$, and $\bar{f}_2 = -1$ were used). As you can see, the responses for both $\bar{x}_1$ and $\bar{x}_2$ are finite at zero frequency (as in the SDOF problem), go to infinity at the resonant frequencies of the system, and die down to zero as the forcing frequency becomes very large (also like the SDOF problem). Figure 4.5 also shows the phase response, which, in this case, is either 0 or $\pi$ rad (that is, a positive or negative response, respectively). Just as for the SDOF problem, the phase shifts $\pi$ rad after each resonant frequency is passed.

The situation is much the same for higher-order systems. For instance, Figure 4.6 shows the response of the first mass in the four-mass system shown in Figure 4.7. With four masses, we now have four natural frequencies, $\omega_1 - \omega_4$. You can see from Figure 4.6 that the response goes

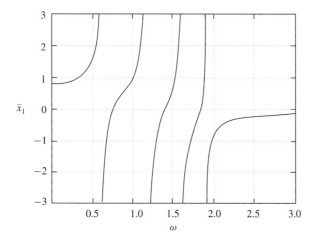

**Figure 4.6**   Response of the first mass of a four-mass system

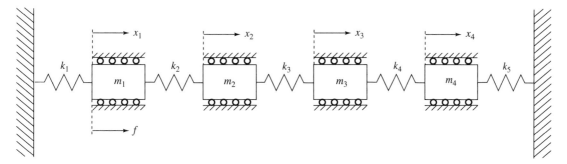

**Figure 4.7**   Forced, 4-DOF spring-mass system

to infinity at each of these frequencies and also that the phase shifts by $\pi$ rad as we pass these frequencies (that is, the response switches sign).

**Example 4.2**

**Problem**   To see how we might use the forced response of a MDOF system, consider Figure 4.8. We can consider this system as approximating an autonomous vehicle, to be used in unmanned explorations of Mars. The two springs represent the front and rear suspensions and the bar represents the body. We'll assume that the disturbance represents the effects of an unbalanced force in the engine. Our design goal will be to ensure that the vibrations are minimized at the location of the main instrumentation. As a simplification, we'll assume that the inertia of the beam doesn't change as the instrumentation is moved. Although we'll work out the governing equations for arbitrary values of the various parameters, for our numerical results we'll use $m = 50$ kg, $l = 1$ m, $\bar{I} = 16.\bar{6}$ kg·m², $k_1 = 30,000$ N/m, $k_2 = 20,000$ N/m, and $f = (100 \text{ N})\cos(\omega t)$.

**Solution**   To solve this problem, we must first obtain the system's equations of motion. A convenient set of coordinates is the vertical translation of the vehicle's center of mass and the rotation

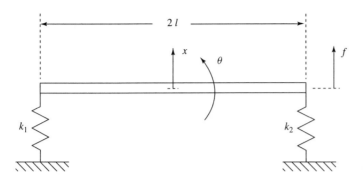

**Figure 4.8**  2-DOF model of an autonomous vehicle

about that center. Summing moments and forces about the center of mass will give us (Figure 4.9)

$$m\ddot{x} = -k_1(x - l\theta) - k_2(x + l\theta) + f \tag{4.3.10}$$

and

$$\bar{I}\ddot{\theta} = -k_2(x + l\theta)l + k_1(x - l\theta)l + fl \tag{4.3.11}$$

Rearranging this yields

$$m\ddot{x} + k_1(x - l\theta) + k_2(x + l\theta) = f \tag{4.3.12}$$

$$\bar{I}\ddot{\theta} + k_2(x + l\theta)l - k_1(x - l\theta)l = fl \tag{4.3.13}$$

or

$$m\ddot{x} + (k_1 + k_2)x + (k_2 - k_1)l\theta = f \tag{4.3.14}$$

$$\bar{I}\ddot{\theta} + (k_1 + k_2)l^2\theta + (k_2 - k_1)lx = fl \tag{4.3.15}$$

Although with more complicated systems we must resort to finding our solutions numerically, we can get a closed-form solution when we have only two equations to deal with. Assuming solutions of $x(t) = \bar{x}\cos(\omega t)$, $\theta(t) = \bar{\theta}\cos(\omega t)$ to go along with the forcing $f = \bar{f}\cos(\omega t)$, (4.3.14) and (4.3.15) will become

$$(k_1 + k_2 - \omega^2 m)\bar{x} + (k_2 - k_1)l\bar{\theta} = \bar{f} \tag{4.3.16}$$

and

$$(k_2 - k_1)l\bar{x} + (k_1 l^2 + k_2 l^2 - \omega^2 \bar{I})\bar{\theta} = \bar{f}l \tag{4.3.17}$$

Solving (4.3.17) for $\bar{x}$ in terms of $\bar{\theta}$ gives

$$\bar{x} = \frac{\bar{f}l - (k_1 l^2 + k_2 l^2 - \bar{I}\omega^2)\bar{\theta}}{(k_2 - k_1)l} \tag{4.3.18}$$

**Figure 4.9**  Free-body diagram for an autonomous vehicle

Substituting (4.3.18) into (4.3.16) and solving for $\bar{\theta}$ yields

$$\bar{\theta} = \frac{\bar{f}l(2k_1 - m\omega^2)}{\Delta(\omega)} \tag{4.3.19}$$

where

$$\Delta(\omega) = (k_1 + k_2 - m\omega^2)(k_1 l^2 + k_2 l^2 - \bar{I}\omega^2) - l^2(k_2 - k_1)^2 \tag{4.3.20}$$

Using (4.3.19) in (4.3.18) gives us

$$\bar{x} = \frac{\bar{f}(2k_1 l^2 - \bar{I}\omega^2)}{\Delta(\omega)} \tag{4.3.21}$$

Now we're in a position to do some design. We'll assume that the instruments are $\alpha l$ m to the right of the bar's center. Thus the displacement is given by $y = x + \alpha l\theta$ for small angles of oscillation, where $\alpha$ runs from $-1$ to $1$. Using this information, together with (4.3.19) and (4.3.21), we can express the displacement of the instruments as

$$\bar{y} = \bar{f}\left(\frac{2k_1 l^2(1 + \alpha) - (\bar{I} + \alpha m l^2)\omega^2}{\Delta(\omega)}\right) \tag{4.3.22}$$

Our minimization problem is now pretty straightforward. Once we're given the frequency of motor vibration, we can use (4.3.22) to determine the placement at which $y$ will be zero. $\bar{y} = 0$ in (4.3.22) means that the numerator must equal zero. Thus we'll have

$$\bar{f}(2k_1 l^2(1 + \alpha) - (\bar{I} + \alpha m l^2)\omega^2) = 0 \tag{4.3.23}$$

or, solving for $\alpha$ in terms of $\omega$,

$$\alpha = \frac{\bar{I}\omega^2 - 2k_1 l^2}{2k_1 l^2 - \omega^2 m l} \tag{4.3.24}$$

If we're told that the forcing frequency is equal to 55 rad/s, then the optimal placement of the instrument package will be $\alpha = .105l$, or .105 m to the right of center.

## Example 4.3

**Problem**    In this example we'll consider another kind of problem. The system will be identical to that of the previous example. The difference in this case is that where we used to have a force applied, we'll now have a sensor mounted. The purpose of the sensor will be to tell us what's happening in the overall system. For specificity, we'll assume that the sensor is an accelerometer that reads vertical accelerations. The question before us is where *not* to place the sensor.

**Solution**    The way to approach this problem is to ask what the worst situation might be. We know that the system will support two modes of vibration. If we want complete knowledge of what's going on, the sensor must be able to respond to both these modes. Therefore, the worst placement of the sensor would be those places on the bar for which no information is transmitted for one or both modes.

We've already calculated the system's natural frequencies in the previous example; $\omega_1$ is equal to to 30.7 rad/s and $\omega_2$ is equal to 55.3 rad/s.

Rewriting (4.3.14) and (4.3.15) in matrix form (neglecting external forcings), we have

$$\begin{bmatrix} m & 0 \\ 0 & \bar{I} \end{bmatrix} \begin{Bmatrix} \ddot{x} \\ \ddot{\theta} \end{Bmatrix} + \begin{bmatrix} k_1 + k_2 & (k_2 - k_1)l \\ (k_2 - k_1)l & (k_1 + k_2)l^2 \end{bmatrix} \begin{Bmatrix} x \\ \theta \end{Bmatrix} = \begin{Bmatrix} 0 \\ 0 \end{Bmatrix} \tag{4.3.25}$$

Thus our eigenvector equations are

$$\begin{bmatrix} k_1 + k_2 - m\omega^2 & (k_2 - k_1)l \\ (k_2 - k_1)l & (k_1 + k_2)l^2 - \bar{I}\omega^2 \end{bmatrix} \begin{Bmatrix} \bar{x} \\ \bar{\theta} \end{Bmatrix} = \begin{Bmatrix} 0 \\ 0 \end{Bmatrix} \tag{4.3.26}$$

Solving for the eigenvectors associated with the two natural frequencies (using the values of the previous example) yields

$$\omega_1 = 30.7 \text{ rad/s}, \quad X_1 = \begin{Bmatrix} .9600 \\ .2799 \end{Bmatrix}, \quad \omega_2 = 55.3 \text{ rad/s}, \quad X_2 = \begin{Bmatrix} -.0967 \\ .9953 \end{Bmatrix} \tag{4.3.27}$$

Now we can ask where on the beam is the deflection equal to zero for each of these modes. The general displacement along the beam is the same as it was for the previous example, $x + \alpha l \theta$. For the first eigenvector this means that the displacement is proportional to $.9600 + .2799\alpha$, where $\alpha$ measures displacement to the right of the midpoint in meters. The displacement is going to be equal to zero when

$$\alpha = -\frac{.9600}{.2799} = -3.430 \text{ m} \tag{4.3.28}$$

Is this a ridiculous answer? After all, it says that the zero point occurs *off* the bar, way to the left. Well, the answer is that it's not ridiculous at all; what it's telling us is that there aren't any points on the bar that *don't* move when the bar is oscillating in the system's first mode. Our accelerometer can be placed anywhere and still pick up the motion from the first mode. The situation, however, is different for the second mode. For this mode, the displacement is proportional to $-.0967 + .9953\alpha$. This is zero when $\alpha = \dfrac{.0967}{.9953} = .0972$. Thus there is a node (point of zero motion) $.0972$ m to the right of center. If we place our accelerometer there, then there is no chance for it to pick up the second mode's motion.

Our conclusion is, therefore, that we should put the accelerometer anywhere but $.0972$ m to the right of center. Of course, physical considerations tell us that the displacement of the second mode is pretty small in that point's surrounding neighborhood, also. Thus we'd want to position the accelerometer a reasonable distance away.

The effect we've just seen is a very important one in vibrational analyses. For complicated systems we'll have a large number of locations for which the response of a particular mode will be very small. Knowing where these vibrational nodes occur is crucial in order to place our sensors intelligently. In the same way, if we want to control a particular mode, then we wouldn't want to apply the force to a node either, since this would keep the force from influencing that mode.

## 4.4 VIBRATION ABSORBERS WITHOUT DAMPING

We've already seen how altering the mass, spring, and damping properties of a SDOF system will alter its response. What this section will discuss is an interesting twist, in which we will control the response of a sinusoidally forced mass by adding more spring and mass elements to the system.

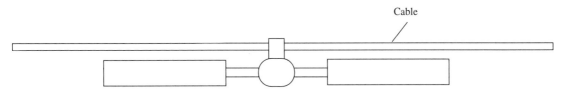

**Figure 4.10**   Stockbridge damper

We won't need to add damping and won't be dissipating more energy to control the response. Instead, we'll distribute the vibrational energy, moving it from the original mass to the new one.

Devices that absorb energy in this way are found in a variety of applications. One common use is to reduce the vibrational response of high-tension power lines. Figure 4.10 shows what a vibration absorber for this application, called a Stockbridge damper, looks like. Keep in mind that, as already mentioned, there is no energy dissipation taking place, in spite of the name. All that is happening is that energy that *was* vibrating the cable will now vibrate the Stockbridge damper.

A simplified picture of a vibration absorber application is shown in Figures 4.11a and b. Figure 4.11a indicates the original system, in which a spring-mass combination ($m_1$ and $k_1$) is acted upon by a sinusoidal force. The vibration absorber consists of the mass $m_2$ and the spring $k_2$, with the total system illustrated in Figure 4.11b. We can picture the original system as representing any one of a variety of situations. For the purposes of illustration, let's presume that it represents an engine on its mount, and the sinusoidal force is the force due to a rotating imbalance in the

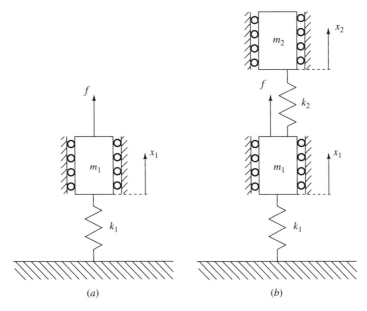

**Figure 4.11**   (*a*) Primary S-M system; (*b*) undamped vibration absorber, primary and secondary S-M systems

engine. The equation of motion for the original system is given by

$$m_1\ddot{x} + k_1 x = \bar{f}\sin(\omega t)$$    (4.4.1)

and the solution is

$$x(t) = \frac{\bar{f}}{(k_1 - m_1\omega^2)}\sin(\omega t)$$    (4.4.2)

To be even more specific, let's assign values to $m_1$, $k_1$, $\bar{f}$, and $\omega$. We'll let $m_1 = 100$ kg, $k_1 = 900$ kg/s², $\bar{f} = 1$ N, and $\omega = 3$ rad/s. What's the result of these choices? Well, since we've chosen the driving frequency to be equal to the system's natural frequency, we're in a resonance condition and the response becomes unbounded. Figure 4.12 shows the displacement response as a function of frequency. As you can see, the response goes to infinity at $\omega = 3$ rad/s.

   Clearly this state of affairs is not a good situation. It is, however, a reasonable one to consider. Even if the engine doesn't usually operate at the natural frequency of the engine/base combination, it may happen in some applications that this frequency must be passed through as the engine comes up to speed. Possibly this is also the frequency at which the engine operates most efficiently. Of course, one solution to the problem might be to redesign the suspension. Unfortunately, the suspension design might well be constrained by other design considerations. Thus, instead we'll investigate how we can mitigate the vibration problem by adding a subsystem ($m_2$ and $k_2$) to the engine. This subsystem will be attached to $m_1$ in the manner shown in Figure 4.11b. What we now have is a 2-DOF problem rather than the original, SDOF one. The $m_1/k_1$ system is referred to as the primary system and the $m_2/k_2$ system is known as the secondary system.

   Since this problem is simply a special case of the more general problem worked out in the previous section, we can substitute our particular values for $m_1$, $k_1$, etc. and obtain the solutions

$$\bar{x}_1 = \frac{(k_2 - m_2\omega^2)\bar{f}_1}{\Delta_1(\omega)}$$    (4.4.3)

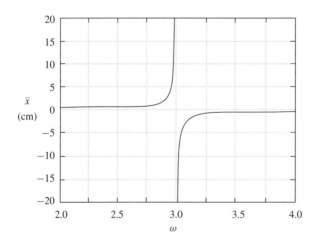

**Figure 4.12** Response of primary mass before the addition of vibration absorber

and

$$\bar{x}_2 = \frac{k_2 \bar{f}_1}{\Delta_1(\omega)} \tag{4.4.4}$$

with $\Delta_1(\omega) \equiv (k_1 + k_2 - m_1\omega^2)(k_2 - m_2\omega^2) - k_2^2$.

We can glean a wealth of observations from this particular case. First we see from (4.4.3) that the response, $\bar{x}_1$, can be zero if the forcing frequency, $\omega$, is equal to $\sqrt{\dfrac{k_2}{m_2}}$. Since we have complete freedom over how we choose $m_2$ and $k_2$, we can easily choose them such that the ratio $\dfrac{k_2}{m_2} = \dfrac{k_1}{m_1}$. Think of what this implies. It means that if our system used to be in a resonant condition because the driving frequency was equal to the SDOF's natural frequency $\sqrt{\dfrac{k_1}{m_1}}$, the newly created 2-DOF system will actually have a zero amplitude response at that frequency. Thus, instead of reducing the amplitude from infinity to something finite, as the addition of damping would have done, we've actually reduced it to zero. Although it isn't obvious right now, adding a degree of freedom, as we've done, will always create one natural frequency above the original natural frequency and one below it. The frequency at which the zero amplitude response occurs will always lie between the two and thus can always be placed where the original natural frequency occurred.

Before moving beyond this point, we should think about what's physically going on here. If the amplitude $\bar{x}_1$ is zero, that means that $m_1$ is completely stationary. Thus we can view our system as not being the forced, 2-DOF problem that it really is, but as the SDOF problem of a mass, connected by a spring to a stationary mass. This is shown in Figure 4.13. In this view, the forcing has disappeared (since it is applied to the stationary mass) and all we're left with is a free-vibration problem.

We can also see why the mass $m_1$ remains stationary. Although it isn't actually fixed in place, we're forcing it at $\omega = \sqrt{\dfrac{k_1}{m_1}}$. As we've mentioned, this would be the natural frequency of oscillation of $m_2$ and $k_2$, if they had been attached to an immovable body. Thus the $m_2$, $k_2$ combination oscillates at just the right frequency to cancel out the external force, and this is exactly what happens. The phase of the absorber force is exactly opposed to that of the external forcing. By oscillating at the correct amplitude, the magnitude of the forcing is also matched. Thus $m_1$ experiences two forces acting on it that are exactly opposed to each other. The end result—no motion. These forces are illustrated in Figure 4.14.

Let's calculate this just to be sure. The external force applied to $m_1$ is $\bar{f} \cos(\omega_n t)$, where $\omega_n^2 = \dfrac{k_1}{m_1} = \dfrac{k_2}{m_2}$. From (4.4.4) we have $\bar{x}_2 = \dfrac{k_2 \bar{f}_1}{\Delta_1(\omega)}$. Since $m_1$ is stationary, the force acting on $m_1$ due to $k_2$ is equal to $k_2\bar{x}_2$. Using $\omega = \omega_n$ to evaluate $\Delta_1(\omega)$ gives us $-\bar{f} \cos(\omega_n t)$ as the force being exerted on $m_1$ by $k_2$. Thus the force due to the external forcing ($\bar{f} \cos(\omega_n t)$) and the

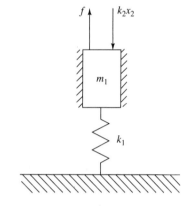

**Figure 4.13**  Physical equivalent to the effect of an absorber spring-mass

**Figure 4.14**  Opposed forces acting on primary mass

moving mass $m_2$ $(-\bar{f}\cos(\omega_n t))$ are exactly equal and opposed, resulting in no overall force acting on $m_1$.

We can gain some additional insight by looking at two particular absorbers. Figure 4.15 shows the response of both the primary and absorber masses for the case $k_2 = 90$ N/m and $m_2 = 10$ kg. What this means is that we've changed the mass of the overall system by 10 percent. As you can see, $\bar{x}_1$ goes through zero at $\omega = 3$ rad/s, as predicted. Furthermore, there is a reasonably broad band of frequencies about 3 rad/s, for which the response level is quite low. Thus, even in the face of fluctuations in the exciting frequency, we'll still have only a small response. Figure 4.15 also shows the response of the absorber mass itself ($\bar{x}_2$). Again, as the equations predict, there is no frequency for which the amplitude is zero.

Although these results look pretty good, you might be wondering if we could do even better. After all, the only requirement for a vibration absorber to work is that $\sqrt{\dfrac{k_2}{m_2}} = 3$ rad/s. Although the 10-percent weight penalty we just used isn't terribly bad, wouldn't it be better to use even less mass? After all, designers are often very concerned about weight, and if the actual mass doesn't matter for a vibration absorber, then why not make it as small as possible?

Figure 4.16 shows what happens when we reduce both $k_2$ and $m_2$ by a factor of 10. In this case, the secondary mass is 100 times smaller than the primary mass, meaning that we've hardly changed the weight of the system at all. But as Figure 4.16 shows, we must pay a price for the weight savings. First off, the two resonant frequencies are closer together than they were for the larger absorber. Instead of a relatively wide range of frequencies for which the absorber is effective,

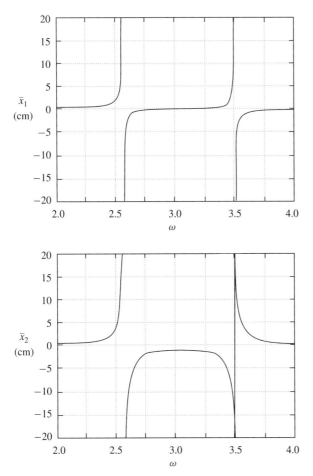

**Figure 4.15**  Response of vibration absorber with absorber mass equal to one-tenth of primary mass

we now have a very tight range. Only very close to 3 rad/s do we have small vibrations for our primary mass. If the excitation frequency should shift a bit, then we could again end up with a vibration problem, since we'll encounter resonance conditions around $\omega = 2.7$ and $\omega = 3.2$ rad/s.

The other downside is seen in the plot of $\bar{x}_2$ versus $\omega$. By decreasing the mass and stiffness we've increased the amplitude of $\bar{x}_2$ by a factor of 10 at $\omega = 3$ rad/s. This is, in fact, a direct consequence of our having reduced the spring stiffness. $k_2$ is the means through which the vibrational force is generated that counters the applied forcing. Since the force generated is equal to the stiffness multiplied by the amplitude of the motion, if you lower the stiffness by a factor of 10, then you must increase the amplitude by the same factor in order to compensate.

These results show us that the familiar engineering dilemma of having to trade off advantages and disadvantages exists for vibration absorbers. To create a wide band of absorption, you need a heavy mass and stiff spring. Along with the wide absorption band, you also have a small,

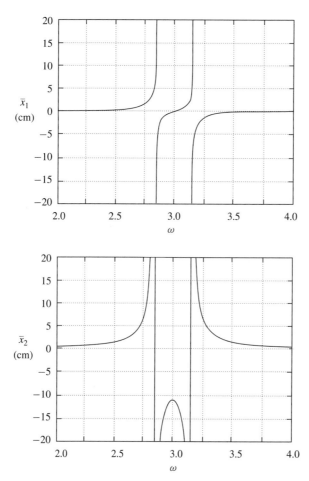

**Figure 4.16**  Response of vibration absorber with absorber mass equal to one one-hundredth of primary mass

secondary-mass motion. Therefore the packaging concerns aren't very significant, while the weight concerns are. If you use a small mass and spring, then you'll have a narrow absorption band, meaning that the absorber isn't terribly robust to changes in the parameters of the system. Furthermore, packaging will now be a problem because of the large deflections that the secondary mass will experience. As a designer, you must balance these competing trends in order to come up with an acceptable overall design.

## 4.5 REAL BEHAVIOR OF A VIBRATION ABSORBER

We've just seen that vibrations can be avoided in a forced mass by the addition of a vibration absorber. However, you might be a bit doubtful. After all, the only reason the second mass could counter the externally applied force was that it was oscillating exactly out-of-phase with this force.

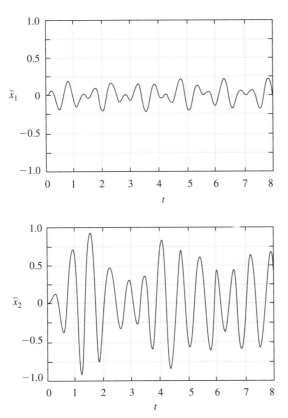

**Figure 4.17**  Actual response of undamped vibration absorber

The obvious question is, then, what if the force is *not* exactly out-of-phase with the second mass? In this case we'll have an unbalanced force on the first mass and it should move. How does the real world get around this problem?

Figure 4.17 shows the actual response (found from numerically integrating the equations of motion) of both the first and second mass of a vibration absorber problem for the system shown in Figure 4.11. Note that the primary mass does not remain stationary. However, it's important to realize that the vibrations are *bounded*. If the force and second mass weren't correctly phase-aligned, we'd actually have a resonance condition and the oscillations would grow. Since we certainly don't have growing oscillations, the force and second mass must be correctly phase-aligned. What causes the vibrations is the fact that, with no damping in the system, the homogeneous solution (that due to initial conditions) never dies out; it rings forever. This is the vibration we're seeing—that due to initial conditions. To avoid this solution, we can carefully choose our initial conditions to obtain the vibration absorption we want. If the initial conditions are chosen so that the spring force is exactly opposing the applied force, then we'll see purely sinusoidal oscillations for $x_2$ and no motion for $x_1$.

Since we can't precisely set up the initial conditions we might need for a real-world implementation, must we assume that vibration absorbers don't actually work, that we'll always have

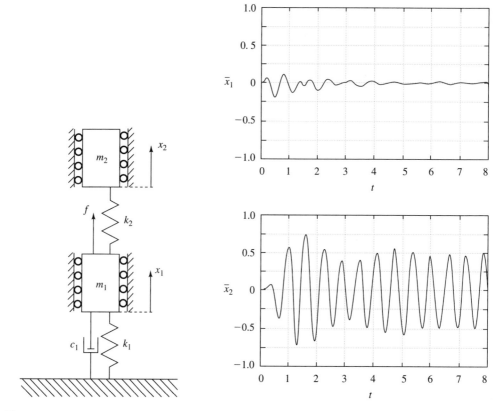

**Figure 4.18** Vibration absorber with one damper

**Figure 4.19** Actual response of damped vibration absorber

transient vibrations due to initial conditions? Luckily, we don't have to assume this at all, and the reason why was just alluded to in the previous paragraph. We'll run into this problem only in the absence of damping. Since purely undamped systems *never* exist in the real world, we really should add some damping to the system and see how it behaves. Figure 4.18 shows the same system as before, except that now we've added a small amount of linear damping to the first mass. A typical response of this system is shown in Figure 4.19. Quite a different result! Now we see that, after a transient phase, the system responds just as predicted, with a stationary first mass and a second mass that oscillates continually in order to cancel the external forcing, just what we'd have seen if the conditions had been perfectly set up.

There are only two problems left to address. First, since we've changed the system by adding damping, how do we know that the system is behaving according to prediction? We only predicted that the *undamped* system would act in the way shown, not a damped one. Second, why is it that damping allowed the system to give us the response we predicted? The first question is most easily answered. Remember our physical discussion of the absorber problem. If the first mass is kept stationary, then we're left with an eigenvalue problem for the second mass. That's how we determined that the correct spring/mass ratio for the absorber should be one for which the natural frequency was equal to the forcing frequency of the external force. Luckily, since the damper is attached between the first mass and ground, it doesn't move if the first mass is stationary. So we've got the same eigenvalue problem as before. The only way the damper is operative is if the first mass is moving; once it stops the damper drops out of the picture. As you can see from looking at Figure 4.19, the first mass is certainly moving at the start of the numerical simulation. Thus the damper is operative and must, in fact, be responsible for bringing about the vibration cancellation that is going on later in the simulation. Therefore, now we're stuck with determining why this happens. This second question will shed a lot of light on the action of damping in vibratory systems and will serve as motivation for the damping analyses to come in later sections.

The best way to approach this second question is to focus our attention on the desired solution (stationary first mass) and ask how perturbations away from this solution will behave. Thus we'll assume that our solution looks like

$$x_1(t) = 0 + \Delta x_1(t) \tag{4.5.1}$$

$$x_2(t) = a \cos(\omega_2 t) + \Delta x_2(t) \tag{4.5.2}$$

for a forcing

$$f(t) = -a k_2 \cos(\omega_2 t) \tag{4.5.3}$$

where $\omega_2 = \sqrt{\dfrac{k_2}{m_2}}$.

The approach we're using is a very important one. Since we're basing our assumed solution on a known, steady-state solution, the added terms ($\Delta x_1$ and $\Delta x_2$) are the motions that occur in addition to the steady-state solution, that is, the transient behavior. Put another way, we want to capture the behavior of the system for $t = 0$ to some future time $t_1$. We already know that a steady-state solution exists for which the first mass is stationary and the second acts to cancel the applied external force. We also know that the actual solution will have a transient and steady-state part. Since we already know what the steady-state part is, we can explicity include it in our assumed solution, leaving the unknown $\Delta x_i$'s to represent the transient solutions. This approach to finding the transient behavior of a system is quite common in nonlinear analyses and will be of equal use here.

Our governing equations of motion are

$$m_1 \ddot{x}_1 + c_1 \dot{x}_1 + (k_1 + k_2)x_1 - k_2 x_2 = f(t) \tag{4.5.4}$$

and

$$m_2 \ddot{x}_2 + k_2 x_2 - k_2 x_1 = 0 \tag{4.5.5}$$

Substituting our assumed solutions into (4.5.4) and (4.5.5) we find

$$m_1 \Delta \ddot{x}_1 + c_1 \Delta \dot{x}_1 + (k_1 + k_2)\Delta x_1 - k_2(a \cos(\omega_2 t) + \Delta x_2) = -a k_2 \cos(\omega_2 t) \tag{4.5.6}$$

and

$$m_2(-A\omega_2^2\cos(\omega_2 t) + \Delta\ddot{x}_2) + k_2(A\cos(\omega_2 t) + \Delta x_2) - k_2\Delta x_1 = 0 \tag{4.5.7}$$

What you can see is that all the steady-state solutions, along with the forcing, cancel out the equations. (For instance, in (4.5.6) the $-k_2 A\cos(\omega_2 t)$ term on the left side cancels out the $-Ak_2\cos(\omega_2 t)$ term on the right.) Thus, once we remove these known terms, we're left with equations that simply involve the $\Delta x_i$'s,

$$m_1\Delta\ddot{x}_1 + c_1\Delta\dot{x}_1 + (k_1 + k_2)\Delta x_1 - k_2\Delta x_2 = 0 \tag{4.5.8}$$

and

$$m_2\Delta\ddot{x}_2 + k_2\Delta x_2 - k_2\Delta x_1 = 0 \tag{4.5.9}$$

Now, instead of the forced problem we previously had to deal with, we have a free-vibration problem. If you examine these equations, you'll see that they're exactly the same equations that you would have derived if you'd been concerned with free vibrations of Figure 4.11 in the absence of any external forcing. Thus we've made the very important observation that the transient motion in a linear, forced system is governed by the *free vibrations* of that system when no forcing is present.

We've already seen in our work with SDOF systems that damping alters the pure sinusoids that occur for undamped problems into damped sinusoids. The same thing occurs with MDOF systems. Rather than begin a detailed look at the general effect of damping in a system, we're going to postpone this topic until we've looked at normal forms in Section 4.7. Thus we'll content ourselves with looking at the numerical response of (4.5.8) and (4.5.9). Figure 4.20 shows the response of these equations to an arbitrarily chosen set of initial conditions. As you can see, the responses simply oscillate and decay. Eventually, both $\Delta x_1$ and $\Delta x_2$ approach zero. What does this mean for our vibration problem? It means that, as $t$ becomes large, the transient solutions $\Delta x_1(t)$ and $\Delta x_2(t)$ die away, leaving the steady-state solution in which the first mass is stationary, that is, the desired solution for a vibration absorber. It's important to realize that there were no approximations involved in this analysis. Thus it is always true that the transient solution will die down as governed by (4.5.8) and (4.5.9), leaving the desired steady-state solution.

Let's recap a bit to realize what we've gotten into. First, we've seen that damping, which always exists in the real world, can be quite helpful, not just because it allows disturbances to a system to die down, but also because it can be the mechanism through which passive vibrational control can be effected. In the case just examined, the steady-state vibrational response of the system with damping was the same as that predicted for the system without damping. We should now be motivated to ask if this is always true (it's not) and to try to understand the different ways in which damping affects the response of MDOF systems. This is (almost) the next item we'll be discussing. However, before jumping into this topic, it will be helpful to look more closely at two more items. The first is to generalize our vibration-absorber observations to the multidimensional case. The second will be to look at the different ways we can express general solutions and, specifically, to see if there's a way to view MDOF systems in a simpler way, that is, as sets of completely independent, SDOF systems. Thus we'll first tackle these issues and then return to the question of damping.

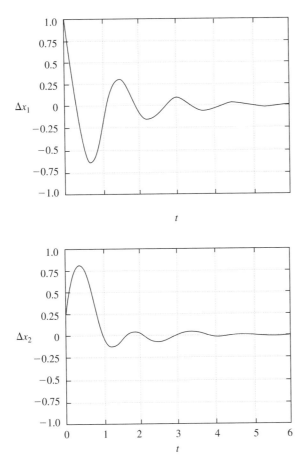

**Figure 4.20**  Transient solution behavior of a damped vibration absorber

## 4.6 ZEROS IN A FORCED RESPONSE

In Section 4.4 we determined what frequencies of excitation will cause the response at the driven mass to equal zero for a 2-DOF system. The problem was stated differently, of course. We said that our original SDOF system had severe vibrations and determined how to reduce the vibrations to zero by the addition of a second mass-spring assembly. The end result, however, is the same as if we'd started with a 2-DOF system and asked what frequency of forcing, when applied to the first mass, would result in that mass being stationary. You'll recall that in the process of solving the problem we noticed that a stationary mass is the same as a rigidly fixed mass. Thus the correct frequency of excitation turned out to be the natural frequency of the remaining mass and springs.

The same thing can be seen in MDOF systems. In an MDOF system, we have $n$ masses. Our question will become, What frequency of force excitation, applied to a particular mass, will cause a particular mass in the system to have no motion? Consider the serial chain of mass and springs shown in Figure 4.21. A sinusoidal force is applied to $m_1$ and what we want to do is determine at

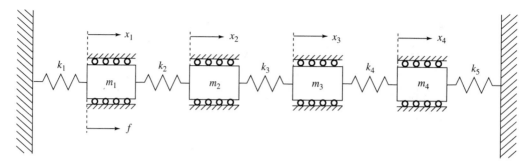

**Figure 4.21**  Serial chain of spring-mass elements, force applied to first mass

what frequencies of excitation $m_1$ will be stationary. Then we'll ask if $m_2$ can also be stationary. After we've examined the different masses, we'll move the force and again see how the individual masses might be stationary, even though a sinusoidally varying force is being applied to the system.

Our first task is to see for what frequencies $m_1$ remains still. Just as in the vibration absorber, we can realize that if $m_1$ is still, it's as if $k_2$ is attached to a rigid wall at the left. We're left with a three-mass system: $k_2 - k_4$ and $m_2 - m_4$. Since this is a three-degree-of-freedom system, it will have three natural frequencies and three eigenvectors. This is the answer we're seeking. $m_1$ can be stationary if the applied force exactly opposes the forces that $k_2$ will transmit (which are due to the free vibrations of the three free masses).

Note that when we had two masses, we were able to get the response of one of the masses to be zero at only one frequency. Furthermore, we saw that the response of the second mass was never zero if the force was applied to the first. In this case we've seen that there are four masses in the system and we can cause the forced mass to have zero response at three frequencies. For both cases, the displacement of the forced mass can be zero at $n - 1$ frequencies, where there are $n$ masses in the system.

What if we apply the force to the second mass, rather than to the first, as shown in Figure 4.22? Are there $n - 1$ frequencies at which the second mass has zero response? If the second mass is stationary, then we have a SDOF system on the left (made up of $k_1$, $k_2$, and $m_1$) and a 2-DOF

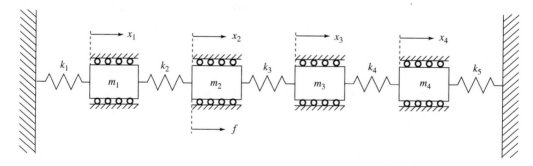

**Figure 4.22**  Serial chain of spring-mass elements force applied to second mass

system on the right (made up of $k_3$, $k_4$, $k_5$, $m_3$, and $m_4$). Thus we'll have one natural frequency from the left and two from the right. Again we've got $n - 1$ frequencies for which we get a zero amplitude response. This finding is a general one. When we have $n$ masses, the number frequencies at which the driven mass is stationary are equal to $n - 1$.

What if we move away from the driven mass? For instance, let's consider the situation of Figure 4.21, where the force is applied to $m_1$. This time, let's ask when $m_2$ is stationary. Since $m_2$ is still, we have two degrees of freedom to the right ($m_3$ and $m_4$) that can oscillate at one of their two natural frequencies. This vibration will cause a force on $m_2$ due to $k_3$. The only way to keep $m_2$ from moving in response is if an equal and opposite force is applied from the left. This force can easily be supplied by appropriately forcing $m_1$. Thus we have 2, or $n - 2$, zeros for $m_2$ when $m_1$ is forced. We saw the same thing for the 2-DOF vibration absorber. When the first mass was forced, the second mass had no zero responses. For this case $n = 2$, therefore $n - 2 = 0$, which tells us that there are no frequencies at which the second mass would be stationary.

If you look at $m_3$ instead of $m_2$ you'll see that the number of frequencies that yield zero responses is equal to $n - 3$, that is, there will be one. Finally, when you force the first mass and look at the mass furthest away ($m_4$ in this case), there will be no frequencies at which the response is zero.

All these deductions can be seen in Figure 4.23. This figure shows the actual response versus forcing frequency for each mass of the system of Figure 4.21. For these plots all the $m_i$'s and $k_i$'s were given unit values and the magnitude of $f$ was also set equal to 1.0. Figure 4.23a shows the response of $m_1$ for a force applied to $m_1$. You can see that the response is equal to zero at three distinct frequencies. Moving from $m_1$ to $m_2$, Figure 4.23b loses one of the zeros. Instead of going through zero near $\omega = 1.4$, the response approaches zero, but then turns around and moves away from zero and out to negative infinity. Figure 4.23c displays only a single zero. Note that this zero occurs near $\omega = 1.4$. Thus we see that the zeros don't drop off in a regular fashion as we move away from the forced mass. $m_2$ (in Figure 4.23b) did not have a zero at the same place that $m_3$ (in Figure 4.23c) had one. In addition, $m_3$ has lost the zeros where $m_2$ had them. Lastly, Figure 4.23d displays no zeros.

These observations give us some insight into the design of our systems. We might well want a zero response somewhere in the system. Perhaps we need to mount a mirror and don't want any vibrations that would disturb it. Or it may be that a sensitive piece of equipment will be mounted in a particular place and we want to minimize any vibrations. The analysis shown above will tell us how many frequencies there are that will permit a vibration absorption to take place.

If we call the forced mass our *actuated* mass and the mass whose output we're observing our *sensed* mass, then we have a sensor-actuator pair. By having the sensor and actuator both be the same mass, we've created what is called a *collocated sensor/actuator pair*. The ability to be collocated is highly desirable in any control system and the reason is pretty clear. What we normally want to do in a control system is keep some part of the system under control, usually implying that the part under control is kept at zero, even in the face of disturbances. Figure 4.23d shows the worst case of noncollocation. In this situation the control force and sensing are located as far from each other as possible; they're separated by the dynamics of the entire system. This

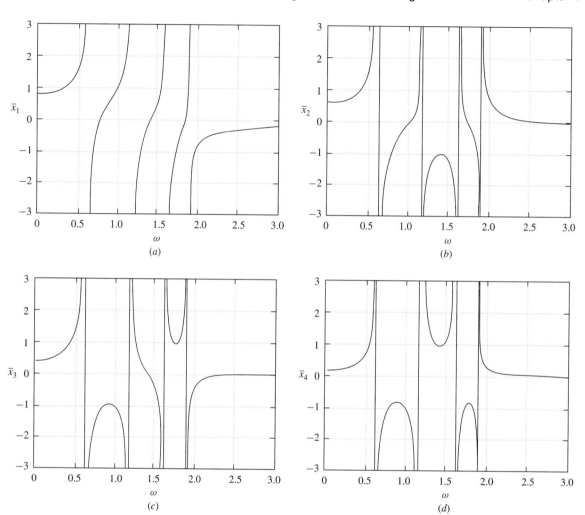

**Figure 4.23** Four graphs showing amplitude responses for different masses in a four-mass system

makes control very difficult, as reflected by the fact that there are no frequencies at which the response is zero.

The physical analogue of this situation is a golfer or a fly fisherman. In each case the person is holding the device (be it a flexible golf club or a really flexible fishing rod), and the part they wish to control is at the other end of the device (the club head or the end of the rod). We could model either device as a string of masses and springs, and, if we assume that the person was watching the tip in order to control the system, we'd have a completely noncollocated setup. (Interesting side note: You'll no doubt be saying—if you're a golfer or fisherman—that people don't actually observe the tip in order to golf or fish. You're right. The reason is that the dynamics and control problems are too complex. They're beyond our abilities to control in any sort of feedback sense. That's why they're done in what's called an *open loop* fashion. We simply learn by rote what the correct motions are and then repeat them mechanically.)

When running a modal test, we'll also have a similar situation in that we have an exciter and sensor that we use to gather the modal data. We'll have either a collocated or noncollocated setup depending upon whether the sensor and exciter act on the same mass. Precise collocation usually is difficult to achieve. An interesting new type of sensor, a *self-sensing* actuator that manages to be a truly collocated system, will be discussed in Chapter 8.

**Example 4.4**

**Problem**  Determine how many zeros to expect in the system shown in Figure 4.22 when the sensed mass is $m_3$. Then determine how many zeros should occur for any combination of forced mass and sensed mass. For this example $m_1 = .75$, $m_2 = .5$, $m_3 = .9$, $m_4 = .55$, $k_1 = 1.1$, $k_2 = 2$, $k_3 = 1$, $k_4 = 1$, and $k_5 = 3.3$. Masses are given in kg and stiffnesses in N/m.

**Solution**  First we'll look to the right of $m_3$. Since $m_3$ is stationary, that implies that the $k_4$-$k_5$-$m_4$ SDOF system could be oscillating at its natural frequency. This will impress a force on $m_3$ by means of $k_4$. If $k_3$ is impressing an equal and opposite force (because of the force $f$) then $m_3$ can remain stationary. That's one frequency. Are there others? Let's look to the left of $m_3$ to find out. For there to be another frequency $m_2$, as well as $m_3$, must be stationary. If $m_2$ wasn't stationary, then its motion would make $m_3$ move, violating our assumption. The only frequency at which $m_2$ could move without moving $m_3$ would be the natural frequency of the $k_4$, $k_5$, $m_4$ system. However, we've already taken that answer into account. Thus we conclude that $m_2$ must be stationary in order for there to be a second zero for $m_3$. Is this possible? Very definitely. The force on $m_2$ can exactly counteract free vibrations of the $k_1$, $k_2$, $m_1$ system. Physically, the $k_1$, $k_2$, $m_1$ system will be oscillating and the external force will exactly cancel the forces due to this oscillation that $k_2$ is impressing on $m_2$. Thus we see that $m_2$ and $m_3$ are both stationary. So, in fact, is $m_4$, since there's no way for it to get excited. This immediately tells us that we've found out something else, namely that this zero for $m_3$ must also be a zero of $m_2$ and $m_4$.

Figure 4.24 shows the forced response for this system. As you can see, we have just two frequencies that correspond to zero motion (around $\omega = 2$ rad/s and $\omega = 2.75$ rad/s). Both zero frequencies that were predicted have shown up.

An interesting fact that can be seen in the previous examples is that the number of zeros in a mass-spring chain is always equal to $n - (s + 1)$, where $n$ is the number of masses and $s$ is the number of masses the sensed mass is away from the driven mass. In our first example the driven mass was $m_1$ and the sensed mass varied from $m_1$ to $m_4$ (Figure 4.23). When $m_1$ was the sensed mass, there was no separation between sensed and forced masses, therefore $s = 0$. The number of zeros was three, which matched the result predicted from evaluating $n - 1$. When the sensed mass was $m_2$, $s = 1$ and $n - (1 + s) = 2$, matching the numerical results. In the example we've just discussed, the separation between the driving and forced mass was again equal to one and $n - (s + 1) = 2$, agreeing with the number of zeros shown in Figure 4.24. This formula also tells us to expect no zeros when the driven and sensed masses are at either end of the chain. In this case $s = n - 1$ and $n - (s + 1)$ is always equal to zero.

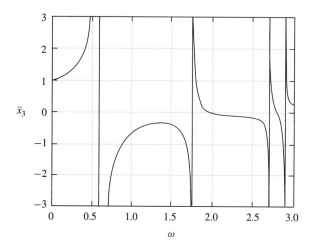

**Figure 4.24** Graph showing zero amplitude for third mass, force applied to second mass

## 4.7 PUTTING PROBLEMS INTO NORMAL FORM

So far, the way we've viewed 2-DOF systems has been reasonably similar to the way we've dealt with SDOFs. We've seen that 2-DOF systems support eigenvector responses, but we haven't particularly exploited this fact. Now we'll do so.

One obvious characteristic of the general 2-DOF systems we've seen so far is the fact that the system equations are coupled. It's impossible to solve for the response of $x_1$ without at the same time solving for $x_2$ (where $x_1$ and $x_2$ are the two coordinates of our system). We even have names for the particular types of coupling that we might run into. For instance, the following equation is *mass coupled*, or *inertially coupled*

$$\begin{bmatrix} 2 & -1 \\ -1 & 4 \end{bmatrix} \begin{Bmatrix} \ddot{x}_1 \\ \ddot{x}_2 \end{Bmatrix} + \begin{bmatrix} 5 & 0 \\ 0 & 8 \end{bmatrix} \begin{Bmatrix} x_1 \\ x_2 \end{Bmatrix} = \begin{Bmatrix} f_1 \\ f_2 \end{Bmatrix} \tag{4.7.1}$$

because the two equations are coupled through the $[M]$ matrix. In a like manner, (4.7.2) is *stiffness coupled* or *spring coupled* because the coupling comes through the $[K]$ matrix.

$$\begin{bmatrix} 1 & 0 \\ 0 & 2 \end{bmatrix} \begin{Bmatrix} \ddot{x}_1 \\ \ddot{x}_2 \end{Bmatrix} + \begin{bmatrix} 7 & -2 \\ -2 & 8 \end{bmatrix} \begin{Bmatrix} x_1 \\ x_2 \end{Bmatrix} = \begin{Bmatrix} f_1 \\ f_2 \end{Bmatrix} \tag{4.7.2}$$

A fully uncoupled set of equations would look like

$$\begin{bmatrix} 1 & 0 \\ 0 & 2 \end{bmatrix} \begin{Bmatrix} \ddot{x}_1 \\ \ddot{x}_2 \end{Bmatrix} + \begin{bmatrix} 7 & 0 \\ 0 & 8 \end{bmatrix} \begin{Bmatrix} x_1 \\ x_2 \end{Bmatrix} = \begin{Bmatrix} f_1 \\ f_2 \end{Bmatrix} \tag{4.7.3}$$

For this case, we see that the equation governing the response of $x_1$ has nothing to do with $x_2$'s behavior. Thus we can solve either equation independently of the other; they're uncoupled.

It's clear that uncoupled equations are nice to have since it's always easier to solve two independent equations than it is to solve a simultaneous set of equations. The question is, Can we

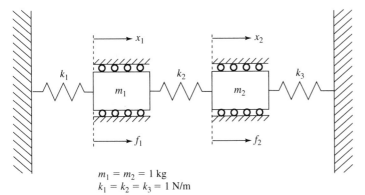

$m_1 = m_2 = 1$ kg
$k_1 = k_2 = k_3 = 1$ N/m

**Figure 4.25**   Forced, 2-DOF system

somehow transform our particular equations into a form in which they're decoupled? Obviously, the answer is yes, or we wouldn't be mentioning it. To take a specific example, let's look at the problem illustrated in Figure 4.25. We'll consider the case for which $m_1 = m_2 = 1$ and $k_1 = k_2 = k_3 = 1$. The governing equations are

$$\begin{bmatrix} 1 & 0 \\ 0 & 1 \end{bmatrix} \begin{Bmatrix} \ddot{x}_1 \\ \ddot{x}_2 \end{Bmatrix} + \begin{bmatrix} 2 & -1 \\ -1 & 2 \end{bmatrix} \begin{Bmatrix} x_1 \\ x_2 \end{Bmatrix} = \begin{Bmatrix} \bar{f}_1 \\ \bar{f}_2 \end{Bmatrix} \cos(\omega t) \tag{4.7.4}$$

This set of equations is stiffness coupled. In order to decouple them, we're going to define a new set of coordinates. Instead of $x_1$ and $x_2$, we'll express the equations of motion in terms of $\eta_1$ and $\eta_2$, where the $\eta$'s are defined as

$$\eta_1 = \tfrac{1}{2}(\sqrt{2}x_1 + \sqrt{2}x_2) \tag{4.7.5}$$

and

$$\eta_2 = \tfrac{1}{2}(\sqrt{2}x_1 - \sqrt{2}x_2) \tag{4.7.6}$$

This can be expressed in matrix form as $H = [V]X$ where

$$H \equiv \begin{Bmatrix} \eta_1 \\ \eta_2 \end{Bmatrix}, \quad X \equiv \begin{Bmatrix} x_1 \\ x_2 \end{Bmatrix}, \quad \text{and} \quad [V] \equiv \frac{1}{\sqrt{2}} \begin{bmatrix} 1 & 1 \\ 1 & -1 \end{bmatrix} \tag{4.7.7}$$

At the moment it isn't clear why in the world we'd want to do this, but we're going to do it anyway. Later, we'll see why this particular transformation was chosen. Before moving on from this point, we should think about what this new coordinate set is all about. $\eta_1$ is simply $\sqrt{2}$ times the mean of $x_1$ and $x_2$, and $\eta_2$ is $\sqrt{2}$ times one-half of the difference between $x_1$ and $x_2$. Thus, if $x_1 = 1$ and $x_2 = 1$, then $\eta_1 = \sqrt{2}$ and $\eta_2 = 0$. Recall that the two eigenvectors for this system are

$$X_1 = \begin{Bmatrix} 1 \\ 1 \end{Bmatrix}, \quad X_2 = \begin{Bmatrix} 1 \\ -1 \end{Bmatrix} \tag{4.7.8}$$

Thus, when only the first eigenvector is involved, $x_1 = x_2 = 1$; therefore $\eta_1 = \sqrt{2}$, and $\eta_2 = 0$. Similarly, when only the second eigenvector is present, $\eta_1 = 0$ and $\eta_2 = \sqrt{2}$, since $x_1 = 1$ and

$x_2 = -1$. Therefore the $\eta$ coordinates are doing a great job of telling us which eigenvector is present in the response. When $\eta_1$ is nonzero and $\eta_2$ equals zero, we know that only the first eigenvector is involved. The same holds when $\eta_1 = 0$ and $\eta_2$ is nonzero, except that it's the second eigenvector that's now involved. And if both eigenvectors are present, the $\eta$'s tell us how much of each is present.

This is not a trivial point. Previously, all we knew were the positions of $m_1$ and $m_2$. We could use this information to determine what was happening with regard to the eigenvectors of the system, but it required some calculation. Now we have an immediate reading as to the eigenvector's participation through the $\eta$ coordinates. Of course in this case we don't know what the individual $x_1$ and $x_2$ responses are, and we'd have to calculate them from our knowledge of the $\eta$'s. We can get this information by multiplying by the inverse of the transformation matrix $V$, so that instead of $H = [V]X$, we have

$$X = [U]H \tag{4.7.9}$$

where

$$[U] = [V]^{-1} = \begin{bmatrix} \frac{1}{\sqrt{2}} & \frac{1}{\sqrt{2}} \\ \frac{1}{\sqrt{2}} & -\frac{1}{\sqrt{2}} \end{bmatrix} \tag{4.7.10}$$

If we go ahead and use (4.7.9) in (4.7.4) we'll get

$$[M][U]\ddot{H} + [K][U]H = F \cos(\omega t) \tag{4.7.11}$$

Now there's one more thing we need to do. Rather than leave our equations in the above form, we're going to premultiply by $[U]^T$ to obtain

$$[U]^T[M][U]\ddot{H} + [U]^T[K][U]H = [U]^T F \cos(\omega t) \tag{4.7.12}$$

We can express this more compactly as

$$[M']\ddot{H} + [K']H = F' \cos(\omega t) \tag{4.7.13}$$

where

$$[M'] \equiv [U]^T[M][U], \quad [K'] \equiv [U]^T[K][U], \quad F' \equiv [U]^T F \tag{4.7.14}$$

If you carry through with the indicated operations, you'll find that (4.7.13) has the following particular matrix entries:

$$[M'] = \begin{bmatrix} 1 & 0 \\ 0 & 1 \end{bmatrix}, \quad [K'] = \begin{bmatrix} 1 & 0 \\ 0 & 3 \end{bmatrix}, \quad F' = \begin{bmatrix} \frac{1}{2}(f_1' + f_2') \\ \frac{1}{2}(f_1' - f_2') \end{bmatrix} \tag{4.7.15}$$

Note the good news: we've decoupled the equations. The equation for $\eta_1$ is completely independent of the equation for $\eta_2$ and vice versa. Furthermore, we can notice some interesting facts. First, the mass matrix $[M']$ is the identity matrix. This isn't a big deal, since it was already equal to the identity matrix when we started. However, what's interesting is that we can work it so that *any* initial mass matrix is transformed into the identity matrix. Obviously this fact will save storage space on a computer if it's true for any system size (and it is). Secondly, we can see that the entries in the $[K']$ matrix are exactly equal to the squares of the natural frequencies. It seems that this coordinate transformation is quite powerful. It makes the mass matrix equal to the identity matrix, produces a diagonal stiffness matrix having the square of each natural frequency on the main diagonal, and, to top it off, gives us a direct readout of the modal character of the response through the $\eta$ coordinates.

Having seen all this good stuff, you're probably interested in why and how this happened and how you can do it yourself. Amazingly, the procedure is pretty simple. Just look at the transformation matrix $[U]$ for a second and ask yourself what's familiar about the columns of this matrix. Yes, that's right! They're simply equal to the eigenvectors of the original system. That's almost all there is to it. We find the eigenvectors, put them side by side to define $[U]$, premultiply by $[U]^T$, and get our decoupled form. There are some additional fine points, which we'll talk about as we rigorously show why this decoupling takes place. If you're completely uninterested in this, you can skip to Section 4.9 now.

## 4.8 ORTHOGONALITY OF SYSTEM EIGENVECTORS

You're probably all familiar with the term *orthogonality* as applied to spatial vectors. It means that the vectors are oriented at right angles to each other. Mathematically, when two vectors are orthogonal, their dot product is zero. Additionally, one way of calculating the dot product of two column vectors $A$ and $B$ is by evaluating $A^T B$. These notions are very close to those that we'll develop in this section. The only twist is that, instead of saying certain vectors are orthogonal only to each other, we'll see that the orthogonality conditions include the $[M]$ and $[K]$ matrices. Thus orthogonality is very tightly tied to the particular system we're examining.

We'll sneak up on these results in a very straightforward way, namely by making a simple examination of our equations and then by drawing some conclusions from them. For clarity, we'll derive our results with a 2-DOF example. Keep in mind, though, that the approach is applicable to a problem with any number of degrees of freedom, as we'll indicate after finishing with the 2-DOF example.

To begin, we'll look at our free vibration problem,

$$[M]\ddot{X} + [K]X = 0 \tag{4.8.1}$$

We know now that the solutions to this problem define our system eigenvectors ($X_1$ and $X_2$) and natural frequencies ($\omega_1$ and $\omega_2$). Thus, writing (4.8.1) in terms of these known solutions gives us

$$-\omega_1^2[M]X_1 + [K]X_1 = 0 \tag{4.8.2}$$

and

$$-\omega_2^2[M]X_2 + [K]X_2 = 0 \tag{4.8.3}$$

Rearranging these equations gives us

$$\omega_1^2[M]X_1 = [K]X_1 \tag{4.8.4}$$

and

$$\omega_2^2[M]X_2 = [K]X_2 \tag{4.8.5}$$

What we've got here are two sets of vector equalities (the vector on the left equals the vector on the right). Clearly, we'll still have an equality if we premultiply (4.8.4) by $X_2^T$ and (4.8.5) by $X_1^T$. It's not obvious yet why we'd want to do this, but there's certainly nothing that says we can't. So now we have

$$\omega_1^2 X_2^T[M]X_1 = X_2^T[K]X_1 \tag{4.8.6}$$

and

$$\omega_2^2 X_1^T[M]X_2 = X_1^T[K]X_2 \tag{4.8.7}$$

Now let's take the transpose of (4.8.7). (Yes, I know. This seems pointless. But it's not, so just hold on for another moment or two.) Transposing (4.8.7) gives us

$$\omega_2^2 X_2^T[M]^T X_1 = X_2^T[K]^T X_1 \tag{4.8.8}$$

Since the $[M]$ and $[K]$ matrices of spring-mass systems are symmetric, this is the same as

$$\omega_2^2 X_2^T[M]X_1 = X_2^T[K]X_1 \tag{4.8.9}$$

Take a second to compare (4.8.6) and (4.8.9). They are identical except for the fact that one involves $\omega_1$ while the other contains $\omega_2$. We're now in a position to be clever. Let's subtract (4.8.9) from (4.8.6). This will give us

$$(\omega_2^2 - \omega_1^2)X_2^T[M]X_1 = 0 \tag{4.8.10}$$

This is interesting! It tells us something fundamental about our eigenvectors. If $\omega_2 \neq \omega_1$ then $X_2^T[M]X_1$ must equal zero (otherwise (4.8.10) wouldn't be satisfied). This is the start of our orthogonality condition. For vibrations problems, the eigenvectors aren't directly orthogonal to each other (i.e., it's not true that $X_1^T X_2 = 0$), but rather they're orthogonal with respect to the mass matrix $[M]$. But wait, there's even more. From (4.8.9) we see that if $X_2^T[M]X_1 = 0$, then $X_2^T[K]X_1$ must also equal zero. Thus the eigenvectors are orthogonal with respect to both the stiffness *and* the mass matrices.

The next thing we need to ask is whether the eigenvectors are also orthogonal to themselves, that is, does $X_1^T[M]X_1 = 0$? One result of linear algebra is that $X^T[A]X$ is never equal to zero if $[A]$ is positive definite, unless $X$ is itself equal to zero. Since $[M]$ is always a positive definite matrix, this means that $X_1^T[M]X_1 \neq 0$. The precise value that $X_1^T[M]X_1$ takes on depends on both $[M]$ and $X_1$. $[M]$ is fixed, but as we've discussed, $X_1$ can only be determined up to a point. All multiples of our chosen $X_1$ are also valid eigenvectors. This actually can be turned to our advantage. Rather than always having this uncertainty about $X_1$, we can demand that it be chosen so that $X_1^T[M]X_1 = 1.0$. This is a type of normalization (known as mass normalization) and, if we choose each eigenvector such that its transpose multiplied by $[M]$ multiplied by the eigenvector equals 1.0, we have what is called an orthonormal set of vectors. The ortho refers to the fact that they are orthogonal and the normal refers to the fact that they're normalized. We'll use a tilde to indicate that the eigenvectors are normalized ($\tilde{X}_i$) and discuss the details of how to do this in the next section.

Next, we need to ask what $\tilde{X}_1^T[K]\tilde{X}_1$ is equal to. Using $\tilde{X}_1$ in (4.8.4) and premultiplying by $\tilde{X}_1^T$ yields

$$\omega_1^2 \tilde{X}_1^T[M]\tilde{X}_1 = \tilde{X}_1^T[K]\tilde{X}_1 \tag{4.8.11}$$

Since $\tilde{X}_1^T[M]\tilde{X}_1 = 1.0$, we see that $\tilde{X}_1^T[K]\tilde{X}_1 = \omega_1^2$. In exactly the same manner, $\tilde{X}_2^T[K]\tilde{X}_2 = \omega_2^2$.

Right about now is a good time to recall that one of our goals was to determine how to get a system into a decoupled form. Well, we've done it. All we need to do is make sure our transformation matrix $[U]$ is made up of our normalized system eigenvectors, that is,

$$[U] = [\tilde{X}_1 \vdots \tilde{X}_2] \tag{4.8.12}$$

Using $X = [U]H$ in (4.8.1) and premultiplying by $[U]^T$ gives us

$$[U]^T[M][U]\ddot{H} + [U]^T[K][U]H = 0 \tag{4.8.13}$$

which, in expanded form, is equal to

$$\begin{bmatrix} \tilde{X}_1^T[M]\tilde{X}_1 & \tilde{X}_1^T[M]\tilde{X}_2 \\ \tilde{X}_2^T[M]\tilde{X}_1 & \tilde{X}_2^T[M]\tilde{X}_2 \end{bmatrix} \begin{Bmatrix} \ddot{\eta}_1 \\ \ddot{\eta}_2 \end{Bmatrix} + \begin{bmatrix} \tilde{X}_1^T[K]\tilde{X}_1 & \tilde{X}_1^T[K]\tilde{X}_2 \\ \tilde{X}_2^T[K]\tilde{X}_1 & \tilde{X}_2^T[K]\tilde{X}_2 \end{bmatrix} \begin{Bmatrix} \eta_1 \\ \eta_2 \end{Bmatrix} = \begin{Bmatrix} 0 \\ 0 \end{Bmatrix} \tag{4.8.14}$$

or, in terms of what we now know these products are equal to,

$$\begin{bmatrix} 1 & 0 \\ 0 & 1 \end{bmatrix} \begin{Bmatrix} \ddot{\eta}_1 \\ \ddot{\eta}_2 \end{Bmatrix} + \begin{bmatrix} \omega_1^2 & 0 \\ 0 & \omega_2^2 \end{bmatrix} \begin{Bmatrix} \eta_1 \\ \eta_2 \end{Bmatrix} = \begin{Bmatrix} 0 \\ 0 \end{Bmatrix} \tag{4.8.15}$$

Although we derived these results using a 2-DOF model, it should be clear that the results are actually general and can be applied to any size system. Thus, for an $n$-DOF system, our orthogonality conditions would read

$$X_i^T[M]X_j = X_i^T[K]X_j = 0, \quad i \neq j \tag{4.8.16}$$

In this case, the matrices $[M]$ and $[K]$ are $n \times n$, $X_i$ is the $i$th eigenvector, and $X_j$ is the $j$th eigenvector. If the eigenvectors are mass normalized then

$$\tilde{X}_i^T[M]\tilde{X}_i = 1.0 \tag{4.8.17}$$

and

$$\tilde{X}_i^T[K]\tilde{X}_i = \omega_i^2 \tag{4.8.18}$$

We've now learned that the eigenvectors of our system are not in general orthogonal to each other, but rather are orthogonal with respect to the $[M]$ and $[K]$ matrices. In order to distinguish between this situation and the more common concept of orthogonal vectors, we'll make some definitions. If we have a set of vectors that are orthogonal with respect to the $[M]$ matrix, we'll call them *mass orthogonal*. A set of vectors that are orthogonal to the $[K]$ matrix are *stiffness orthogonal*. If the set is simultaneously orthogonal to both the $[K]$ and $[M]$ matrix, then they are *system orthogonal*. Finally, if they are orthogonal to themselves, that is,

$$X_i^T X_j = 0 \tag{4.8.19}$$

then they are *self-orthogonal*. Note that if $[M]$ is equal to a multiple of the identity matrix, then the eigenvectors are self-orthogonal, since with $[M] = \alpha[I]$ we have

$$X_i^T [M] X_j = \alpha X_i^T [I] X_j = \alpha X_i^T X_j = 0 \tag{4.8.20}$$

This is a good time to make an observation. The eigenvectors of our system are orthogonal to the $[M]$ and $[K]$ matrices. As you might suspect, this implies that they are also independent. In fact, taken together, the eigenvectors form a very special set. If our system is $n$ dimensional ($n$ physical coordinates needed to describe the system's configuration), then we can obtain any possible configuration by suitably combining the system eigenvectors. A simple example would be to look at a 2-DOF system shown in Figure 4.1. If all the spring constants are identical and the masses are equal, then the eigenvectors will be $X_1 = \{1 \ 1\}^T$ and $X_2 = \{1 \ -1\}^T$. Any position of the two masses can be represented by these vectors. For instance, let's say the first mass has zero displacement and the second is moved to the right by four units. Then

$$\begin{Bmatrix} x_1 \\ x_2 \end{Bmatrix} = \begin{Bmatrix} 0 \\ 4 \end{Bmatrix} = 2 \begin{Bmatrix} 1 \\ 1 \end{Bmatrix} - 2 \begin{Bmatrix} 1 \\ -1 \end{Bmatrix} \tag{4.8.21}$$

The same holds for any system size. A system's configuration can be represented by an appropriate combination of the eigenvectors. What we say in this case is that the system eigenvectors *span* the configuration space of the system. It's worth noting that you don't need to use eigenvectors to span a system's configuration space. In fact, all you need is a set of independent vectors. Since the eigenvectors are independent, they'll do the job, but so will any number of other vector sets. In fact, this will be one of the main points to consider when we introduce approximation techniques in Chapter 6.

We can show some more interesting results by using a little linear algebra. The first thing to realize is that our mass matrix is symmetric and positive definite. One linear algebra result tells us that such a matrix can always be written as the product of two identical matrices,

$$[M] = [M]^{\frac{1}{2}} [M]^{\frac{1}{2}} \tag{4.8.22}$$

If we use this fact in our equations of motion, we'll have

$$[M]^{\frac{1}{2}} [M]^{\frac{1}{2}} \ddot{X} + [K] X = 0 \tag{4.8.23}$$

If we define a new set of coordinates by

$$[M]^{\frac{1}{2}} X = H \tag{4.8.24}$$

then we'll have

$$[M]^{\frac{1}{2}}\ddot{H} + [K][M]^{-\frac{1}{2}}H = 0 \tag{4.8.25}$$

Premultiplying by $[M]^{-\frac{1}{2}}$ leaves us with

$$\ddot{H} + [M]^{-\frac{1}{2}}[K][M]^{-\frac{1}{2}}H = 0 \tag{4.8.26}$$

This shows us that we can always transform our equations into a form for which the mass matrix is equal to the identity matrix and the stiffness matrix is symmetric. We know the transformed stiffness matrix must be symmetric because $[M]^{-\frac{1}{2}}$ and $[K]$ are symmetric and products of symmetric matrices of the form shown above are themselves symmetric.

Note that (4.8.26) is in the form

$$\ddot{X} + [A]X = 0 \tag{4.8.27}$$

where $[A]$ is symmetric. Now let's see what we can do here. We know that the solutions of this problem will support $n$ natural frequencies and eigenvectors. If we put the eigenvectors together into a transformation matrix $[U]$, as suggested earlier, then we can write

$$-[U][\Omega^2] + [A][U] = 0 \tag{4.8.28}$$

where $[\Omega^2]$ is an $n \times n$ diagonal matrix having the squares of the system natural frequencies on the diagonal. $[U]$ is called the *modal matrix* for the system. Pulling the left term onto the other side of the equality gives us

$$[U][\Omega^2] = [A][U] \tag{4.8.29}$$

and premultiplying by $[U]^{-1}$ gives us

$$[\Omega^2] = [U]^{-1}[A][U] \tag{4.8.30}$$

What we'll do now is take the transpose of (4.8.29) and show that there is a very special relationship between $[U]^T$ and $[U]^{-1}$. Taking the transpose of (4.8.29) gives us

$$[\Omega^2]^T[U]^T = [U]^T[A]^T \tag{4.8.31}$$

But remember that both $[\Omega^2]$ and $[A]$ are symmetric. Thus (4.8.31) is equal to

$$[\Omega^2][U]^T = [U]^T[A] \tag{4.8.32}$$

Postmultiplying (4.8.32) by $([U]^T)^{-1}$ gives us

$$[\Omega^2] = [U]^T[A]([U]^T)^{-1} \tag{4.8.33}$$

Now compare (4.8.33) and (4.8.30). What we have are essentially two expressions of a similarity transformation, between $[A]$ and $[\Omega^2]$. Since transformations such as these are unique, we see by comparing terms that $[U]^T$ must be equal to a scalar constant times $[U]^{-1}$. If we use

$$[U]^T = \alpha[U]^{-1} \tag{4.8.34}$$

in (4.8.33), we'll obtain

$$[\Omega^2] = \alpha[U]^{-1}[A][U]\frac{1}{\alpha} \qquad (4.8.35)$$

or

$$[\Omega^2] = [U]^{-1}[A][U] \qquad (4.8.36)$$

Note that this is exactly the same as (4.8.30). So we've found an important fact, namely that the transpose of the modal matrix is, within a scalar multiple, equal to the inverse of the modal matrix when the associated mass matrix is an identity matrix. Knowing this certainly allows us to save time, as calculating a transpose is substantially easier than calculating an inverse.

## 4.9 MORE ON NORMAL FORMS

Now that we've seen how to put our system into normal form, we can go ahead and tidy up the remaining loose ends. To start, let's consider what to do when our problem has an external forcing. In this case, the equations of motion are given by

$$[M]\ddot{X} + [K]X = F \qquad (4.9.1)$$

If we go ahead and use $X = [U]H$ to decouple this set of equations, we'll obtain

$$[M']\ddot{H} + [K']H = F' \qquad (4.9.2)$$

where

$$[M'] \equiv [U]^T[M][U], \quad [K'] \equiv [U]^T[K][U], \quad [F'] \equiv [U]^T F \qquad (4.9.3)$$

This was previously derived in ((4.7.13) and (4.7.14)) and is simply rewritten here for convenience. If the modal matrix $[U]$ has been mass normalized then $[M']$ is equal to the identity matrix and $[K']$ is equal to $[\Omega^2]$. Thus our equations are given by

$$\ddot{H} + [\Omega^2]H = F' \qquad (4.9.4)$$

This is a completely decoupled problem. In this form, our problem is exactly equivalent to $n$, individual, second-order equations. Taking the $i$th one at random, we have

$$\ddot{\eta}_i + \omega_i^2\eta_i = f_i' \qquad (4.9.5)$$

$f_i'$ is simply the $i$th component of the modified forcing vector. We know from previous work that $\eta_i$ represents the degree of participation of the $i$th eigenvector in the response. Thus we can clearly see how the external force affects the various modes of the system.

**Example 4.5**

> **Problem**   Consider the 2-DOF example we recently saw (in which $m_1 = m_2 = 1$ and $k_1 = k_2 = k_3 = 1$), the equations for which are given in (4.7.4). Put the system into normal form and analyze the resulting equations.

**Solution**   We've already seen that, if the modal matrix is given by

$$[U] = \begin{bmatrix} \frac{1}{\sqrt{2}} & \frac{1}{\sqrt{2}} \\ \frac{1}{\sqrt{2}} & -\frac{1}{\sqrt{2}} \end{bmatrix} \tag{4.9.6}$$

then

$$[M'] = \begin{bmatrix} 1 & 0 \\ 0 & 1 \end{bmatrix}, \quad [K'] = \begin{bmatrix} 1 & 0 \\ 0 & 3 \end{bmatrix}, \quad [F'] = \begin{bmatrix} \frac{1}{\sqrt{2}}(f_1 + f_2) \\ \frac{1}{\sqrt{2}}(f_1 - f_2) \end{bmatrix} \tag{4.9.7}$$

The equations are decoupled, in just the form we expected. Let $f_1 = \bar{f}_1 \cos(\omega t)$ and $f_2 = \bar{f}_2 \cos(\omega t)$. The two relevant equations of our normal formal representation are

$$\ddot{\eta}_1 + \eta_1 = \frac{1}{\sqrt{2}}(\bar{f}_1 + \bar{f}_2)\cos(\omega t) \tag{4.9.8}$$

and

$$\ddot{\eta}_2 + 3\eta_2 = \frac{1}{\sqrt{2}}(\bar{f}_1 - \bar{f}_2)\cos(\omega t) \tag{4.9.9}$$

The solutions to these problems are simply

$$\eta_1 = \frac{\frac{1}{\sqrt{2}}(\bar{f}_1 + \bar{f}_2)}{1 - \omega^2}\cos(\omega t) \tag{4.9.10}$$

and

$$\eta_2 = \frac{\frac{1}{\sqrt{2}}(\bar{f}_1 - \bar{f}_2)}{3 - \omega^2}\cos(\omega t) \tag{4.9.11}$$

This seems to make sense. If we realize that the first mode of the system is simply both masses moving together in unison, then the best way to excite it is to have both forces acting in unison, one on each mass. Of course, in general, this doesn't happen. So what counts is how much force is being applied on average to the two of them. That's what the forcing component represents, an average force. Similarly, the second mode consists of both masses moving in an opposite direction. Thus, to best excite this mode, the forces must be opposed to each other.

Let's look at a couple of cases to make sure this is clear. If both forces are acting in the same direction and with the same magnitude, then they're aimed directly at the first mode. In this case ($\bar{f}_1 = \bar{f}_2 = \bar{f}$) our equations become

$$\ddot{\eta}_1 + \eta_1 = \sqrt{2}\bar{f} \tag{4.9.12}$$

and

$$\ddot{\eta}_2 + 3\eta_2 = 0 \tag{4.9.13}$$

Thus only the first equation is being forced while the second is completely unaffected. This means that the first mode will respond and the second mode will not.

Similarly, if the forces are of equal magnitude and opposed ($-\bar{f}_2 = \bar{f}_1 = \bar{f}$), we'll have

$$\ddot{\eta}_1 + \eta_1 = 0 \tag{4.9.14}$$

and

$$\ddot{\eta}_2 + 3\eta_2 = \sqrt{2}\,\bar{f} \qquad (4.9.15)$$

In this case, the second mode is excited while the first is unaffected.

There are two fundamental observations we can draw from all this. The first is that it is possible to excite individual modes of a system and leave others totally unaffected. This fact is often used during modal tests of large systems, such as airliners. For instance, let's say you want to perform a modal test on a large airplane. Something this big is difficult to excite effectively. One way to get a reasonably sized modal response is to place shakers along the wings and body and then vibrate the plane in such a way as to couple into just a single vibrational mode. This will give you a large, well-defined response, just what you need for a good experimental test.

The second observation is the fact that, in general, *all* modes are excited when a system is forced. As we saw in the previous example, only for the special cases of $\bar{f}_1 = \bar{f}_2$ and $\bar{f}_1 = -\bar{f}_2$ did we avoid exciting one mode while getting a response from another. For any other combination of inputs, both modes will be involved in the total response. This fact is also of great use in modal testing. A very common way of running modal tests is to hit the structure with an instrumented hammer at a particular point. Since all modes are (usually) excited by such an input, a full range of spectral data can be acquired from this single test.

## 4.10 LINEAR DAMPING

We've already seen that damping isn't a trivial matter, even for the SDOF case. By approximating the actual damping by a linear damping characteristic, we kept the problem solvable and were able to draw some conclusions about how damped systems behave. And, as Section 2.5 showed, damping can often be very helpful. So let's see how we can add damping to our repertoire.

Unfortunately, the addition of one or more additional degrees of freedom causes some real complications, even if we assume linear damping. To illustrate why this is so, look at Figure 4.26. This figure shows a 2-DOF spring-mass damper system. If the dampers are removed, we're left with our familiar spring-mass example. We know that with no damping the system has the two orthonormalized eigenvectors $\tilde{X}_1 = \frac{1}{\sqrt{2}}\{1 \ \ 1\}^T$ and $\tilde{X}_2 = \frac{1}{\sqrt{2}}\{1 \ \ -1\}^T$. If the system is displaced into either of these configurations, it will oscillate forever in that mode.

It would be nice if, when damping is added, the modal character of the system was preserved. Thus, if the system was displaced into the shape of an eigenvector, we'd expect to find that it remained in that configuration. However, instead of oscillating forever, the oscillations would die down at a rate that was proportional to the damping. Unfortunately, the actual behavior is most often not like this. What happens most often can be illustrated by looking at the behavior of the system shown in Figure 4.26, with the specific damping values $c_1 = 0$, $c_2 = .2$ N/m/s, and $c_3 = .1$ N/m/s. Figure 4.27 shows what the resultant vibrations look like if the damped system is first displaced into the first (undamped) eigenvector's shape. As you can see, the relative displacements of the

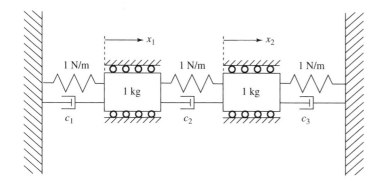

**Figure 4.26**   Damped 2-DOF system

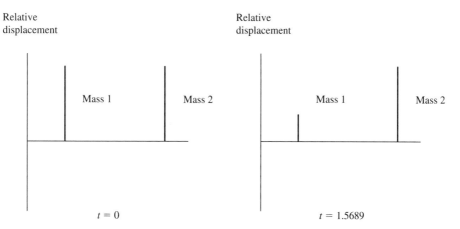

**Figure 4.27**   Relative displacements in nonproportionally damped system

two masses are definitely not remaining the same as they were initially. They start out equal, but at a later time are very different from each other.

Interestingly, although the behavior we've just seen didn't give us the modal response we were looking for, a small change in the damping matrix can remedy the situation. Instead of the damping values we used previously, we can use $c_1 = c_2 = c_3 = .1$ N/m/s. If the two masses are equally displaced (first undamped eigenvector) and then released, we'll see what we had originally hoped for. The relative displacements of the masses will always be equal as the oscillation dies down.

What this means is that some damping matrices preserve the undamped modal character of the solution and others don't. The obvious step, then, is to ask when we can expect the solutions to behave one way or the other.

To do this, let's consider the $n$-DOF problem shown in Figure 4.28. The equations of motion for this particular system are

$$[M]\ddot{X} + [C]\dot{X} + [K]X = F \tag{4.10.1}$$

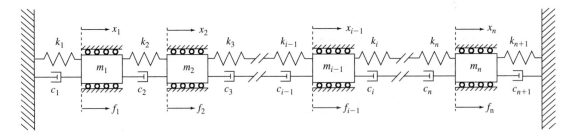

**Figure 4.28**  Forced $n$-DOF, spring-mass-damper system

where

$$[M] \equiv \begin{bmatrix} m_1 & 0 & \cdots & & 0 \\ 0 & m_2 & \cdots & & \vdots \\ \vdots & \cdots & m_{n-1} & 0 & \\ 0 & \cdots & & 0 & m_n \end{bmatrix}$$    (4.10.2)

$$[C] \equiv \begin{bmatrix} c_1 + c_2 & -c_2 & 0 & 0 & \cdots & & 0 \\ -c_2 & c_2 + c_3 & -c_3 & 0 & & & \vdots \\ 0 & -c_3 & & & & & \vdots \\ \vdots & & & & -c_{n-1} & & 0 \\ & & & & c_{n-1} + c_n & -c_n & \\ 0 & \cdots & & 0 & -c_n & c_n + c_{n+1} \end{bmatrix}$$    (4.10.3)

$$[K] \equiv \begin{bmatrix} k_1 + k_2 & -k_2 & 0 & 0 & \cdots & & 0 \\ -k_2 & k_2 + k_3 & -k_3 & 0 & & & \vdots \\ 0 & -k_3 & & & & & \vdots \\ \vdots & & & & -k_{n-1} & & 0 \\ & & & & k_{n-1} + k_n & -k_n & \\ 0 & \cdots & & 0 & -k_n & k_n + k_{n+1} \end{bmatrix}$$    (4.10.4)

and

$$F \equiv \begin{Bmatrix} f_1 \\ f_2 \\ \vdots \\ f_n \end{Bmatrix}$$    (4.10.5)

Although the system shown in Figure 4.28 has the particular $[M]$, $[C]$, and $[K]$ entries listed above (valid for a serial chain of masses), the analysis we'll be undertaking will be valid for any other configuration of masses we might wish to examine. To simplify our analysis, we'll restrict ourselves to a 2-DOF example. As usual when examining the fundamental responses of a system, we'll consider the unforced case. In order to illustrate the effect of damping, we'll keep the $[M]$ and $[K]$ matrices fixed and will vary $[C]$.

The first issue that we must address is the question of how we'll be solving our problem. The equations of motion will be in the form

$$[M]\ddot{X} + [C]\dot{X} + [K]X = 0 \tag{4.10.6}$$

We could certainly attempt to solve this by assuming $X = \bar{X}e^{\lambda t}$ to obtain

$$\left[\lambda^2[M] + \lambda[C] + [K]\right]\bar{X} = \{0\} \tag{4.10.7}$$

Unfortunately, solving this is not particularly easy. For instance, assume that we have a 2-DOF system. With no damping, we'd have to solve a fourth-order characteristic equation, but $\lambda$ would appear only in powers of two ($\lambda^2$, $\lambda^4$, etc.). Thus we'd really have a quadratic in $\lambda^2$, not a fourth-order problem in $\lambda$. But with the addition of damping, when we take the determinant of $\lambda^2[M] + \lambda[C] + [K]$, we'll get an honest-to-goodness, fourth-order equation. In a like manner, any $n$-DOF problem will give us a $2n$-order polynomial. And to make matters even more interesting, this polynomial will have complex solutions. This comes about because of the damping matrix's contribution ($\lambda[C]$). Thus, we'd need to assume a complex solution ($\lambda = \lambda_r + i\lambda_i$) from the outset. We'd have to substitute this into our characteristic polynomial, set it equal to zero, separate all the real parts of the equation and set them equal to zero, while simultaneously setting the imaginary parts equal to zero, also. We'd then need to create a root-finder program that would solve these equations for us.

Clearly, this isn't overly easy and it would be far preferable to use a prepackaged code. Unfortunately, there is no code that currently accepts problems in the form we're using (separate $[M]$, $[C]$, and $[K]$ matrices). Most often, a canned numerical code will want to use an eigenvalue problem in the form

$$\dot{X} = [A]X \tag{4.10.8}$$

or perhaps

$$[A]\dot{X} = [B]X \tag{4.10.9}$$

Certainly this is the way MATLAB expects to see the problem. The basic difference is that our equations are in second-order form (two differentiations with respect to time) as opposed to first-order form (one differentiation with respect to time). Thus, in order to use MATLAB or similar software, we need to get our equations into first-order form.

Those of you who have already taken a controls course will find this very familiar. The standard controls formulation is one that looks like

$$\dot{X} = [A]X + [B]u \tag{4.10.10}$$

This can be contrasted with our familiar forced representation,

$$[M]\ddot{X} + [K]X = F \tag{4.10.11}$$

The common theme behind our solution procedure for both these problems is to use linear algebra. Both our representation and the controls representation are cast as problems involving combinations of matrices and vectors, something that defines a linear algebra problem. The one advantage that those who work with controls formulation have over those who work with vibrations problems is that they've always used the (4.10.10) formulation. We had the advantage as long as we were dealing with undamped problems, but now we're going to have to use the same form as the controls people if we want to get anywhere quickly. In fact, the MATLAB program was originally seized upon by the controls researchers as an excellent tool with which to solve problems, and many applications have been written for controls applications. Thus, all the development work has already been done for us and all we need to do is put our equations in a form MATLAB recognizes. When equations are in a form that looks like

$$[A]\dot{X} = [B]X \tag{4.10.12}$$

we say they are in *first-order form*.

One nice formulation that takes advantage of symmetry is to put our vibration problem into the form

$$[B]\dot{Y} = [A]Y \tag{4.10.13}$$

where

$$[B] \equiv \begin{bmatrix} -[K] & [0] \\ [0] & [M] \end{bmatrix}, \quad [A] \equiv \begin{bmatrix} [0] & -[K] \\ -[K] & -[C] \end{bmatrix}, \quad Y^T \equiv \{x_1 \ x_2 \ \dot{x}_1 \ \dot{x}_2\} \tag{4.10.14}$$

You're certainly free to choose another representation if you wish. The main thing is to represent the system in an $n$-DOF vibration problem as a $2n$th-order set of differential equations. The next section will look more precisely at what sort of solution behavior we can expect.

## 4.11 COMPARISON OF DAMPED EIGENSOLUTIONS

One of the more confusing aspects of eigenanalysis is how to interpret the complex eigenvectors and eigenvalues that result from looking at damped systems. In order to shed some light on this problem, we'll look at a carefully chosen succession of problems that will illustrate what you can expect from typical analyses. All the calculations will be done using MATLAB, so you should feel free to double-check any of the results.

MATLAB has two basic routines that handle eigenvalue problems, both called `eig`. If you type

$$\texttt{eig([A]) (cr)}$$

where (cr) indicates a carriage return, MATLAB will determine the solutions to the problem

$$\dot{X} = [A]X \tag{4.11.1}$$

If you type

$$\text{eig}([A], [B]) \quad (\text{cr})$$

then MATLAB will solve

$$[B]\dot{X} = [A]X \tag{4.11.2}$$

Note that within MATLAB you can call your matrix $m$ or $M$ or $[M]$; it won't matter. Since typing $M$ is quicker than typing $[M]$, it'll be easier to forget the brackets when in MATLAB. However, to remain consistent when referring to a matrix in this discussion, we'll use the bracket notation ($[M]$), but when giving directions for MATLAB use we'll use the simpler $M$ form.

It'll help if you know ahead of time what the series of examples will be illustrating. In Example 4.6 we'll look at an undamped problem. We'll obtain the natural frequencies and eigenvectors in the familiar manner. Then we'll put the system into first-order form and see how this new form gives us our old results in a new way. In Example 4.7 we'll add a special kind of damping to the system, one for which the real displacements of the undamped eigenvectors are identical to those of the damped system. Example 4.8 will show what this same system looks like if it's put into normal form and will demonstrate how the new eigenvectors will reflect the fact that the system is decoupled. In Example 4.9 we'll use a different damping matrix, one for which the undamped system's eigenvectors are no longer the same as the damped system's eigenvectors. In fact, we'll see that there will no longer exist a set of eigenvectors having real displacements. In Example 4.10 we'll try to decouple the system of Example 4.9 using the original (undamped) modal matrix, an attempt that will fail. Finally, Example 4.11 will show how neglecting certain damping entries can bring back the old (undamped) eigenvectors and produce a decoupled form, while at the same time not changing the system eigenvalues by a significant amount.

**Example 4.6**

**Problem**    To start, let's look at a problem that we're already familiar with, an undamped oscillator

$$\begin{bmatrix} 1 & 0 \\ 0 & 2 \end{bmatrix} \begin{Bmatrix} \ddot{x}_1 \\ \ddot{x}_2 \end{Bmatrix} + \begin{bmatrix} 2 & -1 \\ -1 & 2 \end{bmatrix} \begin{Bmatrix} x_1 \\ x_2 \end{Bmatrix} = \begin{Bmatrix} 0 \\ 0 \end{Bmatrix} \tag{4.11.3}$$

Our usual analysis of this equation would be to assume motion of the form $Xe^{i\omega t}$, solve the characteristic equation (quadratic in $\omega^2$), and then determine the eigenvectors. Doing so would lead to

$$\omega_1^2 = .634, \quad X_1 = \begin{Bmatrix} .731 \\ 1.0 \end{Bmatrix} \tag{4.11.4}$$

and

$$\omega_2^2 = 2.366, \quad X_2 = \begin{Bmatrix} 1.0 \\ -.366 \end{Bmatrix} \tag{4.11.5}$$

What would we get if we put our system into symmetric first-order form?

**Solution**   From (4.10.13) we'll have

$$[B] = \begin{bmatrix} -2 & 1 & 0 & 0 \\ 1 & -2 & 0 & 0 \\ 0 & 0 & 1 & 0 \\ 0 & 0 & 0 & 2 \end{bmatrix}$$

(4.11.6)

and

$$[A] = \begin{bmatrix} 0 & 0 & -2 & 1 \\ 0 & 0 & 1 & -2 \\ -2 & 1 & 0 & 0 \\ 1 & -2 & 0 & 0 \end{bmatrix}$$

(4.11.7)

If we then type

$$[v, d] = \text{eig}(A, B) \quad (\text{cr})$$

MATLAB will construct a modal matrix ($v$) and an $n \times n$ matrix ($d$) that contains the eigenvalues of the problem on the diagonal (and zeros as the off-diagonal entries). Note that the $d(1,1)$ will be the eigenvalue that corresponds to the first column of $v$, $d(2,2)$ will correspond to the second column, etc. Keep in mind that these column entries are eigenvector solutions.

Going ahead and typing

$$[v, d] = \text{eig}(A, B) \quad (\text{cr})$$

will get us

$$v = \begin{bmatrix} .5118 & -.4621i & .5118 & -.4621i \\ -.1873 & -.6312i & -.1873 & -.6312i \\ -.7873i & -.3679 & .7873i & .3679 \\ .2882i & -.5026 & -.2882i & .5026 \end{bmatrix}$$

(4.11.8)

and

$$d = \begin{bmatrix} -1.5382i & 0 & 0 & 0 \\ 0 & -.7962i & 0 & 0 \\ 0 & 0 & 1.5382i & 0 \\ 0 & 0 & 0 & .7962i \end{bmatrix}$$

(4.11.9)

Note that we now have four eigenvalues and four eigenvectors. We may immediately notice a few things about the results. To start off, the entries in the $d$ matrix are not in any particular order. First we have $-1.5382i$, followed by $-.7962i$, $1.5382i$, and finally $.7962i$. We can recognize the magnitudes of these as $\omega_1$ or $\omega_2$. The reason for them being imaginary is that the general solution assumed by MATLAB is of the form $Xe^{\lambda t}$. Thus the first entry indicates a time behavior of $e^{-i1.5382t}$, that is, sinusoidal motion.

Why didn't MATLAB order these from smallest to highest, with the positive value first? The reason is because the programmers didn't think it was important to order them in this way. We'll see this kind of thing crop up continually. Just because we might want the answer in a particular way, there's no reason to assume that the software is set up to give us that solution form. In MATLAB's case, the ordering of the eigenvalues is essentially random. Remember this! Canned routines are convenient, but it's *very* important to understand what the subroutines are doing, how they're doing it, and in what form they present the results. Failure to do so will often cause pain and frustration.

Now let's examine the eigenvectors that correspond to the eigenvalues (found in the matrix $v$). We can see that the first eigenvector matches our predictions. We previously found the eigenvector of $\omega_2^2$ to be $\{1. \quad -.366\}^T$. The first two entries of the first column of $v$ are $\{.5118 \quad -.1873\}^T$. If we multiply both terms by 1.9539, we obtain $\{1. \quad -.366\}^T$, matching our original determination. Remember that eigenvectors are only known within a scalar multiple anyway; thus there's no problem with having to multiply MATLAB's results by a scalar. The reason that we need look only at the first two entries is that these correspond to $x_1$ and $x_2$, the physical displacements of our system. The last two entries, $-.7873i$ and $.2882i$, correspond to $\dot{x}_1$ and $\dot{x}_2$. We know that, given $X = \bar{X}e^{\lambda t}$, the velocity will simply be $\dot{X} = \lambda \bar{X}e^{\lambda t}$. Thus the velocity entries should differ from the displacement entries by $\lambda$. For our first eigenvector, the corresponding $\lambda$ ($d(1,1)$) is $-1.5382i$. Thus we'd expect $-1.5382i \times \{.5118 \quad -.1873\}^T$ to equal the third and fourth entries of the first column of $v$, that is, $\{-.7873i \quad .2882i\}^T$. And, luckily for us, it does. Therefore we see that the latter half of each eigenvector is really redundant. Once we know the first half, along with the associated eigenvalue, we can compute the second half.

**Example 4.7**

**Problem**   Now we will consider what happens if we add damping to our problem for which the damping matrix is a linear combination of $[M]$ and $[K]$. For this case, we'll let $[C] = .1[M]+.1[K]$:

$$\begin{bmatrix} 1 & 0 \\ 0 & 2 \end{bmatrix} \begin{Bmatrix} \ddot{x}_1 \\ \ddot{x}_2 \end{Bmatrix} + \begin{bmatrix} .3 & -.1 \\ -.1 & .4 \end{bmatrix} \begin{Bmatrix} \dot{x}_1 \\ \dot{x}_2 \end{Bmatrix} + \begin{bmatrix} 2 & -1 \\ -1 & 2 \end{bmatrix} \begin{Bmatrix} x_1 \\ x_2 \end{Bmatrix} = \begin{Bmatrix} 0 \\ 0 \end{Bmatrix} \tag{4.11.10}$$

What will an eigenanalysis yield for this problem?

**Solution**   Putting the equations into symmetric first-order form gives us

$$[B] = \begin{bmatrix} -2 & 1 & 0 & 0 \\ 1 & -2 & 0 & 0 \\ 0 & 0 & 1 & 0 \\ 0 & 0 & 0 & 2 \end{bmatrix} \tag{4.11.11}$$

and

$$[A] = \begin{bmatrix} 0 & 0 & -2 & 1 \\ 0 & 0 & 1 & -2 \\ -2 & 1 & -.3 & .1 \\ 1 & -2 & .1 & -.4 \end{bmatrix} \tag{4.11.12}$$

with eigensolutions:

$$v = \begin{bmatrix} .5118 + .0041i & -.2126 - .4103i & .4681 + .2070i & .0248 - .4614i \\ -.1873 - .0015i & -.2904 - .5605i & -.1714 - .0758i & .0339 - .6303i \\ -.0799 - .7833i & -.3076 + .2019i & -.3952 + .6809i & .3634 + .0574i \\ .0292 + .2867i & -.4202 + .2758i & .1447 - .2492i & .4965 + .0783i \end{bmatrix} \tag{4.11.13}$$

and

$$
d = \begin{bmatrix} -.1683 - 1.5290i & 0 & 0 & 0 \\ 0 & -.0817 - .7920i & 0 & 0 \\ 0 & 0 & -.1683 + 1.5290i & 0 \\ 0 & 0 & 0 & -.0817 + .7920i \end{bmatrix} \quad (4.11.14)
$$

So, is this good or bad? Well, actually it's good, because the damped solution is behaving in the way we want. Of course, it's not crystal clear that this is so (because of the way MATLAB displays its results), so we'll clarify it a bit.

First off, note that our eigenvalues have changed in an appropriate way. They were previously imaginary. This was good because it indicated undamped oscillatory motion. Now they are complex. The imaginary component changed only slightly (because the damping wasn't large) and we've gotten a new real component. Specifically, the undamped solution $-1.5382i$ has now changed to $-.1683 - 1.529i$. The real component is negative for all four eigenvalues, indicating that the solutions are all going to decay in time. As expected, the roots are complex conjugates of each other. Therefore we can add them together to obtain sine and cosine solutions with an exponentially decaying envelope.

Now, on to the eigenvectors. Note that they are all different. Again, this is due to MATLAB making whatever normalization choices it feels like doing. For clarity, we'll take the complex vector and rotate it in the complex plane so that the first entry of each column is purely real. To do this, we must first express the vectors in magnitude/phase notation. Next, we see what the phase happens to be for the first column entry and then subtract this phase from all entries in the column, including the first one. This will have the effect of rotating the entire vector. To avoid major boredom, we'll look at just one set of eigenvectors, those associated with the $-.1683 \pm 1.529i$ roots.

The easiest way to rotate the vector is to let MATLAB do the work. To do this we must determine the magnitude and phase of the vector. Luckily, MATLAB has two functions called abs and angle that do just this. If our input vector is $X$, then abs($X$) returns the absolute value of each component in the vector and angle($X$) returns the value of the phase associated with each entry. Thus to rotate the vector so that the first entry is real we need simply use angle, record the phase angle (call it $\phi$), and then multiply the vector by $\exp(-i * \phi)$.

Using angle on the first column of $v$ gives us

$$
\left\{ \begin{array}{c} .0080 \\ -3.1336 \\ -1.6724 \\ 1.4692 \end{array} \right\} \quad (4.11.15)
$$

Thus we must remove a phase angle of .008 rad from each entry of our vector. Doing this (by multiplying by $\exp(-.008i)$ gives us

$$
\left\{ \begin{array}{c} .5118 \\ -.1873 \\ .7874e^{-1.6804i} \\ .2882e^{1.4612i} \end{array} \right\} \quad (4.11.16)
$$

Note that the velocity terms (the third and fourth entries) have been expressed in magnitude/phase form in order to illustrate another point. The two entries should be exactly out-of-phase with each other since the displacements are of opposite sign. This is exactly what we see. $-1.6804$ rad is $\pi$ rad different from $1.4612$ rad. Thus the two velocity entries are also exactly out-of-phase with each other.

Notice further that the relative magnitudes of the first two entries are exactly the same as those for the undamped case. Not just close—but exact. This means that we've retained the same displacement shape for our eigenvector! Thus, if we displace the system into the configuration defined by this eigenvector, the masses will stay in this configuration as the overall motion dies down. You can verify that the third entry $(.7874e^{-1.6804i})$ is simply the first one $(.5118)$ multiplied by the eigenvalue $-.1683 - 1.529i$. The same holds for the second and fourth entries. This shows that the third and fourth entries are still just the first and second entries multiplied by the eigenvalue.

Proceeding in the same manner with the third column of $v$ transforms it to

$$\left\{ \begin{array}{c} .5118 \\ -.1873 \\ .7874e^{1.6804i} \\ .2882e^{-1.4612i} \end{array} \right\} \tag{4.11.17}$$

The only difference between this vector and the first is that the imaginary parts of the third and fourth entries have switched signs. This is expected, since the associated eigenvalues ($-.1683 - 1.529i$ and $.1683 + 1.529i$) are complex conjugates of each other. If you wish, you can go ahead and verify that this same type of behavior is exhibited by the remaining two eigenvectors.

The reason that the undamped eigenvector was preserved in the previous example, even when damping was added, was because the damping was of a very special form. Recall that the damping matrix was defined as $[C] = .1[M] + .1[K]$. This is called *proportional damping* for the very obvious reason that $[C]$ is made up of components that are proportional to both the $[M]$ matrix and the $[K]$ matrix. Thus we define a proportionally damped system as one in which the damping matrix can be expressed by

$$[C] = \alpha[M] + \beta[K] \tag{4.11.18}$$

where $\alpha$ and $\beta$ are real, scalar constants.

We can understand what's happening by applying our knowledge of normal forms. To start, let's consider

$$[M]\ddot{X} + [C]\dot{X} + [K]X = 0 \tag{4.11.19}$$

If we go ahead and use normal coordinates ($X = [U]H$) and then premultiply by $[U]^T$, we'll get

$$[U]^T[M][U]\ddot{H} + [U]^T[C][U]\dot{H} + [U]^T[K][U]H = 0 \tag{4.11.20}$$

Assuming that the $[U]$ matrix has been mass normalized, this becomes

$$[I]\ddot{H} + [U]^T[C][U]\dot{H} + [\Omega^2]H = 0 \tag{4.11.21}$$

However, we should now realize that $[C] = \alpha[M] + \beta[K]$. Thus we have

$$[I]\ddot{H} + [U]^T[\alpha[M] + \beta[K]][U]\dot{H} + [\Omega^2]H = 0 \tag{4.11.22}$$

or

$$[I]\ddot{H} + \left[\alpha[U]^T[M][U] + \beta[U]^T[K][U]\right]\dot{H} + [\Omega^2]H = 0 \tag{4.11.23}$$

Since we already know what $[U]^T[M][U]$ and $[U]^T[K][U]$ equal, we can rewrite this equation as

$$[I]\ddot{H} + \left[\alpha[I] + \beta[\Omega^2]\right]\dot{H} + [\Omega^2]H = 0 \tag{4.11.24}$$

Notice that all the matrices are diagonal. Thus we have a set of $n$ completely decoupled equations:

$$\ddot{\eta}_i + (\alpha + \beta\omega_i^2)\dot{\eta}_i + \omega_i^2\eta_i = 0, \qquad i = 1, 2, \ldots n \tag{4.11.25}$$

So there you have it. Using the undamped modal matrix produced a set of decoupled equations, each of which governs the motion of one particular eigenvector. Since there is no coupling between the eigenvectors, we can start the system off in the configuration of a given eigenvector and it will stay in that configuration. The rate of decay will be governed by the particular $\alpha$, $\beta$, and $\omega_i$ parameters.

At this point you may be tempted to ask what would happen with our eigenvalue analysis if we had used the normal form matrices rather than the physical mass, spring, and damping matrices. We'll do just that in the next example.

**Example 4.8**

**Problem**   Determine the results of using normal form matrices rather than physical ones for the system of Example 4.7.

**Solution**   If we carry through the operations shown in (4.11.20), we'll obtain

$$\begin{bmatrix} 1.651 & 0 \\ 0 & 1.118 \end{bmatrix}\begin{Bmatrix} \ddot{x}_1 \\ \ddot{x}_2 \end{Bmatrix} + \begin{bmatrix} .270 & 0 \\ 0 & .376 \end{bmatrix}\begin{Bmatrix} \dot{x}_1 \\ \dot{x}_2 \end{Bmatrix} + \begin{bmatrix} 1.047 & 0 \\ 0 & 2.646 \end{bmatrix}\begin{Bmatrix} x_1 \\ x_2 \end{Bmatrix} = \begin{Bmatrix} 0 \\ 0 \end{Bmatrix} \tag{4.11.26}$$

Putting this into $A$, $B$ form, we'll get

$$[B] = \begin{bmatrix} -1.047 & 0 & 0 & 0 \\ 0 & -2.646 & 0 & 0 \\ 0 & 0 & 1.651 & 0 \\ 0 & 0 & 0 & 1.118 \end{bmatrix} \tag{4.11.27}$$

and

$$[A] = \begin{bmatrix} 0 & 0 & -1.047 & 0 \\ 0 & 0 & 0 & -2.646 \\ -1.047 & 0 & -.270 & 0 \\ 0 & -2.646 & 0 & -.376 \end{bmatrix} \tag{4.11.28}$$

Note that each two-by-two block in the $A$ and $B$ matrices is diagonal, which is expected since all the individual normal coordinate matrices are diagonal. Just to demonstrate that you don't have to mass normalize everything, the precise $[U]$ matrix produced by MATLAB was used in the transformations. Since this matrix wasn't mass normalized, we won't have an identity mass matrix for the $\eta$ coordinate representation. The overall ratios, however, are correct.

If we go ahead and ask MATLAB to analyze (4.11.27) and (4.11.28), we'll obtain

$$
v = \begin{bmatrix} .782 & .782i & 0 & 0 \\ 0 & 0 & .5451 & -.545i \\ -.0639 - .620i & -.620 - .0639i & 0 & 0 \\ 0 & 0 & -.0917 - .833i & .8334 + .0917i \end{bmatrix} \tag{4.11.29}
$$

and

$$
d = \begin{bmatrix} -.0817 - .792i & 0 & 0 & 0 \\ 0 & -.0817 + .792i & 0 & 0 \\ 0 & 0 & -.1683 - 1.529i & 0 \\ 0 & 0 & 0 & -.1683 + 1.529i \end{bmatrix} \tag{4.11.30}
$$

This is right on the money. Note that the displacement entries display complete decoupling, just as they should. Either the first or second entry is nonzero for each eigenvector, while the other is zero. This tells us that our normal coordinates correspond to a single eigenvector, just as we expected. You can also see that MATLAB isn't going to give us a big break, even with such a well-ordered problem. The first and second columns of $v$, which would normally be written in the form

$$
\begin{Bmatrix} a \\ 0 \\ \lambda_1 a \\ 0 \end{Bmatrix} \tag{4.11.31}
$$

and

$$
\begin{Bmatrix} a \\ 0 \\ \lambda_1^* a \\ 0 \end{Bmatrix} \tag{4.11.32}
$$

don't quite fit the bill as the second column of the MATLAB output has been multiplied by $i$. Again, this is because software programs don't often care about how you might want to see the answer; they present it in the way they were programmed to present it.

Unfortunately, it's probably pretty clear to you that the $[C]$ matrices are *not* proportionally damped more often than they are. So it's time to ask what happens to our solutions in the case of nonproportional damping. We'll be looking at three examples. First, we'll analyze the case of nonproportional damping using our physical coordinates. Next, we'll transform them into normal form and see what the solutions look like. Finally, we'll force the equations into a decoupled form in order to see what the effect is on the overall solution.

**Example 4.9**

**Problem**   To start, we'll do an eigenanalysis on the physical system of the previous examples using the damping matrix

$$[C] = \begin{bmatrix} .6 & -.2 \\ -.2 & .3 \end{bmatrix} \tag{4.11.33}$$

The question is to determine what this does to our resulting solution.

**Solution**   Our $A$ and $B$ matrices are

$$[B] = \begin{bmatrix} -2 & 1 & 0 & 0 \\ 1 & -2 & 0 & 0 \\ 0 & 0 & 1 & 0 \\ 0 & 0 & 0 & 2 \end{bmatrix} \tag{4.11.34}$$

and

$$[A] = \begin{bmatrix} 0 & 0 & -2 & 1 \\ 0 & 0 & 1 & -2 \\ -2 & 1 & -.6 & .2 \\ 1 & -2 & .2 & -.3 \end{bmatrix} \tag{4.11.35}$$

The eigenvectors and eigenvalues resulting from using these in `eig` are

$$v = \begin{bmatrix} .512 + .0584i & -.359 - .285i & .297 + .421i & -.0731 - .452i \\ -.172 - .0579i & -.538 - .335i & -.135 - .122i & -.0292 - .633i \\ -.0782 - .786i & -.203 + .304i & -.724 + .316i & .364 - .029i \\ -.0336 + .276i & -.232 + .450i & .225 - .164i & .506 + .0177i \end{bmatrix} \tag{4.11.36}$$

and

$$d = \begin{bmatrix} -.310 - 1.502i & 0 & 0 & 0 \\ 0 & -.0647 - .796i & 0 & 0 \\ 0 & 0 & -.310 + 1.502i & 0 \\ 0 & 0 & 0 & -.0647 + .796i \end{bmatrix} \tag{4.11.37}$$

As in Example 4.7, we can put the eigenvectors into magnitude/phase form and shift each of the vector entries by the appropriate phase amount to cause the first entry to be a real quantity. If our system supported real eigenvectors (as the proportionally damped case did), then both the first and second elements of the eigenvectors must be real.

   Carrying out this procedure on the first column of $v$ gives us

$$X_1 = \begin{Bmatrix} .515 \\ -.177 - .0396i \\ -.156 - .774i \\ -.0046 + .278i \end{Bmatrix} \tag{4.11.38}$$

Bad news. The second entry says that when $x_1 = .515$, $x_2 = -.177 - .0396i$. Since this isn't a real quantity, it's pretty clear that the second mass is going to have a tough time doing this physically.

In fact, it's impossible since we're never going to see imaginary displacement. The reason for the difficulty is that we no longer have distinct, real eigenvectors that will decay smoothly to zero. The convenient view of real eigenvectors that decay at a particular rate is just not applicable in most damped systems. In order to solve for the overall vibrations, we'll have to use *all* the eigenvectors and combine them together with appropriate multipliers (undoubtedly complex ones). The total solution will involve only real displacements, but the individual eigenvectors no longer display purely real characteristics. As shown in Figure 4.27, the overall behavior is certainly real, but there is no discernible eigenvector anymore.

**Example 4.10**

**Problem**   Now we'll try to put the system into normal form. What we should expect to see is that the damping matrix doesn't diagonalize.

**Solution**   For this system, the $A$ and $B$ matrices are given by

$$[B] = \begin{bmatrix} -.786 & 0 & 0 & 0 \\ 0 & -.641 & 0 & 0 \\ 0 & 0 & .332 & 0 \\ 0 & 0 & 0 & 1.01 \end{bmatrix} \tag{4.11.39}$$

and

$$[A] = \begin{bmatrix} 0 & 0 & -.786 & 0 \\ 0 & 0 & 0 & -.641 \\ -.786 & 0 & -.206 & -.0591 \\ 0 & -.641 & -.0591 & -.131 \end{bmatrix} \tag{4.11.40}$$

Keep in mind that to get $[A]$ and $[B]$ we took the *undamped* system eigenvectors, formed the modal matrix $[U]$, used $[U]$ to transform $[M]$, $[C]$, and $[K]$, and then assembled them into the larger $[A]$ and $[B]$ matrices to have first-order system representation. We can see from the lower two-by-two block of $[A]$ that the undamped eigenvectors didn't decouple our $[M]$, $[C]$, and $[K]$. If $[U]$ had diagonalized our matrices, then the off-diagonal entries $-.0591$ would not have shown up.

     If we use `eig` to find the eigenvalues and eigenvectors, we'll obtain

$$v = \begin{bmatrix} .526 + .143i & .147 + .525i & .0618 + .0149i & -.0144 - .0619i \\ .0136 - .0248i & -.0247 + .0138i & .0712 - .776i & .776 - .0646i \\ .0514 - .835i & -.835 + .0585i & .0079 - .0502i & .0502 - .0075i \\ -.0415 - .0127i & -.0131 - .0414i & -.622 - .0065i & .0013 + .622i \end{bmatrix} \tag{4.11.41}$$

and

$$d = \begin{bmatrix} -.310 - 1.502i & 0 & 0 & 0 \\ 0 & -.310 + 1.502i & 0 & 0 \\ 0 & 0 & -.0647 - .796i & 0 \\ 0 & 0 & 0 & -.0647 + .796i \end{bmatrix} \tag{4.11.42}$$

As you can see, the columns in (4.11.41) are definitely not showing decoupling. All entries have finite values; there are no zero entries that indicate nonparticipation of one coordinate while another is active. Thus the eigenvectors (as expected) are not the same as the uncoupled eigenvectors.

The final topic is an interesting one to think about. Clearly, it would be nice if our system were decoupled. The problem is easier to solve this way and decoupling supports the attractive notion of physically identifiable eigenvectors that govern the motion. Given that this is the case, some vibrations engineers wondered what the consequences might be of *forcing* the system to decouple. Since damping is such a difficult concept already, and since our notion of purely viscous damping is already somewhat arguable, why not just throw out the damping terms that are spoiling the equations?

Actually, there are some more sophisticated ways to reorder the damping, but for our purposes it will be sufficient to examine the simplest approach. And it certainly is simple. What you want to do is try to put your system in normal form, the way we did with the $A$ and $B$ matrices of (4.11.39) and (4.11.40). Then you simply cross out all terms that don't support proportional damping, that is, all the off-diagonal terms of the transformed $[C]$ matrix. The next example demonstrates what this will buy us.

**Example 4.11**

**Problem**    Construct the normal form of the equations from the previous example and discard all off-diagonal damping terms. Examine the resulting solution.

**Solution**    Putting the system into normal form gives us

$$[B] = \begin{bmatrix} -.786 & 0 & 0 & 0 \\ 0 & -.641 & 0 & 0 \\ 0 & 0 & .332 & 0 \\ 0 & 0 & 0 & 1.01 \end{bmatrix} \tag{4.11.43}$$

and

$$[A] = \begin{bmatrix} 0 & 0 & -.786 & 0 \\ 0 & 0 & 0 & -.641 \\ -.786 & 0 & -.206 & 0 \\ 0 & -.641 & 0 & -.131 \end{bmatrix} \tag{4.11.44}$$

Note that we simply erased the old $A(3, 4)$ and $A(4, 3)$ entries of (4.11.40). Because of this, the two-by-two submatrices in (4.11.43) and (4.11.44) that correspond to the transformed $[M]$, $[C]$, and $[K]$ matrices are diagonal. If we go ahead and use eig on these, we'll obtain

$$v = \begin{bmatrix} .541 + .0657i & .0657 + .541i & 0 & 0 \\ 0 & 0 & .249 + .742i & -.745 - .249i \\ -.0689 - .836i & -.836 - .0689i & 0 & 0 \\ 0 & 0 & .572 - .246i & .246 - .572i \end{bmatrix} \tag{4.11.45}$$

and

$$d = \begin{bmatrix} -.310 - 1.507i & 0 & 0 & 0 \\ 0 & -.310 + 1.507i & 0 & 0 \\ 0 & 0 & -.0648 - .794i & 0 \\ 0 & 0 & 0 & -.0648 + .794i \end{bmatrix} \tag{4.11.46}$$

Just compare these results with the exact case. The damped natural frequencies have changed hardly at all. For instance, the damped solution $-.310 - 1.502i$ from (4.11.42) is now $-.310 - 1.507i$. This isn't a very large difference. Furthermore, the eigenvectors are completely real again and nicely decoupled. If you compare the two cases, you'll see that crossing out those pesky off-diagonal damping terms didn't really damage our accuracy all that much and allowed us to have a beautifully decoupled system.

The reason for the good results of the last example is that the off-diagonal terms were smaller than those on the diagonal. Now it's possible to come up with specially tailored examples in which it isn't safe to neglect small, off-diagonal damping entries as well as examples of relatively large damping entries that don't cause great problems. But the general rule of thumb is that, when the off-diagonal elements are small compared to the diagonal ones, you won't induce a lot of error by neglecting them. The inverse of this view is also used frequently. Oftentimes researchers will identify the mass and spring matrices, run some modal tests, and then fit the best approximate *proportional* damping that matches the test results.

Just to make sure the analyses we've just gone through are clear, we'll work a final example.

**Example 4.12**

**Problem**    Analyze the system shown in Figure 4.29. All the relevant parameter values are shown on the figure, where the units are kg, N/m/s, and N/m for the masses, dampers, and springs, respectively.

**Solution**    The equations of motion for this system are

$$\begin{bmatrix} 1 & 0 \\ 0 & 2 \end{bmatrix} \begin{Bmatrix} \ddot{x}_1 \\ \ddot{x}_2 \end{Bmatrix} + \begin{bmatrix} .5 & -.1 \\ -.1 & .5 \end{bmatrix} \begin{Bmatrix} \dot{x}_1 \\ \dot{x}_2 \end{Bmatrix} + \begin{bmatrix} 40 & -10 \\ -10 & 30 \end{bmatrix} \begin{Bmatrix} x_1 \\ x_2 \end{Bmatrix} = \begin{Bmatrix} 0 \\ 0 \end{Bmatrix} \tag{4.11.47}$$

This system has proportional damping ($[C] = .1[M] + .1[K]$), so we should obtain purely real displacements for our eigenvectors. Furthermore, the eigenvectors should be the same as those of the undamped case. If we use MATLAB to find the eigenvectors and eigenvalues of the undamped case (using $[v, d] = \text{eig}(K, M)$ (cr)) it will give us

$$v = \begin{bmatrix} .9831 & .3489 \\ -.1830 & .9372 \end{bmatrix} \tag{4.11.48}$$

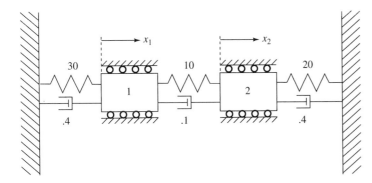

**Figure 4.29**   2-DOF
spring-mass-damper system

and

$$d = \begin{bmatrix} 41.8\dot{6}14 & 0 \\ 0 & 13.1386 \end{bmatrix} \qquad (4.11.49)$$

Thus the first and second natural frequencies will be $\sqrt{13.1386} = 3.62$ rad/s and $\sqrt{41.8614} = 6.47$ rad/s, respectively. Just to make it easy to compare with our damped results, we'll rescale the eigenvectors (the columns of $v$) so that the first entry is equal to 1.0 for both cases. Dividing our first eigenvector by .3489 and our second by .9831 gives us

$$X_1 = \begin{Bmatrix} 1.0000 \\ 2.6861 \end{Bmatrix} \quad \text{and} \quad X_2 = \begin{Bmatrix} 1.0000 \\ -.1861 \end{Bmatrix} \qquad (4.11.50)$$

Now let's look at our damped problem. Putting our system into $[A]$, $[B]$ form yields

$$[B] = \begin{bmatrix} -40 & 10 & 0 & 0 \\ 10 & -30 & 0 & 0 \\ 0 & 0 & 1 & 0 \\ 0 & 0 & 0 & 2 \end{bmatrix} \qquad (4.11.51)$$

and

$$[A] = \begin{bmatrix} 0 & 0 & -40 & 10 \\ 0 & 0 & 10 & -30 \\ -40 & 10 & -.5 & .1 \\ 10 & -30 & .1 & .5 \end{bmatrix} \qquad (4.11.52)$$

Typing $[v, d] = \text{eig}(A, B)$ gives

$$v = \begin{bmatrix} .1502 + .0012i & -.0620 - .0691i & .1443 + .0417i & .0397 - .0839i \\ -.0280 - .0002i & -.1664 - .1855i & -.0269 - .0078i & .1066 - .2253i \\ -.0312 - .9711i & -.2430 + .2325i & -.3070 + .9218i & .2993 + .1534i \\ .0058 + .1808i & -.6529 + .6245i & .0572 - .1716i & .8039 + .4122i \end{bmatrix} \qquad (4.11.53)$$

and

$$d = \begin{bmatrix} -.2593 - 6.4648i & 0 & 0 & 0 \\ 0 & -.1157 - 3.6229i & 0 & 0 \\ 0 & 0 & -.2593 + 6.4648i & 0 \\ 0 & 0 & 0 & -.1157 + 3.6229i \end{bmatrix} \qquad (4.11.54)$$

This looks pretty reasonable. The first two imaginary entries in $d$ are $-6.4648$ and $-3.6229$, which are very close in magnitude to the undamped natural frequencies. The question is whether the associated eigenvectors have real displacement entries. We'll start by looking at the first column of $v$, which is the eigenvector associated with the eigenvalue $.2593 - 6.4648i$. Let's call this eigenvector $V1$. Currently, the displacement entries of $V1$ are complex. Since we know that MATLAB will happily multiply our eigenvectors by complex constants, we must phase shift the entire vector. Typing

$$\text{angle}(V1) \quad \text{(cr)}$$

will give us the output

$$\left\{ \begin{array}{c} .0080 \\ -3.1336 \\ -1.6029 \\ 1.5387 \end{array} \right\} \qquad (4.11.55)$$

Thus we see that the first entry has a phase angle of .0080 rad. We can phase shift the whole vector if we multiply it by

$$e^{-.008i} \qquad (4.11.56)$$

This will rotate the first entry back onto the real axis. If all goes as expected, it should also move the second entry onto the real axis. To do this with MATLAB we must type

$$V1*\exp(-i*.008) \quad \text{(cr)}$$

which will return

$$\left\{ \begin{array}{c} .1502 - 0i \\ -.0280 - 0i \\ -.0389 - .9798i \\ .0072 + .1807i \end{array} \right\} \qquad (4.11.57)$$

The first half of this vector corresponds to the $x_1$ and $x_2$ displacements. As hoped, the entries are both real. If we use the same normalization that we used for the unforced case (i.e., dividing both entries by .1502 so that the first entry equals 1.0), we obtain

$$\left\{ \begin{array}{c} 1.0000 \\ -.1861 \end{array} \right\} \qquad (4.11.58)$$

which matches precisely the second undamped eigenvector (4.11.50).

Just to be complete, we can go ahead and compute the other eigenvector using the second column of $v$ (which we'll call $V2$). angle($V2$) will produce

$$\left\{ \begin{array}{c} -2.302 \\ -2.302 \\ 2.3784 \\ 2.3784 \end{array} \right\} \qquad (4.11.59)$$

Rotating the vector by 2.302 rad ($V2*\exp(2.302*i)$) yields

$$\left\{ \begin{array}{c} .0928 - 0i \\ .2492 + 0i \\ -.0107 - .3362i \\ -.0288 - .9030i \end{array} \right\} \qquad (4.11.60)$$

Finally, dividing the first two rows by .0928 gives us

$$\left\{ \begin{array}{c} 1.0000 \\ 2.6861 \end{array} \right\} \qquad (4.11.61)$$

which matches up with the first undamped eigenvector.

Note that it was easy to see from (4.11.59) that both entries had the same phase for $V2$ (both displacement phases were $-2.302$ rad and both velocity phases were 2.3784 rad). Although it looked less obvious for $V1$ (4.11.55), if you look closely you'll realize that both displacement and velocity phases were equal in this case, also. For instance, the phase for $x_1$ was .0080 rad and the phase for $x_2$ was $-3.1336$ rad. Although these look different, they're saying the same thing. It's just that one phase angle is expressed positively and the other negatively. But both correspond to a phase that's in the first quadrant and equal to .008 rad. The same argument holds for the velocities. Again, this is just the way MATLAB does things. Sometimes you'll get both phases one way and other times it'll be done the other way.

## 4.12 FORCED RESPONSE OF DAMPED SYSTEMS

We've looked at MDOF, forced, undamped systems, and MDOF, unforced, damped systems; all that remains is to examine MDOF, forced, damped systems. The equations of motion are given by

$$[M]\ddot{X} + [C]\dot{X} + [K]X = F \tag{4.12.1}$$

where $F = \{f_1, \quad f_2, \quad \ldots\}^T$. The easiest way to solve this set of equations is to use complex exponential notation. Thus we'll substitute $F = \bar{F}e^{i\omega t}$ and $X = \bar{X}e^{i\omega t}$ into (4.12.1), which, after cancelling out the common $e^{i\omega t}$ factor, gives us

$$\left[-\omega^2[M] + i\omega[C] + [K]\right]\bar{X} = \bar{F} \tag{4.12.2}$$

Solving for $\bar{X}$ yields

$$\bar{X} = \left[-\omega^2[M] + i\omega[C] + [K]\right]^{-1}\bar{F} \tag{4.12.3}$$

As (4.12.3) indicates, there will be a phase shift between the input $\bar{F}$ and the output $\bar{X}$. For instance, if $\bar{F}$ is purely real, then $\bar{X}$ can't be real since $\bar{F}$ is multiplied by the complex matrix $\left[-\omega^2[M] + i\omega[C] + [K]\right]^{-1}$. Just as in the SDOF cases, the response will lag the input. Thus, if we choose $\bar{F}$ to be real and are concerned with the real part of $\bar{F}e^{i\omega t}$, our forcing will be of the form $\bar{F}\cos(\omega t)$ and our output will be of the form $\bar{X}\cos(\omega t - \phi)$, where $\phi$ is greater than zero.

**Example 4.13**

**Problem** Consider Figure 4.28 for $n = 2$. We'll use the particular values $m_1 = 1$ kg, $m_2 = 2$ kg, $c_1 = .1$ N/m/s, $c_2 = .2$ N/m/s, $c_3 = .1$ N/m/s, $k_1 = 4$ N/m, $k_2 = 6$ N/m, and $k_3 = 16$ N/m. The forcing will have the units of newtons and be given by

$$\left\{\begin{matrix} f_1 \\ f_2 \end{matrix}\right\} = \left\{\begin{matrix} 1 \\ -1 \end{matrix}\right\}\cos(3t) \tag{4.12.4}$$

Find the response of the system to this forcing.

**Solution** The forcing is in the form of a cosine wave, that is, the real part of $e^{3it}$. After solving for $\bar{X}$ using the complex exponential approach, we'll need to take the real part to match the solution with the forcing.

For the given values, (4.12.3) becomes

$$\left\{ \begin{array}{c} \bar{x}_1 \\ \bar{x}_2 \end{array} \right\} = \left[ \begin{array}{cc} 1+9i & -6-.6i \\ -6-.6i & 4+.9i \end{array} \right]^{-1} \left\{ \begin{array}{c} 1 \\ -1 \end{array} \right\} \tag{4.12.5}$$

Taking the indicated inverse and multiplying out gives us

$$\left\{ \begin{array}{c} \bar{x}_1 \\ \bar{x}_2 \end{array} \right\} = \left\{ \begin{array}{c} .0604 - .01431i \\ -.1523 + .0219i \end{array} \right\} \tag{4.12.6}$$

Keep in mind that this doesn't mean the response is imaginary. Remember that we're looking for the real part of $\bar{X}e^{3it}$. Thus we need the real part of

$$\left\{ \begin{array}{c} .0604 - .01431i \\ -.1523 + .0219i \end{array} \right\} e^{3it} \tag{4.12.7}$$

or

$$\left\{ \begin{array}{c} x_1(t) \\ x_2(t) \end{array} \right\} = \left\{ \begin{array}{c} .0604 \\ -.1523 \end{array} \right\} \cos(3t) - \left\{ \begin{array}{c} -.0143 \\ .0219 \end{array} \right\} \sin(3t) \tag{4.12.8}$$

Thus the final response for $f_1 = \cos(3t)$, $f_2 = -\cos(3t)$ is given by

$$x_1 = .0604 \cos(3t) + .0143 \sin(3t)$$

and

$$x_2 = -.1523 \cos(3t) - .0219 \sin(3t)$$

Putting this in phase-lag form, we have

$$x_1 = .0621 \cos(3t - 1.34)$$

and

$$x_2 = -.1539 \cos(3t - 1.43)$$

Note that the displacement values are given in meters.

Now that we have an understanding of damping it's time to return to the topic of vibration absorbers (Section 4.4). You'll recall that, for zero damping, we were able to cause the response of a forced main mass to be zero, at the expense of adding a secondary vibrational absorber to the system. For small, secondary, spring-mass systems, we had to endure large excursions of the absorber mass in order to attain vibration cancellation at the main mass. To reduce these excursions, you might think about adding a damper or two to the system. These would actually dissipate energy, rather than simply redirect it to a vibration absorber. However, as we've now seen, damping causes the responses to phase shift. Thus it probably will be impossible to *exactly* cancel the driving force, since the response of the masses will have a phase shift with respect to the input force. So we'll end up with our familiar engineering trade-off—balancing reduced vibration levels of one part of the system at the expense of increasing the vibration levels of another part.

One additional goal of adding damping would be to widen the frequency range over which vibration reduction occurs. As we saw in Section 4.4, the range of vibration reduction can be very

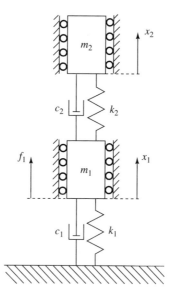

**Figure 4.30**   Damped vibration absorber

narrow and depends directly on how large a secondary spring-mass system is added. Perhaps by adding damping we'll be able to widen this range, thus making the system more robust in the face of disturbance forces that might occur over a wide frequency range.

Figure 4.30 illustrates our basic damped-absorber model. The disturbance force is acting on $m_1$ and $k_2$, and $m_2$ and $c_2$ represent the secondary system. The equations of motion are given by

$$\begin{bmatrix} m_1 & 0 \\ 0 & m_2 \end{bmatrix} \begin{Bmatrix} \ddot{x}_1 \\ \ddot{x}_2 \end{Bmatrix} + \begin{bmatrix} c_1 + c_2 & -c_2 \\ -c_2 & c_2 \end{bmatrix} \begin{Bmatrix} \dot{x}_1 \\ \dot{x}_2 \end{Bmatrix} + \begin{bmatrix} k_1 + k_2 & -k_2 \\ -k_2 & k_2 \end{bmatrix} \begin{Bmatrix} x_1 \\ x_2 \end{Bmatrix} = \begin{Bmatrix} f_1 \\ 0 \end{Bmatrix} \qquad (4.12.9)$$

There's no problem in solving this; it's basically the same system we looked at in the previous section's example. What we'll have to be concerned with here is not how to solve the problem, but rather how the damping elements help (or hurt) our quest for vibration absorption. Here we will show a few examples of how judiciously choosing the absorber parameters can yield good design solutions to our vibration problem. If we solve (4.12.9) by assuming a solution of the form $X = \bar{X}e^{i\omega t}$, as we learned in the previous section, we'll obtain

$$\bar{x}_1 = \frac{\bar{f}_1(k_2 - \omega^2 m_2 + i\omega c_2)}{\Delta(\omega)} \qquad (4.12.10)$$

and

$$\bar{x}_2 = \frac{\bar{f}_1(k_2 + i\omega c_2)}{\Delta(\omega)} \qquad (4.12.11)$$

where $\Delta(\omega)$ is the system's characteristic determinant and is given by

$$\Delta(\omega) = (m_1 + m_2)\omega^4 - i\omega^3 (c_1 m_2 + c_2(m_1 + m_2))$$

$$- \omega^2 (m_2 k_1 + (m_1 + m_2)k_2 + c_1 c_2) + i\omega(c_2 k_1 + c_1 k_2) + k_1 k_2 \qquad (4.12.12)$$

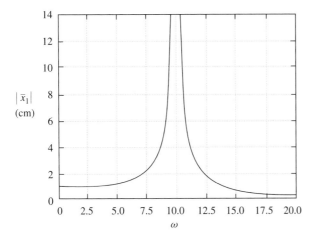

**Figure 4.31**  Amplitude response for a forced spring-mass system

To illustrate how damping affects our solutions, we'll look at a few graphical examples. Figure 4.31 shows the absolute value of the response for our base mass in the absence of any absorber and without damping. The magnitude of the force is equal to 10 N, $m_1 = 10$ kg, and $k_1 = 1000$ N/m. In addition to looking at the amplitude of the response, we can also examine the magnitude of the force that's transmitted to the base of the system. Since the only connection between $m_1$ and ground is the spring $k_1$, the magnitude of the force is given by $|k_1 x_1|$. This response is shown in Figure 4.32.

Now we'll add a vibration absorber to the system without any damping. The parameters will be $m_2 = 1$ kg and $k_2 = 100$ N/m. Plots of the amplitude response for both $x_1$ and $x_2$ are shown in Figure 4.33 while Figure 4.34 shows the magnitude of the transmitted force (again equal to $k_1 x_1$). Here we see all the classic hallmarks of an undamped vibration absorber. At the original system's natural frequency (its resonance frequency, in this case, equal to 10 rad/s) we now have a zero response for

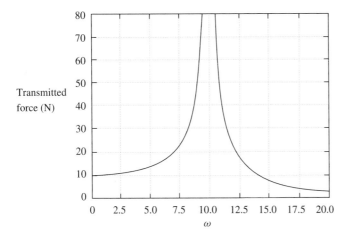

**Figure 4.32**  Transmitted force for a forced spring-mass system

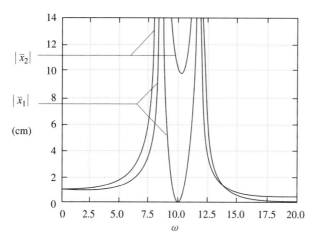

**Figure 4.33** Undamped vibration absorber, amplitude response of both masses

the main mass. However, in order to accomplish this, the secondary mass has a very sizable excursion (just under 10 cm). Thus we've traded off the infinite excursion we had originally for $m_1$ for no motion of $m_1$ and a relatively large motion of $m_2$. The transmitted force is zero at the same frequency that $x_1$ is equal to zero and grows in proportion to $x_1$ for all other frequencies (Figure 4.34).

The picture changes when we add damping. We'll let $c_1 = 2$ N/m/s and $c_2 = 2$ N/m/s. If we view the absorber and main-mass systems as separate systems, these values correspond to $\zeta$'s of $\zeta_1 = .01$ and $\zeta_2 = .1$. Figure 4.35 shows how $|x_1|$ and $|x_2|$ look for this case. Certainly things have changed. Instead of going to zero, $|x_1|$ is now finite near $\omega = 10$ rad/s (the original system's natural frequency). The resonances that used to exist below and above this frequency have been tamed, and only reach amplitudes of about 8 cm and 3 cm, respectively. This is certainly better than the infinite response we saw in Figure 4.33. The motions of the second mass have also been reduced, especially those around the undamped, second resonant frequency (approximately 11.5 rad/s).

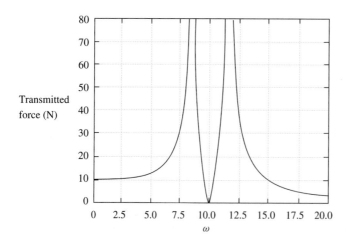

**Figure 4.34** Undamped vibration absorber, transmitted force

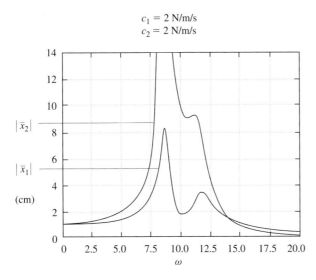

$c_1 = 2$ N/m/s
$c_2 = 2$ N/m/s

**Figure 4.35**  Damped vibration absorber, amplitude response of both masses

If we increase the damping in the secondary mass to 8 N/m/s, the response changes to that of Figure 4.36. The secondary mass's motion has been further reduced, but the maximum amplitude of the first mass hasn't really altered. One big qualitative change is that the two peaks we had previously have now merged into a single peak. If we look at Figure 4.37, we can see how the transmitted force was affected by our changes. $|\bar{f}_{T_1}|$ shows the transmitted force when $c_2$ was equal to 2 N/m/s and $|\bar{f}_{T_2}|$ is the plot for the case of $c_2 = 8$ N/m/s. As you can see, the transmitted force has actually gone up around the original system's natural frequency for the higher value of damping.

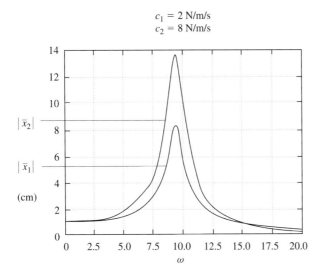

$c_1 = 2$ N/m/s
$c_2 = 8$ N/m/s

**Figure 4.36**  Damped vibration absorber, amplitude response of both masses

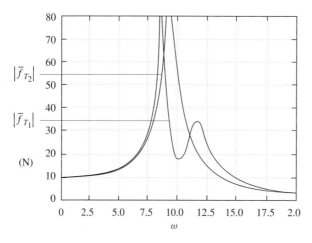

**Figure 4.37** Damped vibration absorber, transmitted force

What's the conclusion? Simply that optimizing a vibration absorber isn't a trivial job. By adding damping to the secondary absorber we can reduce $m_2$'s excursions, which is good from a packaging viewpoint. We may have limited space to work with and excessive oscillations simply can't be permitted. However, we just saw that reducing the level of vibration for this particular case also increased the transmitted forces. One of the goals of vibration absorption is to reduce these forces; indeed, for the undamped case the transmitted force is zero. If the vibrating object is an automobile engine, for example, then any transmitted vibration will excite the chassis, leading to higher vibration levels in the automobile cabin. This is clearly undesirable.

An additional goal in the design process might be to obtain a low-amplitude response level over a wide frequency range. But in doing so we must make sure that we're not unduly increasing the transmitted force or the vibration amplitude of the second mass. Although we won't get into it, we can try to optimize this problem, setting performance goals (such as minimizing the vibration amplitude) subject to constraints (such as keeping the secondary mass below 20 percent of the total mass). Those interested in pursuing this topic can refer to [6] for some ideas on how to proceed.

## 4.13 SYMMETRY OF MASS AND STIFFNESS MATRICES

In all our work so far, the mass and stiffness matrices have invariably been symmetric. This isn't an accident—it's a fundamental property of undamped vibrating structures. However, it's also possible to get a bit confused because you can easily wind up with mass and stiffness matrices that appear nonsymmetric. This might make you doubt yourself since you've come to expect $[K]$ and $[M]$ to be symmetric. Let's show how this can happen. Consider the system of Figure 4.38. We'll only consider vertical motions, thereby limiting the system to two degrees of freedom. Two different coordinate sets are illustrated in Figure 4.38. The $x$, $\theta$ set tracks the vertical motion of the beam's center of mass ($x$) and the beam's rotation ($\theta$). The $x_1$, $x_2$ set gives us a direct reading of the springs' deflections. Either set is a reasonable one from which to find the system's equations of

**Figure 4.38**   2-DOF system, with two illustrated coordinate systems

motion. The kinetic and potential energies in terms of $x$, $\theta$ are given by

$$k_e = \tfrac{1}{2}m\dot{x}^2 + \tfrac{1}{2}\bar{I}\dot{\theta}^2 \tag{4.13.1}$$

and

$$p_e = \tfrac{1}{2}k_2(x + l\theta)^2 + \tfrac{1}{2}k_1(x - l\theta)^2 \tag{4.13.2}$$

An application of Lagrange's equations gives us

$$\begin{bmatrix} m & 0 \\ 0 & \bar{I} \end{bmatrix} \begin{Bmatrix} \ddot{x} \\ \ddot{\theta} \end{Bmatrix} + \begin{bmatrix} k_1 + k_2 & l(k_2 - k_1) \\ l(k_2 - k_1) & l^2(k_1 + k_2) \end{bmatrix} \begin{Bmatrix} x \\ \theta \end{Bmatrix} = \begin{Bmatrix} 0 \\ 0 \end{Bmatrix} \tag{4.13.3}$$

As usual, the $[M]$ and $[K]$ matrices are symmetric. We'll be saying more about these equations in a bit, but for now let's carry on and find what the equations look like in terms of the $x_1$, $x_2$ coordinates. For this case, the energies are given by

$$k_e = \frac{1}{2}m\left(\frac{\dot{x}_1 + \dot{x}_2}{2}\right)^2 + \frac{1}{2}\bar{I}\left(\frac{\dot{x}_2 - \dot{x}_1}{2l}\right)^2 \tag{4.13.4}$$

and

$$p_e = \frac{1}{2}k_1 x_1^2 + \frac{1}{2}k_2 x_2^2 \tag{4.13.5}$$

Applying Lagrange's equations yields

$$\begin{bmatrix} \dfrac{m}{4} + \dfrac{\bar{I}}{4l^2} & \dfrac{m}{4} - \dfrac{\bar{I}}{4l^2} \\ \dfrac{m}{4} - \dfrac{\bar{I}}{4l^2} & \dfrac{m}{4} + \dfrac{\bar{I}}{4l^2} \end{bmatrix} \begin{Bmatrix} \ddot{x}_1 \\ \ddot{x}_2 \end{Bmatrix} + \begin{bmatrix} k_1 & 0 \\ 0 & k_2 \end{bmatrix} \begin{Bmatrix} x_1 \\ x_2 \end{Bmatrix} = \begin{Bmatrix} 0 \\ 0 \end{Bmatrix} \tag{4.13.6}$$

These matrices are also symmetric. So what's the point? Well, you'll note that we used a Lagrangian approach each time to find our equations. Let's see what happens if we apply Newton's laws instead of using the Lagrangian approach. We'll apply $f = ma$ and $\bar{I}\ddot{\theta} = M$, using the center of mass as our reference point. And just to be difficult, we'll use the $x_1$, $x_2$ coordinate set rather than the more immediate $x$, $\theta$ one. $f = ma$ gives us

$$m\left(\frac{\ddot{x}_1 + \ddot{x}_2}{2}\right) = -k_1 x_1 - k_2 x_2 \tag{4.13.7}$$

and $\bar{I}\ddot{\theta} = M$ yields

$$\bar{I}\left(\frac{\ddot{x}_2 - \ddot{x}_1}{2l}\right) = -k_2 l x_2 + k_1 l x_1 \tag{4.13.8}$$

Put this in matrix form and you have

$$\begin{bmatrix} \dfrac{m}{2} & \dfrac{m}{2} \\ -\dfrac{\bar{I}}{2l} & \dfrac{\bar{I}}{2l} \end{bmatrix}\begin{Bmatrix} \ddot{x}_1 \\ \ddot{x}_2 \end{Bmatrix} + \begin{bmatrix} k_1 & k_2 \\ -k_1 l & k_2 l \end{bmatrix}\begin{Bmatrix} x_1 \\ x_2 \end{Bmatrix} = \begin{Bmatrix} 0 \\ 0 \end{Bmatrix} \tag{4.13.9}$$

Now we've got something. You'll note that both the $[M]$ and $[K]$ matrices are decidedly not symmetric. And yet we've already seen that the $x_1$, $x_2$ coordinates do yield symmetric matrices when we use Lagrange's equations. So what's happening? Are (4.13.6) and (4.13.9) really different equations or is something more going on here? Logically, we know that they've got to be telling us the same thing because both Lagrange's equations and Newton's laws are perfectly reasonable ways to go about getting a system's equations of motion; the equations can't fundamentally change just because we use one approach or the other. Thus the change must not be fundamental. In fact, since they must display the same physics, it's reasonable to assume that we can manipulate one set so as to yield the other. We can do this by first dividing the second row of (4.13.9) by $l$ to produce

$$\begin{bmatrix} \dfrac{m}{2} & \dfrac{m}{2} \\ -\dfrac{\bar{I}}{2l^2} & \dfrac{\bar{I}}{2l^2} \end{bmatrix}\begin{Bmatrix} \ddot{x}_1 \\ \ddot{x}_2 \end{Bmatrix} + \begin{bmatrix} k_1 & k_2 \\ -k_1 & k_2 \end{bmatrix}\begin{Bmatrix} x_1 \\ x_2 \end{Bmatrix} = \begin{Bmatrix} 0 \\ 0 \end{Bmatrix} \tag{4.13.10}$$

Next, we'll subtract the second row from the first to produce a new equation and then add the first and second rows together, giving us a second new equation (the new equations are equal to the difference and sum of the foregoing equations, respectively)

$$\begin{bmatrix} \dfrac{m}{2} + \dfrac{\bar{I}}{2l^2} & \dfrac{m}{2} - \dfrac{\bar{I}}{2l^2} \\ \dfrac{m}{2} - \dfrac{\bar{I}}{2l^2} & \dfrac{m}{2} + \dfrac{\bar{I}}{2l^2} \end{bmatrix}\begin{Bmatrix} \ddot{x}_1 \\ \ddot{x}_2 \end{Bmatrix} + \begin{bmatrix} 2k_1 & 0 \\ 0 & 2k_2 \end{bmatrix}\begin{Bmatrix} x_1 \\ x_2 \end{Bmatrix} = \begin{Bmatrix} 0 \\ 0 \end{Bmatrix} \tag{4.13.11}$$

You can see that by dividing both rows of (4.13.11) by 2 we'll exactly replicate (4.13.6). Thus (4.13.6) and (4.13.9) are both giving us the same information; one is simply a linear combination of the other. What we're seeing is simply an example of a coordinate transformation, going from one set of coordinates to another.

But we still haven't addressed the problem of symmetry. Why is one formulation symmetric and the other not? The reason has to do with how the equations were found. Lagrange's equations work with the system's energies, quantities that are quadratic in nature (kinetic energy is proportional to velocity squared, spring potential is proportional to displacement squared, etc.). By going through this route, the mass and stiffness matrix will always be symmetric in the way we've been led to expect. The Newtonian approach, however, deals with force and moment balances and there's nothing that dictates a symmetric form. What *is* dictated is that you can make the equations symmetric by suitably rearranging them. And we can find out the appropriate rearrangement pretty simply.

As far as our current example is concerned, we would have obtained precisely the same equations as those shown in (4.13.9) if we'd started from (4.13.3) and simply used the substitutions

$$x = \frac{x_1 + x_2}{2} \tag{4.13.12}$$

and

$$\theta = \frac{x_2 - x_1}{2l} \tag{4.13.13}$$

that is,

$$\left\{ \begin{array}{c} x \\ \theta \end{array} \right\} = \left[ \begin{array}{cc} \frac{1}{2} & \frac{1}{2} \\ -\frac{1}{2l} & \frac{1}{2l} \end{array} \right] \left\{ \begin{array}{c} x_1 \\ x_2 \end{array} \right\} \tag{4.13.14}$$

We'll call the transformation matrix $[A]$, that is,

$$[A] = \left[ \begin{array}{cc} \frac{1}{2} & \frac{1}{2} \\ -\frac{1}{2l} & \frac{1}{2l} \end{array} \right] \tag{4.13.15}$$

What we have is a matrix relationship between the two coordinates. Operations such as this, however, don't preserve symmetry. Thus, if we start with a symmetric $[K]$ (which we did), switching coordinates will give us the new stiffness matrix $[K]'$, where

$$[K]' = [K][A] \tag{4.13.16}$$

In general, $[K][A]$ will not be symmetric. Luckily, we can regain symmetry by premultiplying by $[A]^T$. That this is so can be seen in the following manner. Assume that a matrix $[K]$ is symmetric. We'll form a new matrix $[C]$ thusly,

$$[C] = [A]^T[K][A] \tag{4.13.17}$$

Transposing both sides of (4.13.17) yields

$$[C]^T = [A]^T[K]^T[A] \tag{4.13.18}$$

Since we've already said that $[K]$ is symmetric, $[K]^T = [K]$, we can rewrite (4.13.18) as

$$[C]^T = [A]^T[K][A] \tag{4.13.19}$$

Equations (4.13.17) and (4.13.19) together imply that $[C]$ is equal to $[C]^T$. Thus we've retained symmetry by using the operation $[A]^T[K][A]$ instead of just sticking with $[K][A]$.

To see if this helps in the problem we're looking at, take the transpose of $[A]$

$$[A]^T = \begin{bmatrix} \frac{1}{2} & -\frac{1}{2l} \\ \frac{1}{2} & \frac{1}{2l} \end{bmatrix} \tag{4.13.20}$$

What does premultiplying by this matrix do to our equations? Look at the first row. It takes one-half of our first equation and then subtracts $\frac{1}{2l}$ multiplied by our second equation. This is exactly what we did when rearranging the equations. The second row takes one-half of the first equation and then adds $\frac{1}{2l}$ multiplied by the second equation. Again, just what we did.

Thus we have our final observation. If we use Lagrange's equations to get our equations of motion, the spring and mass matrices will be symmetric. If we use Newton's laws, they might or might not be symmetric. If they aren't, it's because we've essentially performed half a transformation, the $[K][A]$ part. We now need to premultiply by the appropriate $[A]^T$ in order to get to the symmetric form.

**Example 4.14**

**Problem**   As another example of how we can obtain nonsymmetric mass and stiffness matrices, consider the system shown in Figure 4.39. This is our standard two-mass, three-spring problem. We've seen repeatedly that, if our coordinates tell us the absolute displacement of both $m_1$ and $m_2$, then our spring and mass matrices will be symmetric. However, this time we're not going to use coordinates that tell us how much each mass has moved. Our first coordinate, $x$, does tell us the absolute motion of $m_1$. But the second coordinate, $y$, tells us the *relative* motion between $m_1$ and $m_2$. In Figure 4.39, $y_0$ represents the unstretched length of the physical spring between the two

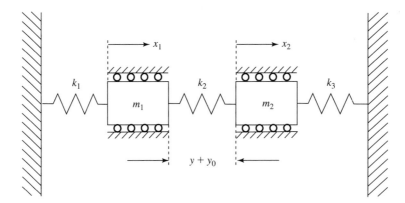

**Figure 4.39**   2-DOF system with coordinates showing absolute and relative motions

masses. Show that using a Newtonian approach to find the equations of motion in terms of these coordinates will yield nonsymmetric $[M]$ and $[K]$ matrices.

**Solution**    If we form the equations of motion using these two coordinates, we'll obtain

$$m_1 \ddot{x} = -k_1 x + k_2 y$$

$$m_2 (\ddot{x} + \ddot{y}) = -k_2 y - k_3 (x + y) \tag{4.13.21}$$

The quantities $\ddot{x} + \ddot{y}$ and $x + y$ represent the absolute acceleration and displacement of $m_2$.
Putting these equations in matrix form gives us

$$\begin{bmatrix} m_1 & 0 \\ m_2 & m_2 \end{bmatrix} \begin{Bmatrix} \ddot{x} \\ \ddot{y} \end{Bmatrix} + \begin{bmatrix} k_1 & -k_2 \\ k_3 & k_2 + k_3 \end{bmatrix} \begin{Bmatrix} x \\ y \end{Bmatrix} = \begin{Bmatrix} 0 \\ 0 \end{Bmatrix} \tag{4.13.22}$$

As you can see, both $[M]$ and $[K]$ are nonsymmetric for these coordinates.

## 4.14  REPEATED FREQUENCIES AND ZERO FREQUENCIES

Although not the norm, we sometimes run into systems that have repeated natural frequencies. In such a system, there will be more than one distinct modal deflection that corresponds to a particular natural frequency. We won't go into great detail with regard to this subject, but we'll indicate what the basic phenomena are and discuss what such a situation means for our analyses.

We'll begin by examining Figure 4.38. We've already found the equations of motion for this system in Section 4.13, which we'll repeat here for convenience:

$$\begin{bmatrix} m & 0 \\ 0 & \bar{I} \end{bmatrix} \begin{Bmatrix} \ddot{x} \\ \ddot{\theta} \end{Bmatrix} + \begin{bmatrix} k_1 + k_2 & l(k_2 - k_1) \\ l(k_2 - k_1) & l^2(k_1 + k_2) \end{bmatrix} \begin{Bmatrix} x \\ \theta \end{Bmatrix} = \begin{Bmatrix} 0 \\ 0 \end{Bmatrix} \tag{4.14.1}$$

We'll now set both spring constants equal to $k$,

$$\begin{bmatrix} m & 0 \\ 0 & \bar{I} \end{bmatrix} \begin{Bmatrix} \ddot{x} \\ \ddot{\theta} \end{Bmatrix} + \begin{bmatrix} 2k & 0 \\ 0 & 2kl^2 \end{bmatrix} \begin{Bmatrix} x \\ \theta \end{Bmatrix} = \begin{Bmatrix} 0 \\ 0 \end{Bmatrix} \tag{4.14.2}$$

What we see is that the equations are now decoupled. Since they are decoupled, that means that $x$ and $\theta$ must be the system's normal coordinates. Oscillations in the $x$ direction represent one mode and oscillations of $\theta$ represent the other mode. The natural frequency of the $x$ mode is given by

$$\omega_1^2 = \frac{2k}{m} \tag{4.14.3}$$

and the second natural frequency (associated with $\theta$) is given by

$$\omega_2^2 = \frac{2kl^2}{\bar{I}} \tag{4.14.4}$$

If we divide the first equation by $m$ and the second by $\bar{I}$, then our equations will be

$$\begin{bmatrix} 1 & 0 \\ 0 & 1 \end{bmatrix} \begin{Bmatrix} \ddot{x} \\ \ddot{\theta} \end{Bmatrix} + \begin{bmatrix} \omega_1^2 & 0 \\ 0 & \omega_2^2 \end{bmatrix} \begin{Bmatrix} x \\ \theta \end{Bmatrix} = \begin{Bmatrix} 0 \\ 0 \end{Bmatrix} \tag{4.14.5}$$

Since we're already in normal form, the corresponding eigenvectors are $\{1 \quad 0\}^T$ (for $\omega_1$) and $\{0 \quad 1\}^T$ (for $\omega_2$).

Now here's the question. What happens if the parameters work out so that

$$\frac{2k}{m} = \frac{2kl^2}{\bar{I}} \tag{4.14.6}$$

In this case we'd have a repeated natural frequency. The equations of motion would be given by

$$\begin{bmatrix} 1 & 0 \\ 0 & 1 \end{bmatrix} \begin{Bmatrix} \ddot{x} \\ \ddot{\theta} \end{Bmatrix} + \begin{bmatrix} \omega_1^2 & 0 \\ 0 & \omega_1^2 \end{bmatrix} \begin{Bmatrix} x \\ \theta \end{Bmatrix} = \begin{Bmatrix} 0 \\ 0 \end{Bmatrix} \tag{4.14.7}$$

What happens if we try to analyze (4.14.7)? If we assume harmonic motion and set the determinant of $[K] - \omega^2[M]$ equal to zero, we'll obtain

$$(\omega_1^2 - \omega^2)(\omega_1^2 - \omega^2) = 0 \tag{4.14.8}$$

This tells us that both natural frequencies are equal to $\omega_1$. No problem there. But what about when we try to find the eigenvectors? If we assume an eigensolution of

$$\begin{Bmatrix} x \\ \theta \end{Bmatrix} = \begin{Bmatrix} \bar{x} \\ \bar{\theta} \end{Bmatrix} e^{i\omega_1 t} \tag{4.14.9}$$

and substitute this into (4.14.7) we'll get

$$\begin{bmatrix} 0 & 0 \\ 0 & 0 \end{bmatrix} \begin{Bmatrix} \bar{x} \\ \bar{\theta} \end{Bmatrix} = \begin{Bmatrix} 0 \\ 0 \end{Bmatrix} \tag{4.14.10}$$

This is a weird one. How can we figure out the eigenvectors if the entire matrix is composed of zeros? Are there any eigenvectors anymore? If we think logically about the problem, it seems that there must be eigenvectors. Why? Well, our problem only occurred when $\omega_2$ was made to equal $\omega_1$. As long as they aren't equal, no matter how close they might be to each other, there are still distinct eigenvectors, $\{1 \quad 0\}^T$ and $\{0 \quad 1\}^T$. It doesn't make sense for them to suddenly disappear just because $\omega_2$ has changed from being almost equal to $\omega_1$ to being exactly equal to $\omega_1$. Let's presume that both of these are still eigenvectors of the problem. Do they satisfy (4.14.10)? Yes, they definitely do. And are they independent of each other? They certainly are. So they *are* eigenvectors for the system. The interesting point is that they are no longer uniquely defined. For instance, what if we looked at $\{1 \quad 1\}^T$ and $\{1 \quad -1\}^T$. These also satisfy (4.14.10) and are independent. So are $\{2 \quad 1\}^T$ and $\{1 \quad 2\}^T$. In fact, so are an infinite number of choices.

What's happened is that the problem has become indeterminate. Since both the original eigenvectors have the same natural frequency, any initial displacement of the system will oscillate at $\omega_1$.

**Example 4.15**

Say that $l$ is 1 m for the system shown in Figure 4.38 and you give the system initial conditions of $x_1 = 1$ cm and $x_2 = 0$. Find the response of the system.

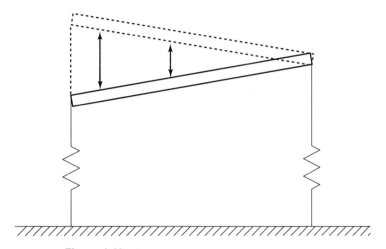

**Figure 4.40**   System with repeated natural frequencies

**Solution**   In this case we have

$$x_1 = x - l\theta = .01 \tag{4.14.11}$$

and

$$x_2 = x + l\theta = 0 \tag{4.14.12}$$

The solution to this is $x = .5$ cm and $\theta = -.005$ rad. Since the translational and rotational frequencies are the same, the system stays in this configuration, as shown in Figure 4.40. Thus it could be thought that this initial deflection is itself an eigenvector. In the same way, any initial deflection will simply oscillate at $\omega_1$, as if it is an eigenvector.

Since our system is no longer telling us what the eigenvectors should be when we have repeated roots, we must figure them out for ourselves. In a case such as the one we've just seen, it would be easiest to simply choose $\{1 \quad 0\}^T$ and $\{0 \quad 1\}^T$ as our eigenvectors. This way there'd be no discontinuity in our eigenvectors for the cases of $\omega_1 = \omega_2$ and $\omega_1 \neq \omega_2$. In general, you'll want to have $n$ independent eigenvectors for an $n$-DOF problem. If you have a repeated root, then you should construct two vectors that satisfy the problem and that are independent of the other (well-determined) eigenvectors, as well as being independent of each other.

Up to this point, all the systems we've been analyzing have been attached to ground in some way. Although this is a pretty common occurrence in real life, it's also reasonably common to have systems that aren't connected to the ground. Airplanes and spacecraft come to mind immediately, as do boats and submarines. The complicating factor when dealing with systems of this sort is that, according to Newton, once these systems are set in motion they'll remain in motion until acted on by some force. This behavior gives rise to *zero frequencies*, that is, natural frequencies that are equal to zero.

**Figure 4.41**  2-DOF system with a rigid body mode

To see how these come about, we'll consider the system shown in Figure 4.41. The equations of motion for this system are given by:

$$\begin{bmatrix} m & 0 \\ 0 & m \end{bmatrix} \begin{Bmatrix} \ddot{x}_1 \\ \ddot{x}_2 \end{Bmatrix} + \begin{bmatrix} k & -k \\ -k & k \end{bmatrix} \begin{Bmatrix} x_1 \\ x_2 \end{Bmatrix} = \begin{Bmatrix} 0 \\ 0 \end{Bmatrix} \tag{4.14.13}$$

If you consider the $[K]$ matrix you'll soon notice that it's different in a fundamental way from those we've seen so far, namely in that its determinant is equal to zero. Thus it isn't invertible. Note that $[M]$ is always invertible since the mass elements can't be equal to zero (unless you've set your problem up incorrectly). To see what this lack of invertibility does to the problem, we'll assume a solution of $\bar{X}e^{i\omega t}$ and substitute this into (4.14.13), leading to

$$\begin{bmatrix} k - m\omega^2 & -k \\ -k & k - m\omega^2 \end{bmatrix} \begin{Bmatrix} \bar{x}_1 \\ \bar{x}_2 \end{Bmatrix} = \begin{Bmatrix} 0 \\ 0 \end{Bmatrix} \tag{4.14.14}$$

Setting the determinant of the matrix to zero gives us

$$m^2\omega^4 - 2mk\omega^2 = 0 \tag{4.14.15}$$

or, factoring this into two parts and dividing through by $m$,

$$\omega^2(m\omega^2 - 2k) = 0 \tag{4.14.16}$$

The two solutions are therefore

$$\omega_1^2 = 0 \tag{4.14.17}$$

and

$$\omega_2^2 = \frac{2k}{m} \tag{4.14.18}$$

The second solution corresponds to a *breathing mode* of the system. Both masses move out of phase with each other at a frequency of $\sqrt{\dfrac{2k}{m}}$. You can see this by solving (4.14.14) for $\omega^2 = \dfrac{2k}{m}$, which will give us the eigenvector solution $\{1 \quad -1\}^T$. The more relevant solution for us is the one corresponding to $\omega_1 = 0$. If we solve (4.14.14) for this case, we'll obtain the associated eigenvector $\{1 \quad 1\}$. This looks just like the first eigenvector of our familiar three-spring, two-mass problem, when both masses had equal values, as did all the springs (4.7.8). However, the difference between these two cases is that now the frequency of the solution is zero. This mode represents both masses moving off to the right or left, each being displaced the same amount as the other. Given an initial velocity, they'll keep moving until something stops them. Since there is no relative motion, this type of response is referred to as a rigid-body mode, since this is the same response you'd see if the springs' values were infinite (making the system a rigid body).

You'll note that this particular example only allowed rigid-body translation in one direction. Since a general rigid body has six degrees of freedom (three translations and three rotations), you should expect the equations of motion for such systems, satellites, aircraft, etc., to have six zero-frequency modes corresponding to these degrees of freedom. Although this complicates the analysis when we are interested in controlling these systems, it doesn't really affect a general vibrational analysis. A modal analyzer will simply record these as zero-frequency modes and the transfer functions will show a peak as the driving frequency approaches zero.

## 4.15 INFLUENCE COEFFICIENTS

To finish up this chapter, we'll look at something called *influence coefficients*. The best way to let you know what these are is to examine them in a one-dimensional setting, that is, a SDOF spring-mass system (Figure 4.42). For such a system, we know that to displace mass $m$ by an amount $x$ requires a force $f$,

$$f = kx \tag{4.15.1}$$

Similarly, if we apply a force $f$ to the mass, it will deflect thusly

$$x = \tfrac{1}{k} f \tag{4.15.2}$$

If we define $a$ as the inverse of $k$, then we have

$$x = af \tag{4.15.3}$$

$k$ is our *stiffness influence coefficient* and $a$ is our *flexibility influence coefficient*. As you can see, the flexibility influence coefficient is simply the inverse of the stiffness influence coefficient. Thus,

**Figure 4.42**    SDOF influence coefficients

if we have one, we can find the other. To obtain either one experimentally, we'd simply have to apply a force to the mass and observe its deflection, or deflect it and have a force transducer handy that would tell us the force needed to cause that deflection. If you think about it, this is a nice way to determine the spring stiffness without needing to know the mass. We could place a known mass on our current mass, which would translate into a known force being exerted on the spring ($f = mg$). We could then measure the deflection and deduce that $k$ is simply $\frac{f}{x}$.

Now that we understand what's what with the single-degree-of-freedom case, it's time to move onto the multiple-degree-of-freedom case. The same general characteristics will hold. This time we're not going to be finding $k$, but rather $[K]$, the system's stiffness matrix. We'll be calling this the stiffness influence coefficient matrix and all the entries of $[K]$ will be the individual stiffness influence coefficients. We'll also be able to find $[A]$, which will be equal to $[K]^{-1}$. Not surprisingly, the elements of $[A]$ will be the flexibility influence coefficients and $[A]$ itself will be called the flexibility influence coefficient matrix. Just as in the single-degree-of-freedom case, knowing $[K]$ and how the $n$ masses are displaced will allow us to determine the forces acting on the masses. Or, in a reversed sense, knowing what the applied forces are will allow us to calculate the displacements.

For clarity, let's look at a serial chain of masses (Figure 4.43). We'll be solving for the flexibility influence coefficients, $a_{i,j}$. Assume we apply a force to $m_j$. We know that all the masses will (in general) move into a new static configuration as a result of this force. We'll call $a_{i,j}$ the flexibility influence coefficient between mass $i$ and mass $j$. It lets us determine how much the mass $m_i$ will deflect if the force $f_j$ is applied to $m_j$. The deflection due to $f_j$ will simply be $a_{i,j}f_j$. Each time we put a new force on another mass, all the masses will again move to a new static configuration. But since the system is linear, we can just add up all the contributions to find the final deflection. Thus

$$x_i = a_{i,1}f_1 + a_{i,2}f_2 + a_{i,3}f_3 + \cdots + a_{i,n}f_n \qquad (4.15.4)$$

This is simply saying that the deflection at $m_i$ will be equal to the deflection caused by the first force acting on the first mass plus the deflection caused by the second force acting on the second mass, etc.

So far, so good. We can even realize that since we're summing over all the masses, this can be written as

$$x_i = \sum_{j=1}^{n} a_{i,j}f_j \qquad (4.15.5)$$

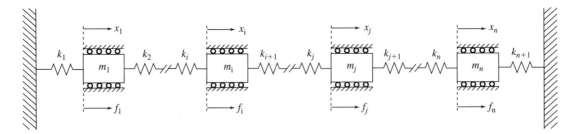

**Figure 4.43**   Serial mass-spring system

This looks very much like summation notation for a matrix multiplied by a vector. Therefore all we need do is define

$$X \equiv \{x_1 \ x_2 \ \dots \ x_n\}^T, \quad F \equiv \{f_1 \ f_2 \ \dots \ f_n\}^T \tag{4.15.6}$$

and

$$[A] \equiv \begin{bmatrix} a_{1,1} & a_{1,2} & a_{1,3} & \cdots \\ a_{2,1} & a_{2,2} & a_{2,3} & \cdots \\ a_{3,1} & a_{3,2} & a_{3,3} & \cdots \\ \vdots & & & a_{n,n} \end{bmatrix} \tag{4.15.7}$$

and we can write (4.15.5) as

$$X = [A]F \tag{4.15.8}$$

In this form, our multiple-degree-of-freedom result looks like the single-degree-of-freedom $x = af$ (with $a = \frac{1}{k}$). Let's see how it all works out in reality, and then let's look at stiffness influence coefficients.

**Example 4.16**

**Problem**   Determine the flexibility influence coefficients for the system shown in Figure 4.44.

**Solution**   First we'll apply a force $f_1$ to the first mass and let $f_2$ equal zero. Since the second mass is unrestrained, the change in displacement for $m_1$ is just

$$x_1^{f_1} = \frac{1}{k_1} f_1 \tag{4.15.9}$$

The displacement is indicated by $x_1^{f_1}$, and not simply $x_1$, to indicate that it's the displacement of the first mass due to the force $f_1$. $x_2$ moves the same amount as $x_1$, due to the fact that there's no

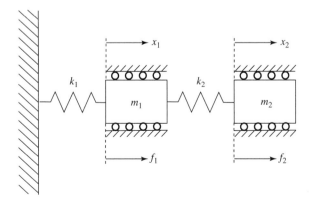

**Figure 4.44**   2-DOF flexibility influence coefficient example

additional spring affecting it, so

$$x_2^{f_1} = \frac{1}{k_1} f_1 \tag{4.15.10}$$

Now let's apply zero force to the first mass and $f_2$ to the second. We can see that the deflection of $m_2$ will depend upon the equivalent spring between it and ground, and the motion of $m_1$ will come from a static force balance. Beginning with $m_2$, the displacement is going to be equal to

$$x_2^{f_2} = \frac{1}{k_{eq}} f_2 \tag{4.15.11}$$

where $k_{eq} = \dfrac{k_1 k_2}{k_1 + k_2}$ (recall our work involving parallel and series springs from Chapter 1). Thus the deflection is

$$x_2^{f_2} = \frac{k_1 + k_2}{k_1 k_2} \tag{4.15.12}$$

To find $x_1^{f_2}$, we'll need a force balance at $m_1$,

$$k_1 x_1^{f_2} = \left( x_2^{f_2} - x_1^{f_2} \right) k_2 \tag{4.15.13}$$

which, upon substitution of $x_2$ and solving for $x_1$, gives us

$$x_1^{f_2} = \frac{1}{k_1} f_2 \tag{4.15.14}$$

Adding together the results of the two cases ($f_1$ finite, $f_2 = 0$, and $f_1 = 0$, $f_2$ finite) gives us

$$\left\{ \begin{array}{c} x_1 \\ x_2 \end{array} \right\} = \left[ \begin{array}{cc} \dfrac{1}{k_1} & \dfrac{1}{k_1} \\ \dfrac{1}{k_1} & \dfrac{k_1 + k_2}{k_1 k_2} \end{array} \right] \left\{ \begin{array}{c} f_1 \\ f_2 \end{array} \right\} \tag{4.15.15}$$

where $x_1 = x_1^{f_1} + x_1^{f_2}$ and $x_2 = x_2^{f_1} + x_2^{f_2}$. Thus we see that

$$[A] = \left[ \begin{array}{cc} \dfrac{1}{k_1} & \dfrac{1}{k_1} \\ \dfrac{1}{k_1} & \dfrac{k_1 + k_2}{k_1 k_2} \end{array} \right] \tag{4.15.16}$$

The first thing we notice about the preceding example is that $a_{1,2} = a_{2,1}$. Is this expected? Well, you'll recall that the stiffness matrix $[K]$ is symmetric. And when we started this section the claim was made that $[A]$ is equal to the inverse of $[K]$. Thus, if this is true, it seems like we should expect symmetric flexibility influence coefficient matrices. And, assuming that $[K]$ is invertible, this will in fact be the case.

Having suggested that $[A]$ is equal to $[K]^{-1}$, it might be worth checking this out for the example we've just gone over. For this example, we can solve for the inverse exactly and find

$$\begin{bmatrix} \dfrac{1}{k_1} & \dfrac{1}{k_1} \\[2ex] \dfrac{1}{k_1} & \dfrac{k_1 + k_2}{k_1 k_2} \end{bmatrix}^{-1} = \begin{bmatrix} k_1 + k_2 & -k_2 \\ -k_2 & k_2 \end{bmatrix} \tag{4.15.17}$$

You can verify this result by multiplying the two matrices together and seeing that the result is indeed the identity matrix. As promised, the inverse we just found is exactly equal to the system's stiffness matrix. So now it's time to look at how we can find the stiffness influence coefficient matrix and see if it's the same as the stiffness matrix.

We'll again look at Figure 4.43. This time, we'll enforce given displacements of the masses and ask what forces are necessary to hold these positions. The stiffness influence coefficient $k_{i,j}$ will tell us what force is needed to hold the $j$th mass at a deflection of $x_j$ according to

$$f_i = k_{i,j} x_j \tag{4.15.18}$$

Note that for this to hold, the deflections of all other masses must be held at zero. This is the analogue of what we did when we determined the flexibility influence coefficients. For these, we applied a particular force to one mass and no force to the others. Now we're deflecting one mass and keeping all the others fixed. Just as in the previous case, we can express the complete stiffness-influence-coefficient information by defining

$$X \equiv \{x_1 \ x_2 \ \dots \ x_n\}, \quad F \equiv \{f_1 \ f_2 \ \dots \ f_n\} \tag{4.15.19}$$

and

$$[K] \equiv \begin{bmatrix} k_{1,1} & k_{1,2} & k_{1,3} & \cdots \\ k_{2,1} & k_{2,2} & k_{2,3} & \\ k_{3,1} & k_{3,2} & k_{3,3} & \\ \vdots & & & k_{n,n} \end{bmatrix} \tag{4.15.20}$$

and writing

$$F = [K]X \tag{4.15.21}$$

**Example 4.17**

**Problem**    Again consider the system of Figure 4.44. Determine the stiffness influence coefficients.

**Solution**    Actually, it's easy enough to solve this problem immediately by moving both $m_1$ and $m_2$ and determining the total force needed at $m_1$ and $m_2$ to hold those positions, but we'll do it the longer way just to be complete. First, we'll move $m_1$ by $x_1$ and hold $m_2$ fixed. In this case the springs are pushing back with a force equal to $k_1 + k_2$, and so our applied force must exactly counter this. Thus $f_1 = k_1 + k_2$. The $k_2$ spring is trying to push $m_2$ to the right with a force equal to $k_2 x_1$, and so the counteracting force must be $f_2 = -k_2 x_1$. Thus we have

$$\left\{ \begin{array}{c} f_1^{x_1} \\ f_2^{x_1} \end{array} \right\} = \left[ \begin{array}{cc} k_1 + k_2 & 0 \\ -k_2 & 0 \end{array} \right] \left\{ \begin{array}{c} x_1 \\ 0 \end{array} \right\} \qquad (4.15.22)$$

Next, we fix $x_1$ at zero and let the second mass move by $x_2$. Now the second spring is pulling $m_1$ to the right with a force equal to $k_2 x_2$, implying that $f_1 = -k_2 x_2$, and is also pulling $m_2$ to the left with the same force, leading to $f_2 = k_2 x_2$. This gives us

$$\left\{ \begin{array}{c} f_1^{x_2} \\ f_2^{x_2} \end{array} \right\} = \left[ \begin{array}{cc} 0 & -k_2 \\ 0 & k_2 \end{array} \right] \left\{ \begin{array}{c} 0 \\ x_2 \end{array} \right\} \qquad (4.15.23)$$

Adding together (4.15.22) and (4.15.23) (and letting $f_1 = f_1^{x_1} + f_1^{x_2}$, $f_2 = f_2^{x_1} + f_2^{x_2}$) gives us

$$\left\{ \begin{array}{c} f_1 \\ f_2 \end{array} \right\} = \left[ \begin{array}{cc} k_1 + k_2 & -k_2 \\ -k_2 & k_2 \end{array} \right] \left\{ \begin{array}{c} x_1 \\ x_2 \end{array} \right\} \qquad (4.15.24)$$

And, as we can see, the flexibility-influence coefficient matrix is exactly equal to the system's stiffness matrix (not to mention equal to the inverse of $[A]$).

Now it's high time we answered the question, What's the point? Why do we care about influence coefficients? Well, think about this for a second. Let's say you've been given the assignment of performing an analysis on some system. You've got a good idea of what the dominant masses in the system are, but you're not so sure about the stiffnesses running around the system. How do you figure it out? Well, one way is to create a detailed model of the system, based on the stiffnesses of the material that make up the structure, and then analyze this model. But to know if you're right, you'd like to find an easy way to check your results. Or you might not want to go to the extensive trouble of building up a detailed model; you want to get an idea of the stiffnesses from some simple experiments. Well, the influence coefficients help you do this. What you need to do in order to find $[K]$ directly is to deflect the structure in a known manner and record the forces needed to do so. Then you can just read off $[K]$ as we did in the previous example. The difficulty with this approach is that it isn't easy to deflect some continuous structure in a set of known manners and then to figure out the forces needed; it would require large clamps that could move the structure in the way you want and would also require some good force transducers.

The simpler way to get the same information is to go after the flexibility-influence-coefficient matrix and then invert it to find $[K]$. Experimentally obtaining $[A]$ isn't nearly as hard as directly finding $[K]$. All you need do is put a predetermined load at a point of the structure corresponding to where you're placing a lumped-mass approximation and then measure how the structure deflects. If the loading is vertical, you need only put a known mass on the structure and let gravity supply the force. By moving the force to all the lumped-mass positions, and reading off all the different deflections, you can build up your $[A]$ matrix. Inverting this then gives you $[K]$.

There's even another reason to be interested in the influence coefficients, although we won't be going into it in this book. As the number of masses in your system becomes larger, the system becomes more and more of a continuum, until finally we arrive at a fully continuous system. At this point, we'll no longer have influence coefficients, but rather will have influence functions, functions that tell us how the system deflects due to any distribution of forces on the object. This is a very useful piece of information to have when trying to determine the response of a continuous system to a continuous loading.

## 4.16 HOMEWORK PROBLEMS

### Section 4.2

1. You've been asked to add realism to a new cartoon. The action of interest is when Dopey Dog, having attached himself by means of a spring to a rocky outcropping, is dismayed to see the outcropping come apart from the ledge and fall. Before the break Dopey was in static equilibrium below the ledge. What will the behavior of the Dopey/spring/ledge system be during the fall? Treat the spring as massless, with a spring constant equal to 1000 N/m. Dopey's mass is 25 kg, the ledge's mass is 250 kg, and the unstretched spring is 1 m long.

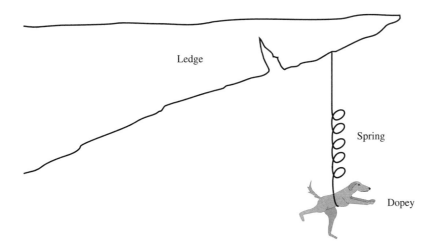

2. Referring to Problem 1, how would the answers change if we ignore the dynamic effect of Dopey on the ledge and treat the ledge as being infinitely massive?

3. Again consider the system of Problem 1. If the connection between the ledge and Dopey is viewed as an inextensible, massless string, what will the dynamic response of the system be after the ledge breaks? Relate this behavior to the case of a spring connection between Dopey and the ledge (Problem 1).

4. Find the natural frequencies and eigenvectors for the system shown in Figure 4.1. $m_1 = 10$ kg, $m_2 = 20$ kg, $k_1 = 100$ N/m, $k_2 = 100$ N/m, and $k_3 = 50$ N/m.

5. The system illustrated here is a double pendulum. It can also be looked at as a model of a two-link robotic manipulator. Find the equations of motion about the system's stable equilibrium position. Once you've found them, linearize the equations by assuming that all the angular deflections are small. Then calculate the system's natural frequencies and eigenvectors for the given parameter values. $m_1 = 1$ kg, $m_2 = 2$ kg, $l_1 = 1$ m, $l_2 = 1.5$ m, and $g = 9.81$ m/s$^2$.

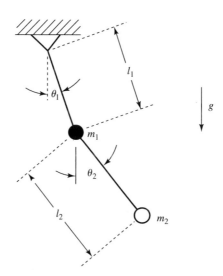

6. A nonuniform beam has a length $l$, mass $m$, and a moment of inertia about $x = \frac{l}{2}$ equal to $\frac{ml^2}{4}$. The bar is supported at each end by a spring with spring constant equal to $k$. Analyze the natural frequencies and eigenvectors equal to this system and comment on why they are unusual. What do the system's modes look like?

7. Derive the linearized equation of motion for the 2-DOF system shown. Find the natural frequencies and eigenvectors. $m_1 = 20$ kg, $m_2 = 1$ kg, $k_1 = 10$ N/m, and $l = 2$ m.

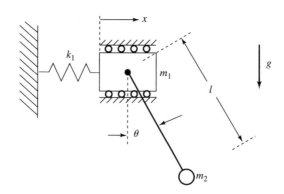

**8.** Consider the illustrated system. All the mass is concentrated in three places; the rest of the rigid bar is massless. Find the system's equations of motion. Determine the eigenvectors and natural frequencies. $m_1 = 1$ kg, $m_2 = 1$ kg, $m_3 = 1$ kg, $k = 2$ N/m, and $l = 1$ m.

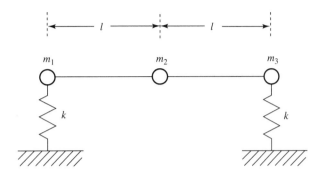

**9.** Analyze the system shown in Problem 8 for $m_1 = 2$ kg, $m_2 = 1$ kg, and $m_3 = 1$ kg and comment on how (and why) the natural frequencies and eigenvectors differ from those of Problem 8. $k = 2$ N/m and $l = 1$ m.

**10.** For a given $n$-DOF system

$$[M]\ddot{X} + [K]X = 0$$

is it possible to have $[K]X_i = \{0\}$? ($X_i$ is the $i$th eigenvector of the system.) If so, under what circumstances? What would this imply about $[K]$?

**11.** You're told that the three eigenvectors for a three-mass vibratory system, corresponding to three distinct natural frequencies, are

$$X_1 = \begin{Bmatrix} 2 \\ 1 \\ 1 \end{Bmatrix}, \quad X_2 = \begin{Bmatrix} 1 \\ 0 \\ -1 \end{Bmatrix}, \quad X_3 = \begin{Bmatrix} 3 \\ 1 \\ 0 \end{Bmatrix}$$

Is this believable? Why or why not?

**12.** For the system shown in Figure 4.7, determine which modal deflections make physical sense and which don't. Explain how you came to your conclusions.

$$X_1 = \begin{Bmatrix} 0 \\ 1 \\ 1 \\ 0 \end{Bmatrix}, \quad X_2 = \begin{Bmatrix} 1 \\ 1 \\ 1 \\ 1 \end{Bmatrix}, \quad X_3 = \begin{Bmatrix} 2 \\ 1 \\ -1 \\ -2 \end{Bmatrix}$$

**13.** Can you use the fact that

$$[K]^{-1}[M] = \begin{bmatrix} 4 & 3 & 6 \\ 1 & 3 & 1 \\ 1 & 1 & 3 \end{bmatrix}$$

to determine if $X = \{3 \ -4 \ 1\}^T$ is an eigenvector of the illustrated system?

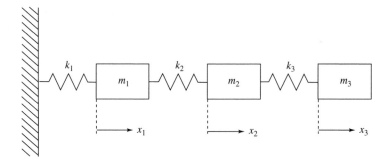

14. Vitally important data with regard to the stiffness and mass of a TOP SECRET device have been lost. The inventor blew himself, and much of his notes, up in the latest experiment. All that was recovered from his journal is the illustrated fragment. From this information, can you deduce the rest of the stiffness matrix? Assume that $[K]$ is symmetric.

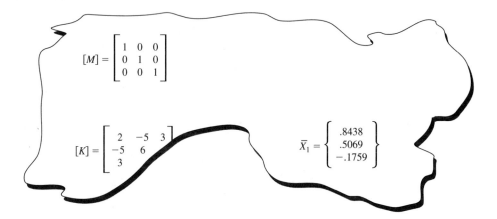

$$[M] = \begin{bmatrix} 1 & 0 & 0 \\ 0 & 1 & 0 \\ 0 & 0 & 1 \end{bmatrix}$$

$$[K] = \begin{bmatrix} 2 & -5 & 3 \\ -5 & 6 & \\ 3 & & \end{bmatrix} \qquad \bar{X}_1 = \begin{Bmatrix} .8438 \\ .5069 \\ -.1759 \end{Bmatrix}$$

15. Solve the following eigenvalue problem and write down the general solution for arbitrary initial conditions.

$$[M]\ddot{X} + [K]X = 0$$

$$[M] = \begin{bmatrix} 1 & 0 \\ 0 & 1 \end{bmatrix}; \quad [K] = \begin{bmatrix} 2 & -2 \\ -1 & 3 \end{bmatrix}; \quad X = \begin{Bmatrix} x_1 \\ x_2 \end{Bmatrix}$$

16. Problem 39 in Chapter 1 looked at the motion of a constrained particle in a tube. Now we're going to go beyond this model by adding a second degree of freedom. The figure here shows the system we'll be considering. If the springs are identical (as shown) then one possible use of this model would be to approximate the first mode of vibration for a water tower of the sort shown. If the spring constant of each spring is equal to $\frac{k}{2}$ and the mass is given by $m$, determine the response of the system to an initial displacement. Is the motion stable? What will the free vibrations look like?

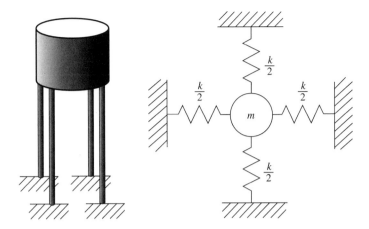

**17.** Consider the system of Problem 16. This time, the spring stiffnesses are not identical, as shown below. Now what will the response be to an initial displacement? Is the motion stable? What do the motions look like?

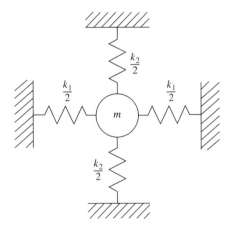

**18.** In this problem, we'll be moving a bit beyond where we were in Problem 16. We'll be considering the same system, but this time we'll let the entire system rotate at an angular rate equal to $\Omega$, as shown. We'll require that $k_1 = k_2$. This system will demonstrate the fundamental response characteristics that we'd see if we examined a rotating disk, such as a computer hard disk. It can also be used as a model for rotating vibration problems, such as those found in pumps, etc. Once again, we should determine the response of the system to initial displacements and/or velocities. Comment on the motion's stability and compare the results to those of Problem 16.

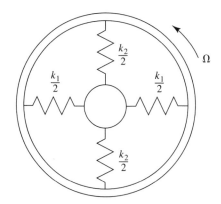

19. The system of Problem 18 is now further refined by dropping the requirement that $k_1 = k_2$. This is the most general problem of all those we've looked at so far. If $k_y$ goes to infinity, we've got Problem 39 in Chapter 1. If $k_1 = k_2$ then we've got Problem 18 of this chapter. This model is applicable to spinning disk problems, such as computer hard drives. It is also a simplified model for a helicopter on its landing gear. The fact that $k_1$ isn't necessarily equal to $k_2$ reflects the different stiffnesses a helicopter would experience in pitch versus roll motions. It could also model turbomachinery, in which the bearing support stiffness isn't identical in all directions. Just as in Problem 18, determine the system's response to different initial conditions. Comment on the stability of the response and compare the results to those of Problems 39 in Chapter 1 and Problem 18 of this chapter (if they've been assigned).

20. In the illustrated system the three rods have rotational stiffness ($k_{\theta_1}$, $k_{\theta_2}$, and $k_{\theta_3}$), but should be treated as massless (zero rotational inertia). Given the values for $k_{\theta_i}$ and $I_i$, find the system's natural frequencies and associated eigenvectors. $k_{\theta_1} = 4.0 \times 10^5$ N·m, $k_{\theta_2} = 10.0 \times 10^5$ N·m, $k_{\theta_3} = 5.0 \times 10^5$ N·m, $I_1 = 10$ kg·m$^2$, and $I_2 = 20$ kg · m$^2$.

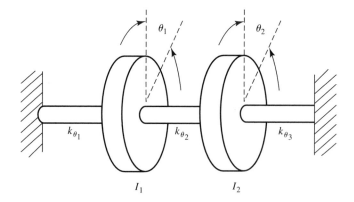

21. Consider the system of Problem 20. Determine how the natural frequencies and eigenvectors change as $I_2$ is varied from 0 to 60 kg·m$^2$. Plot the variation of the natural frequencies as a function of $I_2$. What do the eigenvectors look like as $I_2$ approaches 0 and $\infty$?

22. Consider the system of Problem 20. Determine how the natural frequencies and eigenvectors change as $k_{\theta_2}$ is varied from 0 to $10 \times 10^5$ N·m. Plot the variation of the natural frequencies as a function of $k_{\theta_2}$. What do the eigenvectors look like as $k_{\theta_2}$ approaches $\infty$ and 0?

23. Consider the system of Problem 20. Determine the eigenvectors and natural frequencies if $k_{\theta_1} = 0$.

24. The following scrap of information was found in an inventor's wastebasket. What can you deduce about the $[M]$ and $[K]$ matrix and the eigenvectors and natural frequencies of the problem she was working on?

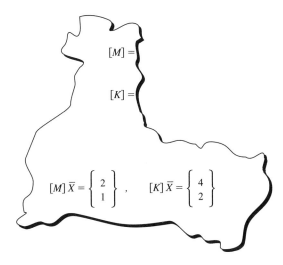

$$[M] = $$

$$[K] = $$

$$[M]\,\bar{X} = \begin{Bmatrix} 2 \\ 1 \end{Bmatrix}, \quad [K]\,\bar{X} = \begin{Bmatrix} 4 \\ 2 \end{Bmatrix}$$

25. Consider the system shown in Problem 20. Let $k_{\theta_3} = 0$. Determine the equations of motion. The rotational inertias are $I_1$ and $I_2$ and the angular stiffnesses equal $k_{\theta_1}$ and $k_{\theta_2}$. What are the system natural frequencies? $k_{\theta_1} = 4000$ N · m, $k_{\theta_2} = 6000$ N · m, $I_1 = 1$ kg · m$^2$, and $I_2 = 2$ kg · m$^2$.

26. Determine the natural frequencies and eigenvectors of the following system: $a = b = .5$ m, $m_1 = m_2 = 1$ kg, $k = 1$ N/m, and $g = 5$ m/s$^2$. (Obviously the system isn't on Earth.)

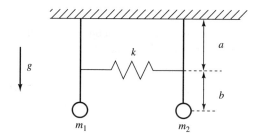

**27.** Find the natural frequencies and the associated eigenvectors for

$$\begin{bmatrix} 2 & -1 \\ -1 & 3 \end{bmatrix} \begin{Bmatrix} \ddot{x}_1 \\ \ddot{x}_2 \end{Bmatrix} + \begin{bmatrix} 4 & -1 \\ -1 & 4 \end{bmatrix} \begin{Bmatrix} x_1 \\ x_2 \end{Bmatrix} = \begin{Bmatrix} 0 \\ 0 \end{Bmatrix}$$

**28.** Illustrated are three snapshots of the vibrational response of a 2-DOF system undergoing free vibration. One possible explanation of the snapshots is that we're seeing the response corresponding to a particular eigenvector, $X_1 = \begin{Bmatrix} 1 \\ 2 \end{Bmatrix}$, which has a natural frequency of $\pi$ rad/s. What is another explanation?

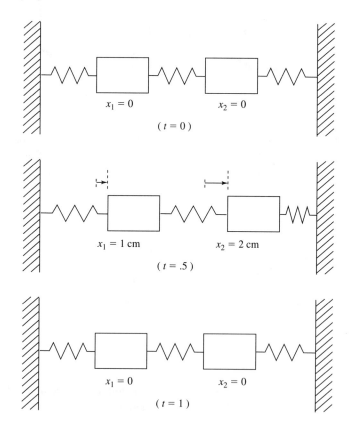

**29.** Consider the system shown in Figure 4.38. Calculate the kinetic and potential energies using the $x$ and $\theta$ coordinates and then use Lagrange's equations to obtain equations of motion. Comment on the couplings in the energies and the equations of motion. Next, determine the kinetic and potential energies using $x_1, x_2$ coordinates and use Lagrange's equations to find the equations of motion. Again, comment on the couplings in both the energies and the equations of motion. Now take the equations of motion found for the $x$ and $\theta$ coordinates and transform them into equations involving $x_1$ and $x_2$ by using the substitutions $x_1 + x_2 = 2x$ and $x_2 - x_1 = 2l\theta$.

 Are the resulting mass and stiffness matrices symmetric? Do the equations look like those you found using Lagrange's equations? Explain how you can transform from this set of equations into the symmetric form you found from Lagrange's equations.

**30.** Consider the system shown below. For one of the system eigenvectors, $m_2$ is completely stationary. Determine for which natural frequency this response occurs and explain physically why it happens. $m_1 = 10$ kg, $m_2 = 20$ kg, $m_3 = 1$ kg, $k_1 = 10$ N/m, $k_2 = 20$ N/m, $k_3 = 2$ N/m, and $k_4 = 1$ N/m.

**31.** Find the linearized equations of motion for the illustrated system (small $\theta$, only linear terms retained). Calculate the natural frequencies and eigenvectors for the given parameter values. ($m_1$ is a uniform bar hinged off $m_2$ and of length $l$.) $m_1 = 2$ kg, $l = .5$ m, $m_2 = 3$ kg, $k_1 = 100$ N/m, $k_2 = 200$ N/m, and $a = .3$ m.

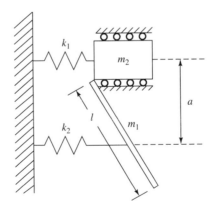

**32.** Determine whether or not

$$X_i = \begin{Bmatrix} 1 \\ 1 \\ 1 \\ 1 \\ 1 \end{Bmatrix}$$

is an eigenvector for the system having the following $[M]$ and $[K]$ matrices:

$$[M] = \begin{bmatrix} 1 & 0 & 0 & 0 \\ 0 & 2 & 0 & 0 \\ 0 & 0 & 3 & 0 \\ 0 & 0 & 0 & 1 \end{bmatrix}; \quad [K] = \begin{bmatrix} 3 & -1 & -1 & -1 \\ -1 & 3 & -2 & 0 \\ -1 & -2 & 4 & -1 \\ -1 & 0 & -1 & 2 \end{bmatrix}$$

*without* solving the eigenvalue problem

$$| [K] - \omega^2 [M] | = 0$$

## Section 4.3

**33.** To analyze the motions of an automobile on its tires, we approximate a car by a bar of uniform mass (1000 kg) with a moment of inertia equal to $1333.\overline{3}$ kg · m². The bar is 4 m long. The tire/suspension is approximated by a pair of springs, with spring constant equal to 11,697 N/m² for each one. If the road profile is sinusoidal, with an amplitude of 20 cm and a wavelength of 10 m, at what speed will the car experience resonance? With what mode is this resonance associated?

**34.** How does the answer for Problem 33 alter if the wavelength of the road is shortened to 8 m?

**35.** For what road wavelength will the car in Problem 33 experience only purely translational motion (no rotation)?

**36.** Consider the system shown. This system is more involved than that of Problem 33 because the front and rear springs of the vehicle model differ from each other. Find the equations of motion, determine the natural frequencies, and comment on whether it is possible to find a vehicle speed for which rotational motions are not present.

The car body is treated as a uniform bar with linear density equal to 250 kg/m. A lumped mass of 150 kg is located at the back end of the bar, approximating a rear-engine configuration. $k_1 = 15,000$ N/m, $k_2 = 10,000$ N/m, the road profile's amplitude is 2 cm, and the road profile's wavelength is 8 m.

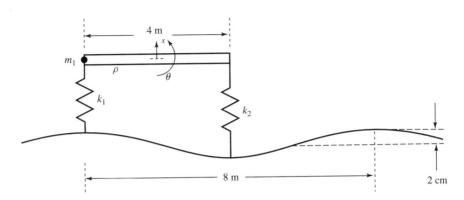

**37.** From the following data, can you deduce the number of degrees of freedom of the tested system and the type of sensor used to acquire the data? The mass was direct forced.

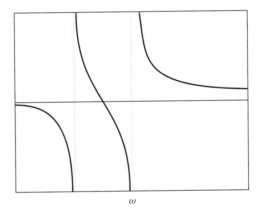

**38.** Could the following be a physically correct transfer function for the displacement response for the system shown in Figure 4.4? Why?

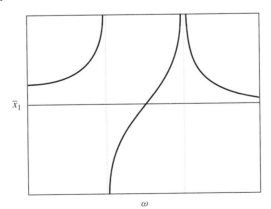

**39.** Explain in detail how you would solve for the steady-state response of Figure 4.4 for the illustrated force inputs. Write down the equations of motion and indicate the solution procedure you would follow. Could you use numerical integration to verify your solution? (Don't actually determine the Fourier coefficients.)

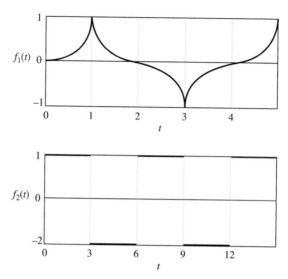

**40.** Some structures can be analyzed as rigid bodies with flexible ground constraints when subjected to loadings due to earthquakes. The figure below illustrates such a model. Assume the horizontal motion of the ground is give by $x = \bar{x} \sin(\omega t)$ and derive the equations governing the structure's response. Treat the structure as a uniform bar and neglect gravity.

**41.** During the filming of a movie, a bus was launched through the air, supposedly clearing a gap in a highway overpass. In order to protect the stunt driver, bungie cords were employed to restrain his body and isolate him from the shock of the impact. This situation can be modeled as shown, where $m_1$ is the stunt driver, $k_1$ is the equivalent spring constant of all the restraining bungie cords, $m_2$ is the mass of the bus, and $k_2$ is the bus's suspension stiffness. If $m_1 = 75$ kg, $m_2 = 10,000$ kg, $k_1 = 73,500$ N/m, and $k_2 = 3.3 \times 10^6$ N/m, determine the natural frequencies and associated eigenvectors for the system. Next, determine the relative motion of $m_1$ with respect to $m_2$, if the bus is dropped from a height of 2 m.

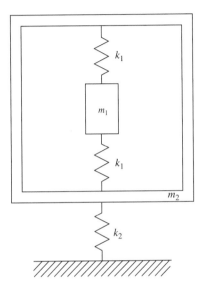

**42.** Find the amplitude of $\theta$ as a function of the forcing amplitude $\bar{f}$ and $\omega$ for the illustrated system. Also, determine an appropriate $m$ and $k_2$ combination so that when $\omega = 10$ rad/s, the amplitude of $\theta = 0$. $f = \bar{f}\cos(\omega t)$.

**43.** Consider the system shown in Figure 4.4. $f_1 = 0$ and $f_2 = \bar{f} \cos(\omega t)$. Which of the illustrated transfer functions correspond to $\bar{x}_1/\bar{f}$ and which to $\bar{x}_2/\bar{f}$. Why?

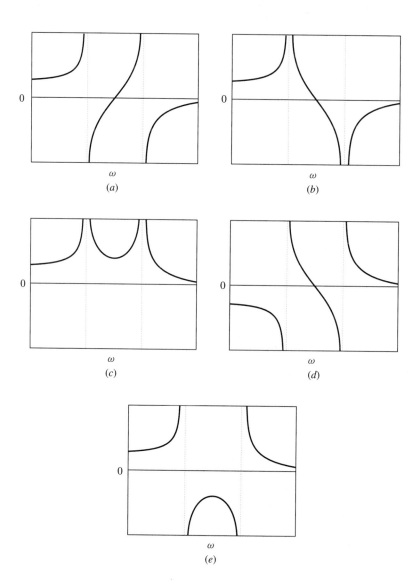

(a)

(b)

(c)

(d)

(e)

**44.** Consider the system of Figure 4.28. Let $n = 3$, $k_1 = 10$ N/m, $k_2 = 20$ N/m, $k_3 = 30$ N/m, and $k_4$ isn't present. $m_1 = 4$ kg, $m_2 = 2$ kg, and $m_3 = 5$ kg, $f_1$ isn't present, $f_2 = 10\cos(1.2t)$, and $f_3 = 30\sin(5.45t)$ (units in newtons). Solve for $x_1(t)$ and plot $x_1(t)$ versus $t$ for 10 s.

## Section 4.4

**45.** Consider the system shown in Figure 4.4. Let $k_3 = 0$, $f_1 = 0$, and $f_2 = \bar{f} \cos(\omega t)$. Can you come up with appropriate values for $k_1$ so that $x_2$ is stationary at $\omega = 60$ rad/s? $m_1 = 20$ kg, $m_2 = 10$ kg, $k_2 = 10{,}000$ N/m.

**46.** Consider the system of Problem 45. Is it possible to come up with appropriate values for $m_1$ and $k_1$ so that $m_1$ is stationary at $\omega = 100$ rads?

**47.** Consider the system shown in Figure 4.11*b*. Determine the smallest secondary mass that you could use if your design constraint is that the magnitude of $x_2$'s motion must be no more than 10 cm. $m_1 = 1000$ kg, $k_1 = 36{,}000$ N/m, the magnitude of the forcing is .7 N, and the forcing frequency is 6 rad/s.

**48.** Consider Problem 47. What we'll now assume is that the forcing frequency increases by only one-half of 1 percent. What effect will this have on the response of both $x_1$ and $x_2$?

**49.** In this problem we'll see what happens for a particular vibration absorber of the type shown in Figure 4.11*b* in two different situations. $m_1 = 10$ kg and $k_1 = 4000$ N/m. We'll consider a vibration absorber that's just as massive as the primary mass, giving us $m_2 = 10$ kg and $k_1 = 4000$ N/m. Determine the amplitude of the primary and secondary masses when $\omega = 20$ rad/s and the forcing magnitude is equal to 10 N. Then remove $k_2$ and simply join $m_2$ and $m_1$. What is the resulting magnitude of oscillation? For this case, was there any benefit to using the spring $k_2$ if the packaging goal was to minimize the overall vibration amplitude of the system?

**50.** You are given the new washing machine to analyze. The mass $m_1$ represents the mass of the washing machine, $m_0$ represents the largest bunch of clothes likely to be found in the washer, $r$ represents the radius of the washer's drum, and $k_1$ represents the spring stiffness of the washer's supporting feet. The spin cycle operates at 14 rad/s. Design a spring-mass vibration absorber to be added to the washer so that the amplitude of vibration of the system is less than 3 mm over a range of 4 rad/s centered at the operating frequency. $r = .05$ m, $m_0 = 1$ kg, $m_1 = 9$ kg, and $k_1 = 2250$ N/m.

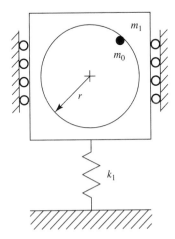

**51.** The figure here shows a transfer function that corresponds to the system of Figure 4.4. What variables might this transfer function be measuring? For instance, it might be the transfer function between $f_1$ and $x_2$ or between $f_2$ and $\dot{x}_2$, etc. Justify your answers.

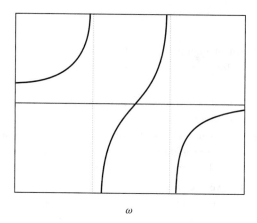

$\omega$

## Section 4.6

**52.** Consider the system of Figure 4.7, but with the force applied to the third mass and not the first. For what $\omega$'s will $x_2$ be equal to zero?

**53.** For what frequencies will $x_3 = 0$ for the illustrated system?

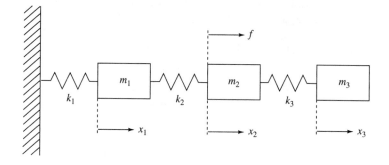

**54.** Is there a finite frequency $\omega$ for which both $m_2$ and $m_4$ are stationary for the illustrated system? $\bar{f} \neq 0$.

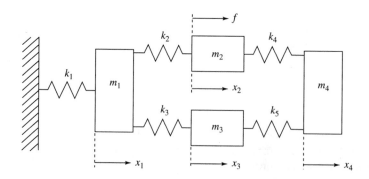

**55.** In an *n*-DOF–lumped system, what is the maximum number of frequencies at which a mass can be forced for which the response of that or some other mass of the system is zero? What is the minimum number? Explain the differences between the two cases.

**56.** What restrictions are there on the values of the $m_1$ and $k_i$ if we wish $m_1$ to remain stationary when a force $f = \bar{f} \cos(30t)$ is applied to $m_4$?

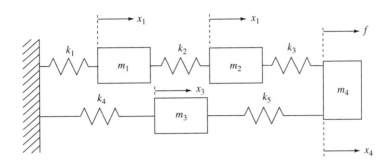

**57.** We're given a forced, spring-supported bar. A forcing frequency is needed that will cause $x$ (the vertical displacement of the bar's center of mass) to be zero at the same time that $\theta$ (the bar's rotation) is nonzero, that is, we'll see only pure rotation about the system's center of mass. (Don't consider any initial condition induced vibrations; look only at the forced solution.) You will discover that such a solution, although seemingly easy to find, isn't attainable. Why? $m = 4$ kg, $f = \bar{f} \cos(\omega t)$, $\bar{I} = 1$ kg $\cdot$ m$^2$, $k_1 = 500$ N/m, $k_2 = 700$ N/m, and $l = 1$ m.

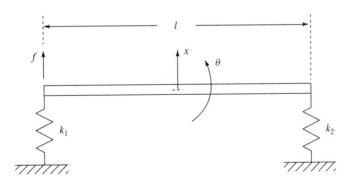

**58.** In Problem 57 we tried to find a forcing frequency for which $x = 0$ and $\theta \neq 0$ and weren't able to do so. Modify the system so that the force doesn't act on the left end of the bar, but rather is applied .2 m to the left of the midpoint. Can a frequency be found for this problem that satisfies $x = 0$ and $\theta \neq 0$? How does this problem differ from Problem 56 (i.e., why can we achieve $x = 0$, $\theta \neq 0$ in this case, but not previously)? $m = 4$ kg, $f = \bar{f} \cos(\omega t)$, $\bar{I} = 1$ kg $\cdot$ m$^2$, $k_1 = 500$ N/m, and $k_2 = 700$ N/m.

## Sections 4.7 and 4.8

**59.** When will the eigenvectors of an $n$-DOF system be geometrically orthogonal to one another?

**60.** Consider the system shown in Figure 4.4. Let $m_1 = 1$ kg, $m_2 = 5$ kg, $k_1 = 100$ N/m, $k_2 = 200$ N/m, and $k_3 = 50$ N/m. $f_1(t) = 1.8318\cos(10t)$ and $f_2(t) = -1.3145\cos(10t)$. (Forces given in newtons.) Put the system into normal form and show that under the given forcing the system responds in the second mode (eigenvector associated with the highest natural frequency).

**61.** Demonstrate that the system modes are orthogonal with respect to the $[M]$ and $[K]$ matrices for the following system:

$$\begin{bmatrix} 1 & 0 & 0 \\ 0 & 2 & 0 \\ 0 & 0 & 4 \end{bmatrix}\begin{Bmatrix} \ddot{x}_1 \\ \ddot{x}_2 \\ \ddot{x}_3 \end{Bmatrix} + \begin{bmatrix} 50 & -5 & 0 \\ -5 & 40 & -2 \\ 0 & -2 & 10 \end{bmatrix}\begin{Bmatrix} x_1 \\ x_2 \\ x_3 \end{Bmatrix} = \begin{Bmatrix} 0 \\ 0 \\ 0 \end{Bmatrix}$$

## Section 4.9

**62.** Decouple the following equations of motion:

$$\begin{bmatrix} 1 & 0 & 0 \\ 0 & 2 & 0 \\ 0 & 0 & 4 \end{bmatrix}\begin{Bmatrix} \ddot{x}_1 \\ \ddot{x}_2 \\ \ddot{x}_3 \end{Bmatrix} + \begin{bmatrix} 100 & -15 & 5 \\ -15 & 20 & -8 \\ 5 & -8 & 10 \end{bmatrix}\begin{Bmatrix} x_1 \\ x_2 \\ x_3 \end{Bmatrix} = \begin{Bmatrix} 1 \\ 0 \\ -1 \end{Bmatrix}$$

**63.** Put the following system into decoupled form:

$$\begin{bmatrix} 10 & 0 & 8 \\ 0 & 12 & 0 \\ 8 & 0 & 34 \end{bmatrix}\begin{Bmatrix} \ddot{x}_1 \\ \ddot{x}_2 \\ \ddot{x}_3 \end{Bmatrix} + \begin{bmatrix} 10 & 0 & 0 \\ 0 & 200 & 0 \\ 0 & 0 & 16 \end{bmatrix}\begin{Bmatrix} x_1 \\ x_2 \\ x_3 \end{Bmatrix} = \begin{Bmatrix} 1 \\ 0 \\ 0 \end{Bmatrix}$$

**64.** Put the following system into normal form (use a mass normalized modal matrix):

$$\begin{bmatrix} 10 & 0 & 18 \\ 0 & 12 & 0 \\ 18 & 0 & 34 \end{bmatrix}\begin{Bmatrix} \ddot{x}_1 \\ \ddot{x}_2 \\ \ddot{x}_3 \end{Bmatrix} + \begin{bmatrix} 100 & -40 & 0 \\ -40 & 200 & -100 \\ 0 & -100 & 250 \end{bmatrix}\begin{Bmatrix} x_1 \\ x_2 \\ x_3 \end{Bmatrix} = \begin{Bmatrix} -1 \\ 3 \\ 5 \end{Bmatrix}$$

**65.** Find the eigenmodes and natural frequencies for the following system. Comment on the results.

$$\begin{bmatrix} 2 & 0 & 0 \\ 0 & 3 & 0 \\ 0 & 0 & 1 \end{bmatrix}\begin{Bmatrix} \ddot{x}_1 \\ \ddot{x}_2 \\ \ddot{x}_3 \end{Bmatrix} + \begin{bmatrix} 1 & -1 & 0 \\ -1 & 2 & -1 \\ 0 & -1 & 1 \end{bmatrix}\begin{Bmatrix} x_1 \\ x_2 \\ x_3 \end{Bmatrix} = \begin{Bmatrix} 0 \\ 0 \\ 0 \end{Bmatrix}$$

**66.** Find the eigenvectors and natural frequencies for the following system. Form the modal matrix and use it to put the system into normal form.

$$\begin{bmatrix} 10 & -20 & 6 \\ -20 & 59 & -34 \\ 6 & -34 & 70 \end{bmatrix}\begin{Bmatrix} \ddot{x}_1 \\ \ddot{x}_2 \\ \ddot{x}_3 \end{Bmatrix} + \begin{bmatrix} 30 & -41 & 14 \\ -41 & 84 & -65 \\ 14 & -65 & 142 \end{bmatrix}\begin{Bmatrix} x_1 \\ x_2 \\ x_3 \end{Bmatrix} = \begin{Bmatrix} 0 \\ 0 \\ 0 \end{Bmatrix}$$

**67.** Given the equations of motion shown below, put the system into a mass normalized form.

$$2\ddot{x}_1 - \ddot{x}_2 + 4x_1 - x_2 = \sin(t) + \cos(3t)$$

$$3\ddot{x}_2 - \ddot{x}_1 + 5x_2 - x_1 = -2\cos(3t)$$

**68.** Given the system illustrated here, determine the magnitude of the first mode's response. Make sure to mass normalize the system modes.

$$\begin{bmatrix} 1 & 0 \\ 0 & 1 \end{bmatrix} \begin{Bmatrix} \ddot{x}_1 \\ \ddot{x}_2 \end{Bmatrix} + \begin{bmatrix} 2 & -1 \\ -1 & 2 \end{bmatrix} \begin{Bmatrix} x_1 \\ x_2 \end{Bmatrix} = \begin{Bmatrix} 2 \\ 3 \end{Bmatrix} \cos(2t)$$

**69.** What is the physical interpretation of normal coordinates?

## Sections 4.10 and 4.11

**70.** Given the following matrices:

$$[M] = \begin{bmatrix} 2 & 0 & 0 \\ 0 & 1 & 0 \\ 0 & 0 & 2 \end{bmatrix}, \quad [K] = \begin{bmatrix} 3 & -2 & 0 \\ -2 & 6 & -1 \\ 0 & -1 & 3 \end{bmatrix}$$

and the fact that the $\omega_i^2$ for the system are equal to 1.0, 1.5, and 6.5, determine the complex eigenvalues that result from adding linear damping to the system for which the damping matrix is such that $[C] = .1[M] + .2[K]$. Deduce the correct result *without* actually solving the eigenvalue problem for the damped system.

**71.** The first eigenvector for an unknown 3-DOF system is given below. The first three entries correspond to the displacement of the three masses and the last three correspond to their velocities. The undamped natural frequency for this mode is equal to 1. Given that the system damping is of the proportional type, determine the damping factor $\zeta$ for the illustrated eigenvector.

$$X_1 = \begin{Bmatrix} 1+i \\ 2+2i \\ 1+i \\ -1.4+.2i \\ -2.8+.4i \\ -1.4+.2i \end{Bmatrix}$$

**72.** Take Problem 30 and add proportional damping of the form

$$[C] = .05[M] + .1[K]$$

Find the eigenvectors and natural frequencies. Mass normalize the eigenvectors and phase shift them so that the first entry is real. Are the other displacement entries real also (as would be expected)? Construct the normal form equations using the mass normalized eigenvectors.

## Section 4.12

**73.** Determine the forced response of the the following system.

$$
\begin{bmatrix} 2 & 0 \\ 0 & 2 \end{bmatrix} \begin{Bmatrix} \ddot{x}_1 \\ \ddot{x}_2 \end{Bmatrix} + \begin{bmatrix} .4 & -.1 \\ -.1 & .2 \end{bmatrix} \begin{Bmatrix} \dot{x}_1 \\ \dot{x}_2 \end{Bmatrix} + \begin{bmatrix} 20 & -4 \\ -4 & 20 \end{bmatrix} \begin{Bmatrix} x_1 \\ x_2 \end{Bmatrix} = \begin{Bmatrix} 40 \\ 0 \end{Bmatrix} \sin(5t)
$$

**74.** Determine the forced response of the following system.

$$
\begin{bmatrix} 2 & 0 & 0 \\ 0 & 4 & 0 \\ 0 & 0 & 10 \end{bmatrix} \begin{Bmatrix} \ddot{x}_1 \\ \ddot{x}_2 \\ \ddot{x}_3 \end{Bmatrix} + \begin{bmatrix} .6 & -.1 & 0 \\ -.1 & .8 & -.2 \\ 0 & -.2 & 1 \end{bmatrix} \begin{Bmatrix} \dot{x}_1 \\ \dot{x}_2 \\ \dot{x}_3 \end{Bmatrix}
$$
$$
+ \begin{bmatrix} 70 & -40 & 0 \\ -40 & 90 & -10 \\ 0 & -10 & 140 \end{bmatrix} \begin{Bmatrix} x_1 \\ x_2 \\ x_3 \end{Bmatrix} = \begin{Bmatrix} 30 \\ 30 \\ 20 \end{Bmatrix} \sin(8t)
$$

**75.** Determine the forced response of the system shown.

$$
\begin{bmatrix} 12 & 0 & 0 \\ 0 & 40 & 0 \\ 0 & 0 & 1 \end{bmatrix} \begin{Bmatrix} \ddot{x}_1 \\ \ddot{x}_2 \\ \ddot{x}_3 \end{Bmatrix} + \begin{bmatrix} 2. & -.9 & 0 \\ -.9 & 1.8 & -.2 \\ 0 & -.2 & 1.6 \end{bmatrix} \begin{Bmatrix} \dot{x}_1 \\ \dot{x}_2 \\ \dot{x}_3 \end{Bmatrix}
$$
$$
+ \begin{bmatrix} 45 & -20 & 0 \\ -20 & 78 & -9 \\ 0 & -9 & 1125 \end{bmatrix} \begin{Bmatrix} x_1 \\ x_2 \\ x_3 \end{Bmatrix} = \begin{Bmatrix} 2000 \\ -1000 \\ 2000 \end{Bmatrix} \cos(20t)
$$

## Section 4.14

**76.** Take Problem 30, set $k_0 = 0$, and use MATLAB to calculate the system eigenvectors and natural frequencies. Do you obtain the expected rigid-body mode? Do the elastic frequencies go up or down from the case where $k_0 = 1000$ N/m? Do the frequency changes make sense?

**77.** We've seen that the $[K]$ and $[M]$ matrix can be nonsymmetric if the equations are derived from a Newtonian framework for the appropriate set of coordinates. Realizing this, is the illustrated $[K]$ matrix possible for the system illustrated below? Vertical translation and rotation of the rigid bar, as well as vertical translation of $m_1$, is allowed.

$$
K = \begin{bmatrix} 2 & -1 & 3 \\ 4 & 7 & -2 \\ 2 & -1 & 3 \end{bmatrix}
$$

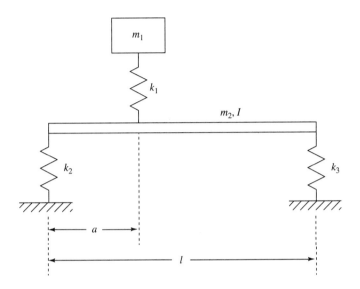

78. Is $X_i = \{1\ 2\ 2\ 1\}^T$, $\omega_i = 0$, a possible eigenvector natural-frequency pair for the system of Figure 4.7 if $k_1 = k_5 = 0$? $m_1 = 10$ kg, $m_2 = 5$ kg, $m_3 = 7$ kg, $m_4 = 9$ kg, $k_1 = 100$ N/m, $k_2 = 200$ N/m, and $k_3 = 400$ N/m.

79. Consider the system of Figure 4.1. Let $k_1 = k_3 = 0$, $k_2 = k$, and $m_1 = m_2 = m$. $k = 2$ N/m and $m = 1$ kg. Find the equations of motion and determine the natural frequencies and eigenvectors. Can you put this system into normal form?

80. Determine the natural frequencies and eigenvectors associated with the system shown below. $m = 1$ kg, $k = 2$ N/m, $l = 1$ m, $\bar{I} = 1$ kg $\cdot$ m$^2$. Vertical displacements and rotations are allowed.

## Section 4.15

81. Consider the system shown below. Determine the displacement of masses one and two for $f_1 = 1$ N, $f_2 = 3$ N, and $f_3 = -1$ N. $k_1 = 1$ N/m, $k_2 = 2$ N/m, $k_3 = 4$ N/m, $k_4 = 1$ N/m, and $k_4 = 3$ N/m.

**82.** Determine the displacement of the three masses for the system shown in Problem 81 if $f_1 = .1$ N and all other forces are zero.

**83.** What forces must be applied to $x_1$, $x_2$, and $x_3$ in order to induce displacements of $x_1 = .4$ cm, $x_2 = -.2$ cm, and $x_3 = 0$ for the system shown below? $k_1 = 12,000$ N/m, $k_2 = 9000$ N/m, $k_3 = 1000$ N/m, and $k_4 = 5000$ N/m.

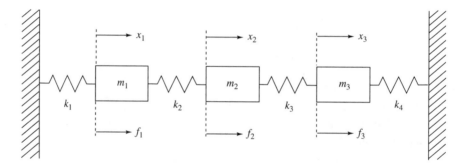

**84.** What forces must be applied to $x_1$, $x_2$, and $x_3$ in order to induce displacements of $x_1 = 1$ cm, $x_2 = 2$ cm, and $x_3 = 1$ cm for the system shown below? $k_1 = 2000$ N/m, $k_2 = 10,000$ N/m, $k_3 = 1500$ N/m, and $k_4 = 400$ N/m.

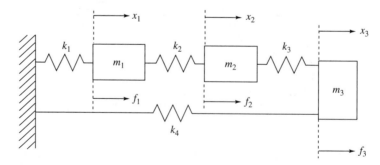

**85.** Consider the system of Problem 83. Let $k_4 = 0$. Determine what the relationship is between the forces applied at $m_1$ and $m_2$ if the goal is to keep $m_3$ from moving with the constraint that $f_3 = 0$.

# 5

## Distributed Systems

WHOOPS!

## 5.1 INTRODUCTION

Up to now the systems we've been examining have been composed of easily identified mass and spring elements. The masses and springs always appeared discretely, never in a distributed way. But in the real world, such idealized structures never exist. The fact of the matter is that real-world structures are *always* continuous. Nobody can construct a true point mass; all real bodies have finite dimensions and thus are characterized by their density or mass per unit volume. Similarly, ideal springs exist only in our minds. In the simple experiments we conducted in Chapters 1, 2, and 4 we treated the physical spring as a pure spring. But it is clear that any physical spring always has mass, a mass that we neglected in our analysis.

So, if we can never construct a system made up of just masses and springs, why did we study them? In the first place, although we can never have systems whose mass and spring elements occur in a precisely discrete manner, we can often approximate real systems as being discrete without losing much accuracy. In addition, the mathematics associated with discrete systems is a bit easier to deal with. And finally, we'll be introducing methods that explicitly include continuous responses, yet put the overall system in a discrete framework.

The point is that we may find ourselves wanting to analyze realistic systems having continuous distributions of mass, elasticity, and damping. However, except in some special cases, we cannot solve continuous problems exactly. Thus we will find ourselves forced to consider approximate solutions. We will see that the approximate solutions fall into two distinct classes: those that discretize the original system into a number of discrete mass and stiffness elements (surprise— back to lumped elements!) and those that approximate the system's response by a finite number of mode shapes. We have already seen that MDOF systems possess a finite number of modes, that is, free vibration configurations of the mass displacements that occur at the system's natural frequencies. In this chapter we'll see that continuous systems also have modes, but in this case they have an infinite number of them. It so happens that the modal approach (that of approximating the system's response through a finite number of mode shapes) is very widely used and forms the basis of modal analysis, the topic of Chapter 8.

Although there are several types of systems that can be solved exactly, we'll look at only two of them in this chapter and look at others in Appendix A (which the interested readers can peruse at their leisure). The underlying mathematical analysis is exactly the same for all these problems; hence, once you've solved one you can pretty much solve the rest, as well. The two examples to be presented in this chapter are longitudinal vibrations of a bar and transverse vibrations of a beam. The bar example will also be used later (in Chapter 6) to illustrate how we can approximate a continuous system by using a lumped one. It is interesting that the mathematical form of the bar equations is exactly the same as those that govern the vibrations of a tensioned string, which are identical to those that govern the torsional motions of a circular rod, which happen to be precisely the same as those governing the behavior of acoustic waves in a one-dimensional duct. Thus, once you've solved one of these problems, you've solved them all; the only thing that changes is the meaning of the dependent variable (transverse deflection of a string, rotation of the circular rod, etc.) and the meaning of the particular physical quantities (tension in the string, torsional stiffness of the rod, etc.).

The reason we'll also be looking at transverse motions of a beam is that this problem is very, very common in practical applications—almost all structures contain beams or beamlike elements. Since you'll be practicing engineering in the real world, it will be helpful to have some knowledge of beam behavior. It would be nice if the beam equations were also exactly the same as those of the string, rod, bar, etc., but unfortunately they're not. Luckily, the differences aren't too bad, as we'll soon see.

Before we get involved with some actual continuous systems, you should think about what the differences are between a vector and a function, since these are the relevant quantities for discrete and continuous problems, respectively. Let's consider position as the variable that we're measuring. In this case, a vector is simply a string of numbers. The position of the numbers within the vector tells us where on the structure we are and the numbers themselves tell us how much motion is taking place at a particular position.

Take, for example, the deflection of a string. A simple way to represent the string's position would be to use three pieces of data: the deflections on each end and the deflection in the middle.

**Figure 5.1**  Pictorial representation of a 3-element vector

A result for a string that's tied down on both its ends would be

$$\left\{ \begin{array}{c} 0 \\ 1 \\ 0 \end{array} \right\} \tag{5.1.1}$$

This vector tells us that the leftmost end's deflection (the first entry) is zero, the middle deflection is 1 (the second entry) and the right end's deflection (the third entry) is zero. This is illustrated in Figure 5.1, where the middle deflection is taken to be 1 cm and the string's length is taken to be 1 m.

Now we can ask ourselves what the difference would be if, instead of having only 3 positions along the string, we had 101. If these were again uniformly distributed, then the second entry in (5.1.1) would become the 51st entry in our new position vector. It would still be the middle entry, but now we'd have many additional entries. A plot of the new vector is shown in Figure 5.2. You'll note that we have a lot more data about what's going on at other points, but our knowledge regarding what's happening at $x = .5$ m is the same as it was. If we had an interest in the behavior at $x = .25$ m and were using only the 3 entry vector, we'd have had to estimate the answer by

**Figure 5.2**  Pictorial representation of a 101-element vector

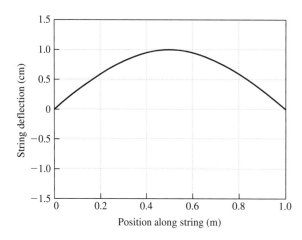

**Figure 5.3**  Pictorial representation of a 1001-element vector

using the first and second entries and averaging them, giving us an estimate of .5 cm. With the greater precision that the 101 entry vector gives us, we can simply read the result as the value associated with the 25th entry of the vector, .707 cm.

You'll notice that the plot of the vector shown in Figure 5.2 looks an awful lot like half a sine wave. If we used a 1001 entry vector instead of a 101 entry one, the plot would look like that shown in Figure 5.3. This is now virtually indistinguishable from half a sine. What's happening here? The answer is, our vector is "trying" to become a function. As we add more and more entries, it "looks" closer and closer to a continuous function. That's all a function really is! It's a very long vector. In fact, it's a vector in which the number of entries has gone to infinity. Pretty neat, eh? This means that, as far as our computers are concerned, there is no difference between vectors and functions since the computers can only deal with finite strings of information. Mathematically there's a difference (since mathematics has no problems with infinity), but practically we'll find that large vectors and functions are handled in very similar ways.

You should keep this little demonstration in mind as you encounter continuous systems in this and following chapters. All of the continuous operations can really be viewed as discrete operations in which all the operations go to their limits. For example, summation of vector entries will mutate into integrations as we go from discrete to continuous descriptions. The distance between two entries of a position vector will go to a differential distance $dx$ as the vector becomes bigger and bigger and finally becomes a function. If you have a problem following what's being done on the continuous problem, try thinking about it in a discrete form and see if that makes it clearer.

## 5.2 FREE VIBRATION OF A BAR (ROD, STRING, ETC.)

As the title says, this section looks at bar vibrations, among other applications. Appendix A shows the equation of motion for several systems, all of which are identical in form although different in the specific parameters used. We'll perform only a full derivation of the equations of motion for a bar, realizing that the approach for all the other applications is essentially the same.

**Figure 5.4**   Schematic of a bar

To begin studying bar vibrations, we must derive the equations of motion for our system. The bar is of length $l$, and we identify locations along the bar with the coordinate $x$ (Figure 5.4). Since the bar is deformable, individual elements will move from their initial positions if a force is applied. The amount of movement will be different at different points along the bar and so we should use $\xi(x, t)$ to indicate these deformations. However, this would quickly become tiring since we'll have to write a lot of $\xi(x, t)$'s in the coming pages. Thus we'll simply write $\xi$. Just keep in mind that this refers to a function that is both position and time dependent. Also, be clear that the bar's motion is horizontal in this case; it's motion is *along* the bar (stretching and compressing of the bar).

Please make sure the distinction between $x$ and $\xi$ is clear in your mind. $x$ indicates a position *on the bar*. If you were asked to put a pencil mark on the bar 5 cm from the left, then the pencil mark is at $x = 5$ cm. $\xi(x, t)$ indicates *how much that pencil mark moves* if the bar is vibrating. The $x$ and $t$ dependences indicate that the deformation depends on the location along the bar, as well as on time.

From your deformable bodies course, you will recall that the relation between an applied force, $f$, and the resultant deformation of the bar is

$$f = EA(x)\xi_x \tag{5.2.1}$$

Thus the spatial rate of change of deformation is proportional to the applied force.

If we separate a small element from the bar, as shown in Figure 5.5, and assume that the motion occurs purely in the longitudinal direction, then we can find the equation of motion by balancing the inertial and applied forces.

$$m(x)\ddot{\xi}dx = \left[EA(x) + (EA(x))_x dx\right]\left[\xi_x + \xi_{xx}dx\right] - EA(x)\xi_x \tag{5.2.2}$$

Note that $m(x)$ denotes the mass per unit length. Multiplying this equation out and canceling the $EA(x)\xi_x$ terms yields

$$m(x)\ddot{\xi}dx = \left(EA(x)_x\right)\xi_x dx + EA(x)\xi_{xx}dx + \left(EA(x)\right)_x\xi_{xx}(dx)^2 \tag{5.2.3}$$

We have a common factor of $dx$ and canceling it out leaves us with

$$m(x)\ddot{\xi} = \left(EA(x)_x\right)\xi_x + EA(x)\xi_{xx} + \left(EA(x)\right)_x\xi_{xx}dx \tag{5.2.4}$$

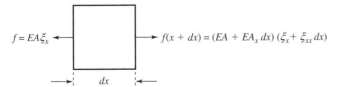

**Figure 5.5**   Differential element of a bar

Keep in mind that the analysis we are doing is an infinitesimal one, that is, the element we've taken out of the bar is only of infinitesimal length. If we take the limit of $dx$ (the length of the element) going to zero, then the $(EA(x))_x \xi_{xx} dx$ term will go to zero. Thus we can eliminate this term and are left with

$$m(x)\ddot{\xi} = EA(x)_x \xi_x + EA(x)\xi_{xx} \qquad (5.2.5)$$

which can be written more compactly as

$$m(x)\ddot{\xi} = \frac{\partial}{\partial x}(EA(x)\xi_x) \qquad (5.2.6)$$

To keep the analysis simple we will assume that $m(x)$ and $EA(x)$ are constant. Thus, even though in general they might vary along the beam, in our case they do not. The basic reason for using this simplifying assumption is that letting these terms vary along the beam moves this from the "simple" category and into the "just about impossible" category. Luckily, Chapter 6 shows how to deal with these more difficult cases. Note that the second term of (5.2.5) disappears under this assumption, since the the spatial variation has been set to zero.

Dropping the $x$ dependence from our notation and dividing by $m$ we have

$$\ddot{\xi} = \frac{EA}{m}\xi_{xx} \qquad (5.2.7)$$

At this point we need to invoke a new mathematical technique in order to get any farther. The technique we'll use is called the separation of variables method. You may have seen this approach in a math class and it may have seemed a bit obscure. This example should clear up any confusion.

Simply put, the method requires us to assume that $\xi(x, t)$ can be written in the form

$$\xi(x, t) = \mathbf{x}(x)\mathbf{t}(t) \qquad (5.2.8)$$

where $\mathbf{x}(x)$ and $\mathbf{t}(t)$ are as yet undetermined functions of position and time. Although this is often a very reasonable assumption, it should be stressed that it is only an assumption at this point. We'll have to carry on and see if we eventually get a reasonable answer. If we do, then we'll know the assumption was justified. As a quick demonstration that this assumption doesn't always have to be satisfied, just try to put $xt + x^2$ into this form.

Substituting (5.2.8) into (5.2.7) yields

$$\frac{EA}{m}\mathbf{x}_{xx}\mathbf{t} = \mathbf{x}\ddot{\mathbf{t}} \qquad (5.2.9)$$

If we divide by $\mathbf{xt}$, we'll obtain

$$\frac{EA}{m}\frac{\mathbf{x}_{xx}}{\mathbf{x}} = \frac{\ddot{\mathbf{t}}}{\mathbf{t}} \qquad (5.2.10)$$

From this result we see that the term on the left (which depends only on position) must be equal to the term on the right (which depends only on time). This is the point that is invariably the most confusing. Ask yourself how two quantities can be equal when each depends upon a different dependent variable. Clearly, this is a tough question. For instance, the term on the left might conceivably be equal to $\sin(2x)$ and the term on the right equal to $t^3$. But then they *can't* be equal to each other for arbitrarily varying values of $x$ or $t$. In fact, the only way for functions that depend on completely different variables to equal each other is if they don't vary at all, that is, if they're equal to a constant. Thus we need to require that

$$\frac{EA}{m}\frac{\mathbf{x}_{xx}}{\mathbf{x}} = c \tag{5.2.11}$$

and

$$\frac{\ddot{\mathbf{t}}}{\mathbf{t}} = c \tag{5.2.12}$$

Although we're calling the constant $c$, in a moment we're going to rename it $-\omega^2$. However, since it's not yet clear why we should do this, we'll hold off for a second.

Taking (5.2.12) first, we multiply by $\mathbf{t}$ and pull both terms to the same side of the equation,

$$\ddot{\mathbf{t}} - c\mathbf{t} = 0 \tag{5.2.13}$$

Now we can see that if we had used $-\omega^2$ instead of $c$, (5.2.13) would be in the form of an oscillator. Since we've very familiar with the oscillator equation, let's do this. Replace $c$ with $-\omega^2$ to obtain

$$\ddot{\mathbf{t}} + \omega^2\mathbf{t} = 0 \tag{5.2.14}$$

which, as we know, has solutions

$$\mathbf{t}(t) = a_1 \cos(\omega t) + a_2 \sin(\omega t) \tag{5.2.15}$$

What this tells us is that the bar undergoes sinusoidal motion at a frequency $\omega$. As far as we know right now, any $\omega$ is as good as any other; there's nothing to indicate that the value of $\omega$ isn't completely arbitrary. However, as we'll see next, the actual choices for $\omega$ are quite restricted and are closely tied to the physical configuration of the system.

Since there's nothing more to be learned from (5.2.12), we'll move on to (5.2.11). Once we've substituted $-\omega^2$ for $c$, we see that multiplying (5.2.11) by $\frac{m}{EA}\mathbf{x}$ and pulling all terms to the left-hand side leaves us with

$$\mathbf{x}_{xx} + \frac{m\omega^2}{EA}\mathbf{x} = 0 \tag{5.2.16}$$

which is also in the form of an oscillator equation. The only difference from the previous example is that the dependent variable is now $x$, not $t$. Just to keep the notation simple, let's define a new variable, $\beta$,

$$\beta \equiv \omega\sqrt{\frac{m}{EA}} \tag{5.2.17}$$

so that (5.2.16) becomes

$$\ddot{\mathbf{x}}_{xx} + \beta^2\mathbf{x} = 0 \tag{5.2.18}$$

**Figure 5.6**  Bar with restrained ends

with solutions

$$\mathbf{x}(x) = b_1 \cos(\beta x) + b_2 \sin(\beta x) \tag{5.2.19}$$

At this point we would seem to be stuck. We have to determine what $\beta$ is equal to (which will tell us $\omega$ from (5.2.17)), but we don't seem to have enough information to do this. The reason for this is that we can't find the precise form of the solution until we've fully pinned down the physics of the problem. And we haven't done that yet. Perhaps you've already noticed that no mention has been made of the boundary conditions at the bar ends. Is the rod fixed to a wall? Is it free to move about? Until we decide on what's happening at the bar ends, we'll be unable to progress further. And this makes sense, since it seems obvious that the bar's response should be different for different boundary conditions. So let's pick fixed-fixed conditions, the situation we'd encounter if the rod were built into rigid walls on both ends (Figure 5.6). In this case the ends can't move longitudinally, so

$$\xi(0, t) = 0 \tag{5.2.20}$$

and

$$\xi(l, t) = 0 \tag{5.2.21}$$

Since $\xi(x, t) = \mathbf{x}(x)\mathbf{t}(t)$, we see that these boundary conditions imply that

$$\mathbf{x}(0) = \mathbf{x}(l) = 0 \tag{5.2.22}$$

Evaluating $\mathbf{x}(x)$ at $x = 0$ leaves us with $\mathbf{x}(0) = b_1$, since the sine term goes to zero and the cosine to 1. Since our boundary condition states that $\mathbf{x}(0) = 0$, we see that $b_1$ must be equal to zero. Therefore we are left with

$$\mathbf{x}(x) = b_2 \sin(\beta x) \tag{5.2.23}$$

Evaluating this expression at $x = l$ and using the remaining boundary condition yields

$$\mathbf{x}(l) = b_2 \sin(\beta l) = 0 \tag{5.2.24}$$

There are two ways (5.2.24) can be satisfied. The first is to set $b_2$ to zero. The only problem is that this would make $\mathbf{x}$ equal to zero, and so $\xi(x, t)$ would equal zero. This is clearly the trivial solution—no motion at all. The other way in which (5.2.24) can be satisfied is if $\sin(\beta l) = 0$. If we look at a plot of $\sin(\beta l)$ versus $\beta l$ (Figure 5.7), we see that this condition occurs for an infinite number of $\beta$ values. If we identify these values as $\beta_1 l, \beta_2 l, \ldots$, we have

$$\beta_n = \frac{n\pi}{l}, \quad n = 1, 2, \ldots, \infty \tag{5.2.25}$$

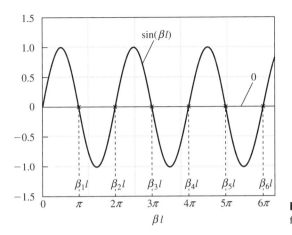

**Figure 5.7** $\beta_n l$ values for a fixed-fixed bar

Thus our spatial solution is

$$\mathbf{x}(x) = b_2 \sin(\beta_n x) \tag{5.2.26}$$

If we recall that $\beta = \omega \sqrt{\dfrac{m}{EA}}$, we see that we can now find the possible frequencies of oscillation for our system, that is,

$$\omega_n = \beta_n \sqrt{\dfrac{EA}{m}} \tag{5.2.27}$$

We have an infinite set of $\beta_n$ values and an associated infinite set of $\omega_n$'s. We'll call $\sin(\beta_n x)$ the modes of our system and $n$ the mode number.

Let's examine some of these solutions more closely. The first value of $\beta$ (represented as $\beta_1$) is equal to $\frac{\pi}{l}$. The mode shape associated with this $\beta$ value is

$$\mathbf{x}(x) = \sin\left(\dfrac{\pi x}{l}\right) \tag{5.2.28}$$

This mode is shown in Figure 5.8. Note that it indicates zero deflection at the bar ends and a maximal deflection in the middle of the bar. The overall time-varying behavior of this mode is given by

$$\xi(x, t) = a_1 \sin\left(\dfrac{\pi x}{l}\right) \cos(\omega_1 t + \phi_1) \tag{5.2.29}$$

where $\omega_1 = \sqrt{\dfrac{EA}{m}} \beta_1$. Note that we've included the constants $a_1$ and $\phi_1$. These are necessary in order to match the particular initial conditions of the problem.

Figure 5.9 shows what the bar deformations would be like over one oscillation cycle. The period of the oscillation is $T \left( T = \dfrac{2\pi}{\omega_1} \right)$. For simplicity, $a_1$ is taken to be 1.0 and $\phi_1$ is set to zero. It is clear from this plot that the deflected shape is always in the form of a half sine wave; only

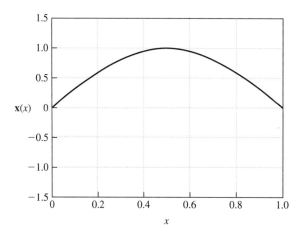

**Figure 5.8** First mode of a fixed-fixed bar

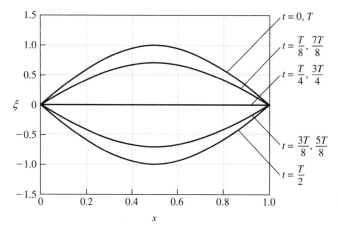

**Figure 5.9** Time behavior of fixed-fixed bar's first mode

the magnitude changes. As time progresses from zero, the deflected positions reduce in magnitude, going to zero at $t = \dfrac{\pi}{2\omega_1}$, reaching a maximum negative deflection at $t = \dfrac{\pi}{\omega_1}$, and returning to the original deflections at $t = \dfrac{2\pi}{\omega_1}$.

    This is an exact analogue of the discrete eigenvector behavior first seen in Chapter 4. If the bar is deflected into the first mode and then released, it will continue to oscillate with the same continuous modal deflection, and at a frequency equal to $\omega_1$, just as discrete systems can oscillate in particular modal vectors. The difference between this continuous system and those of Chapter 4 is that in this case we have an infinite number of mode shapes. This makes sense. A continuous system, such as the bar we're looking at, can be thought of as a collection of an infinite number of infinitesimal mass and spring elements. Since the number of mode shapes depends upon the number of masses in a discrete system, we should expect an infinite number of modes when we've got an infinite number of masses. In fact, part of the motivation behind the approximate methods (discussed in Chapter 6) is to try to approximate the infinite number of mass elements of a continuous structure by a finite number of masses.

The remaining mode shapes for our bar are simply sines with more and more included segments (Figure 5.10). As the mode shape becomes more "wiggly," the frequency of oscillation also rises. This is a characteristic of vibrating systems, more wiggles in the mode shape implies a higher frequency of oscillation.

Now that all our mode shapes are pinned down, we can determine the system's response to a set of initial conditions. The general solution is given by a summation of all the possible modal responses:

$$\xi(x, t) = \sum_{n=1}^{\infty} a_n \sin\left(\frac{n\pi x}{l}\right) \cos(\omega_n t + \phi_n) \tag{5.2.30}$$

The particular values of the $a_n$ and $\phi_n$ will depend on the particular initial conditions. In general, we'll have some initial displacement, $f(x)$, and velocity, $g(x)$, for all points of the bar. Thus, evaluating (5.2.30) at $t = 0$ will give us the two equations

$$f(x) = \sum_{n=1}^{\infty} a_n \sin\left(\frac{n\pi x}{l}\right) \cos(\phi_n) \tag{5.2.31}$$

and

$$g(x) = -\sum_{n=1}^{\infty} a_n \omega_n \sin\left(\frac{n\pi x}{l}\right) \sin(\phi_n) \tag{5.2.32}$$

We'll explore further the meaning of these summations by looking at a particular set of initial conditions. The most immediately understandable initial conditions are those in which the system under consideration has been displaced into some initial configuration and then released. In this case, all the $\phi_n$'s will be equal to zero. This can be seen from an examination of (5.2.32). If all the velocity of all points along the bar are zero at $t = 0$, then having all the $\phi_n$'s equal to zero makes $g(x)$ equal zero since all the individual terms in the summation will be zero. We're then left with

$$f(x) = \sum_{n=1}^{\infty} a_n \sin\left(\frac{n\pi x}{l}\right) \tag{5.2.33}$$

Note that the $\cos(\phi_n)$ terms are equal to 1.0 since all the $\phi_n$ are zero.

If we look at (5.2.33) we see that it is *exactly* in the form of a Fourier series! The $\sin(n\pi x)$ terms are precisely the harmonics we would use if we were trying to construct a Fourier series to approximate a physical displacement that had to start at zero (at $x = 0$) and end at zero (at $x = l$). The $a_n$ are the unknown Fourier coefficients. And, since we now know how to deal with Fourier series problems, we are fully capable of determining the $a_n$ for this problem.

Although this example was a bit easier to study than the most general problem, there are no real difficulties with the fully general problem (for which $f(x)$ and $g(x)$ are both nonzero) other than an increased work load.

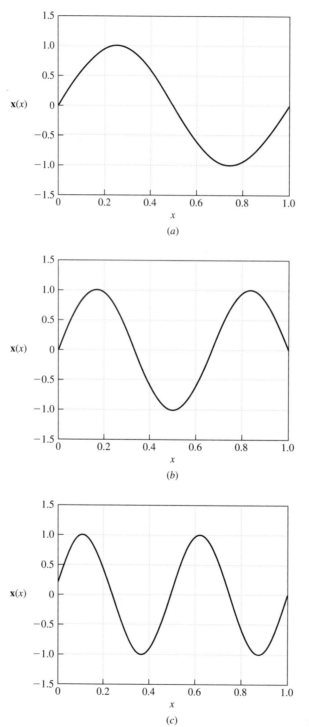

**Figure 5.10** Second, third, and fourth modes of a fixed-fixed bar

Let's be clear about what we've got here. A quick glance at Appendix A (come on, a quick glance won't kill you!) shows us that, as advertised at the start of the chapter, we now know the free responses for a circular rod, a tensioned string, and one-dimensional duct acoustics. So let's say you were interested in building a guitar. You could now analytically predict the necessary length, materials, and tension needed for the strings in order to get the particular pitch you're after. You can also evaluate the differences in pitch that will exist between different string materials. Thus you have a guide to whether you'd want to use nylon or steel for your strings. Or let's say you've decided to go into the pipe organ business. You can now figure out how long your pipes should be in order for them to give you the right note. Or maybe you want to determine the value of $EA$ experimentally. If so, you can set up a bar experiment like the one described here and use your transducers to observe the response of the system to an initial excitation, record the frequencies of the response, and by correlating these with your theoretical results (which involve $EA$) deduce what $EA$ must be for your test specimen.

The bottom line is that this stuff isn't trivial by any means, and you're in a good situation to do a lot of engineering using just the results we've gotten so far. However, our life won't be fully complete until we've looked at a beam. So, on to the next section we go.

## 5.3 FREE VIBRATION OF A BEAM

In the previous section we've seen how to analyze the free vibration of a uniform bar. If the bar materials are known exactly and are completely uniform, then our results are theoretically exact also. Now we'll look at a more complicated question, namely how the transverse vibrations of a beam behave.

Pretty much everyone reading this book has already had a course in deformable bodies (that's the good news), and pretty much everyone has forgotten just what was covered in the course (that's the bad news). If you're one of the fortunate few that can recall what the governing equations of a beam are, congratulations. You can skip to Example 5.1. But for the rest of you, follow along for a bit as we rediscover the wonderful world of beams.

A system that we know how to handle that supports transverse vibrations (as the beam does) is a tensioned string. From Appendix A we see that the governing equations are just like those of a bar, something we're familiar with. Basically, the difference between a beam and a string is that a string is limp and floppy (unless it's under tension), whereas a beam is stiff even without any tensile load. It's this additional characteristic, an internal bending stiffness, that makes beam equations more complex. What we'll derive here is the complete equation of motion, including applied forces, but we won't deal with the effect of forcing immediately.

Figure 5.11 shows what our beam looks like and also shows an infinitesimal element taken out of the beam. We shall be deriving the simplest beam model, namely the Bernoulli-Euler beam model, in this section. There are two basic simplifications that are involved in this model, the first of which has to do with the shape of the deformed element. You'll note that we've drawn the beam

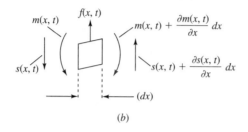

(a)

(b)

**Figure 5.11**  Differential analyses of a beam

element as being a parallelogram. Actually, the drawing would be more accurate if we drew it as a curved element. In this way we'd be including the shear deformation of the beam. Thus, we've reached our first approximation. We're going to assume that the deformations due to bending are much larger than those due to shear; thus we'll be neglecting shear deformation effects. A more complete model, called a Timoshenko beam model, includes this additional effect. As long as the beam doesn't wiggle too much, this approximation is a good one. Thus it will start to break down as we go to higher and higher modes of vibration.

The next step in determining the equations of motion for the beam is to simply apply what you've learned in your dynamics class to the beam element. We're only allowing vertical motions of the beam, so we'll derive only the equation of motion for the vertical direction. Also, the element could conceivably rotate, which would require us to use one of the several formulae that deal with moments and rotation. At this point we'll invoke our second assumption. We'll assume that the rotational inertia of our beam element is negligibly small compared to the translational inertia. As long as our beam is much longer than it is high, this will be a reasonable approximation. However, as the beam becomes "stubbier," we'll start to see error creep in because of our assumption. Once again, Timoshenko beam theory will include this effect. As a rule of thumb, as long as the length is about 10 times greater than the height, neglecting rotational inertia will be a reasonable assumption.

Applying a force balance to the differential element $dx$ of Figure 5.11 will give us

$$f(x, t)dx + \left(s(x) + \frac{\partial s(x)}{\partial x}dx\right) - s(x) = \rho(x)\ddot{u}dx \qquad (5.3.1)$$

where $s(x)$ represents the shear force within the beam and $\rho(x)$ is the mass per unit length along the beam. Canceling out $s(x)$ and dividing through by the common factor of $dx$ leaves us with

$$f(x, t) + \frac{\partial s(x)}{\partial x} = \rho(x)\ddot{u} \qquad (5.3.2)$$

Next, let's look at the rotational part of the problem. Summing moments (in the $z$ direction) about the center of mass of our differential element will give us

$$m(x) + \frac{\partial m(x)}{\partial x}dx - m(x) + \left(s(x) + \frac{\partial s(x)}{\partial x}dx\right)\frac{dx}{2} + s(x)\frac{dx}{2} = I(x)\ddot{\theta} \tag{5.3.3}$$

where $\ddot{\theta}$ is the rotational acceleration of the element and $m(x)$ is the moment acting on the element and $I$ is the moment of area of the element. But recall that we're neglecting rotational inertias. Thus $I\ddot{\theta}dx = 0$. Canceling out the $m$'s, dividing by $dx$, and taking the limit of $dx$ going to zero then gives us

$$\frac{\partial m(x)}{\partial x} = -s(x) \tag{5.3.4}$$

Using (5.3.4) in (5.3.2) yields

$$\rho(x)\ddot{u} = f(x,t) - \frac{\partial^2 m(x)}{\partial x^2} \tag{5.3.5}$$

Now we need only one more piece of information, namely the relationship between $m$ and $u$. Flipping back through whatever text you used in your deformable bodies course, you'll find that this relationship is

$$m(x) = EI(x)\frac{\partial^2 u(x)}{\partial x^2} \tag{5.3.6}$$

where $EI(x)$ is the bending stiffness of the beam. Combining (5.3.6) and (5.3.5) gives us our final beam equation,

$$\rho(x)\ddot{u} = -(EI(x)u_{xx})_{xx} + f(x,t) \tag{5.3.7}$$

The difference between this equation and that governing the bar is that instead of two partial differentiations with respect to $x$, we have four. If we ignore this detail, we'll see that it's simply an equation that equates an inertial force (in this case $\rho\ddot{u}$) with the forces due to the internal workings of the beam (those involving $EI$) and any externally applied loads $f(x,t)$.

Now that we have the equation of motion, we can set $f(x,t)$ to zero and solve for the free vibration of the beam. Just as we did for the bar problem, we're going to use separation of variables, that is, we'll assume that the part of the motion depending upon time can be separated from that part depending on position. Letting $u(x,t) = \mathbf{x}(x)\mathbf{t}(t)$ and dividing by $\rho\mathbf{xt}$, (5.3.7) becomes

$$\frac{\ddot{\mathbf{t}}}{\mathbf{t}} = -\frac{(EI(x)\mathbf{x}_{xx})_{xx}}{\rho(x)\mathbf{x}} \tag{5.3.8}$$

(Note that all the spatially dependent components are now grouped together.)

Just as we did in the bar example, we'll assume that the physical properties of the beam are constant over the beam. Thus (5.3.8) simplifies to

$$\frac{\ddot{\mathbf{t}}}{\mathbf{t}} = -\frac{EI\mathbf{x}_{xxxx}}{\rho\mathbf{x}} \tag{5.3.9}$$

Once again, the only way a time-dependent quantity can equal a spatially varying quantity is if they're both equal to the same constant. Since we now expect the time-dependent behavior to be in the form of an oscillator, we can immediately set the constant equal to $-\omega^2$, giving us the two equations

$$\frac{\ddot{\mathbf{t}}}{\mathbf{t}} = -\omega^2 \tag{5.3.10}$$

and

$$-\frac{EI\mathbf{x}_{xxxx}}{\rho\mathbf{x}} = -\omega^2 \tag{5.3.11}$$

Equation (5.3.10) can be rearranged to give us the same oscillator equation that we found with the bar example,

$$\ddot{\mathbf{t}} + \omega^2\mathbf{t} = 0 \tag{5.3.12}$$

The new discovery is the equation we obtain from (5.3.11),

$$\frac{EI}{\rho}\mathbf{x}_{xxxx} - \omega^2\mathbf{x} = 0 \tag{5.3.13}$$

If we rearrange this a bit we can get it into a form that resembles an oscillator equation,

$$\mathbf{x}_{xxxx} - \frac{\omega^2\rho}{EI}\mathbf{x} = 0 \tag{5.3.14}$$

with the difference being that we have four differentiations with respect to $x$ instead of two and we have a minus sign instead of a plus on the second term.

If we're to get any farther, we're going to have to figure out what the solution to (5.3.14) might be. Before we do this, let's clean up the notation just a bit. Rather than writing $\frac{\omega^2\rho}{EI}$ all the time, we'll replace it with $\beta^4$. You'll recall we did something similar in the bar example. Thus we're left with

$$\mathbf{x}_{xxxx} - \beta^4\mathbf{x} = 0 \tag{5.3.15}$$

The solutions must be functions that, when differentiated four times, give us back the original function (multiplied by some constant). This is because we must subtract $\beta^4\mathbf{x}$ from it, which is simply the original function multiplied by $\beta^4$. Luckily, there are four well-known functions that operate this way: sin, cos, sinh, and cosh! So even though the equation looks more complicated that a simple oscillator, it has the exact same solutions (sin and cos), as well as giving us two new solutions (sinh and cosh). More specifically, our general solution is equal to

$$\mathbf{x} = b_1\cos(\beta x) + b_2\sin(\beta x) + b_3\cosh(\beta x) + b_4\sinh(\beta x) \tag{5.3.16}$$

You'll note that this solution, when differentiated four times, is exactly canceled out by $\beta^4$ multiplied by itself, thus satisfying (5.3.15). For those who don't recall what sinh and cosh are, they're simply given by

$$\sinh(x) = \frac{e^x - e^{-x}}{2} \tag{5.3.17}$$

and

$$\cosh(x) = \frac{e^x + e^{-x}}{2} \tag{5.3.18}$$

Thus both of them get very large for large values of $x$. Only for small $x$ is there much difference between them, with $\sinh(x)$ going to zero as $x$ goes to zero and cosh going to one.

What we have here is simply a richer choice of possible mode shapes than we had with the bar problem. The sin and cos terms give us the wiggle in our solutions, while the sinh and cosh terms give us solutions that get large quickly as $x$ increases. By putting these components together, we can find the solutions for beam vibrations having any possible set of boundary conditions. And now, having mentioned boundary conditions, we should probably move on and examine this subject.

Because the beam has bending stiffness, its boundary conditions are more complicated than those found in the bar example. In that case, only one boundary condition was needed at each end of the bar in order to fully describe the physical setup of the problem. But now we're going to need *two* boundary conditions at each end. Figures 5.14a–5.14c show some of the possible boundary conditions commonly seen associated with beam vibration problems. The first represents a clamped (or cantilevered) boundary condition (Figure 5.12a). For this we visualize the beam as being immovably locked into a wall. Thus the appropriate boundary conditions are $u(0, t) = \frac{\partial u(0, t)}{\partial x} = 0$, that is, the displacement and slope at the clamped end are equal to zero. The next possibility is a pinned boundary condition (Figure 5.12b). In this case we're constraining the displacement to be zero ($u(0, t) = 0$), but we allow finite slopes. What aren't allowed are any moments. This is one of the results that you should have seen in your deformable bodies course or perhaps in a dynamics course. If you didn't see it there, you know it now—pinned joints imply zero moments. Thus, $\frac{\partial^2 u(0, t)}{\partial x^2} = 0$. The final example is that of the free boundary condition (Figure 5.12c). In this case we're not constraining the displacement or slope; both are in general nonzero. But the moment and shear at this point must equal zero, that is, $\frac{\partial^2 u(0, t)}{\partial x^2} = 0$ and $\frac{\partial^3 u(0, t)}{\partial x^3} = 0$.

If we take a moment and look at both the string and the beam problem, we can immediately notice something important. For the string problem, the dependent variable is differentiated twice with respect to $x$ ($y_{xx}$). Also, the problem has two boundary conditions, with one applied at each end: two differentiations and two boundary conditions. In the beam problem we see that the greatest number of differentiations is four ($u_{xxxx}$) and the problem has four boundary conditions, with two at each end: four differentiations and four boundary conditions. There seems to be a pattern here, namely that the number of boundary conditions equals the highest number of differentiations of the dependent variable. Since half the boundary conditions are applied to one end and half to the other, you'll often see a continuous problem characterized as being of order $2p$, where $2p$ is the highest number of differentiations and $p$ is, therefore, the number of boundary conditions that must

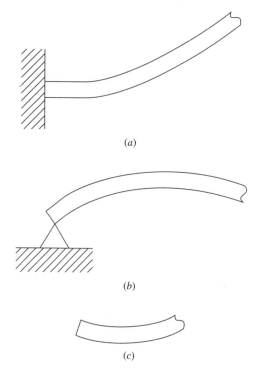

(a)

(b)

(c)

**Figure 5.12** (a) Clamped boundary condition; (b) pinned boundary condition; (c) free boundary condition

be applied at each end. (Keep in mind that we're talking about one-dimensional problems here; the same general concept holds for plates and other higher-dimensional problems.)

We should also take this time to give names to the different kinds of boundary conditions we've just encountered. When a boundary condition involves the geometric configuration of the structure, such as the displacement or the slope at an endpoint, we call it a *geometric boundary condition*. The other type of boundary condition is one that involves the shear force or the moment at an endpoint. These are called *natural boundary conditions*. A pinned end of a beam, therefore, involves both types; the geometric boundary condition is the one that says $u = 0$ and the natural boundary condition is the one that says $u_{xx} = 0$. The distinction between boundary conditions will be important when we look at approximation techniques in Chapter 6.

## Example 5.1

**Problem**   What we'd like to determine in this example is the response of a free-free beam, that is, one for which both the left and right ends are unrestrained. Thus this might be considered the simplest model for the behavior of an airplane or satellite.

**Solution**   We must keep in mind that this is an unrestrained support condition (like it's floating in space), so we'll have rigid-body motion as well as vibrations. The analysis here is concerned solely with the vibrations due to the internal stiffness of the beam. Any rigid-body modes can be solved for separately if needed, a problem we'll consider after we've looked at the flexible modes of the system. For transverse vibrations of the sort we're examining, they'll correspond to rigid translation in the $u$ direction and rigid rotation.

All we must do now is apply the boundary conditions

$$\frac{\partial^2 u(0, t)}{\partial x^2} = \frac{\partial^2 u(l, t)}{\partial x^2} = \frac{\partial^3 u(0, t)}{\partial x^3} = \frac{\partial^3 u(l, t)}{\partial x^3} = 0 \qquad (5.3.19)$$

to (5.3.16). Differentiating (5.3.16) the appropriate number of times, applying the boundary conditions, and using the fact that $\cos(0) = 1$, $\sin(0) = 0$, $\cosh(0) = 1$, and $\sinh(0) = 0$ gives us

$$\beta^2(-b_1 + b_3) = 0 \qquad (5.3.20)$$

$$\beta^3(-b_2 + b_4) = 0 \qquad (5.3.21)$$

$$\beta^2\left(-b_1\cos(\beta l) - b_2\sin(\beta l) + b_3\cosh(\beta l) + b_4\sinh(\beta l)\right) = 0 \qquad (5.3.22)$$

and

$$\beta^3\left(b_1\sin(\beta l) - b_2\cos(\beta l) + b_3\sinh(\beta l) + b_4\cosh(\beta l)\right) = 0 \qquad (5.3.23)$$

Equations (5.3.20) and (5.3.21) tell us that $b_1 = b_3$ and $b_2 = b_4$ (since letting $\beta = 0$ is clearly an uninteresting solution). Using this bit of knowledge and factoring out the $\beta$ terms lets us rewrite (5.3.22) and (5.3.23) as

$$b_1\left(\cosh(\beta l) - \cos(\beta l)\right) + b_2\left(\sinh(\beta l) - \sin(\beta l)\right) = 0 \qquad (5.3.24)$$

$$b_1\left(\sin(\beta l) + \sinh(\beta l)\right) + b_2\left(\cosh(\beta l) - \cos(\beta l)\right) = 0 \qquad (5.3.25)$$

or

$$\begin{bmatrix} \left(\cosh(\beta l) - \cos(\beta l)\right) & \left(\sinh(\beta l) - \sin(\beta l)\right) \\ \left(\sin(\beta l) + \sinh(\beta l)\right) & \left(\cosh(\beta l) - \cos(\beta l)\right) \end{bmatrix} \begin{Bmatrix} b_1 \\ b_2 \end{Bmatrix} = \begin{Bmatrix} 0 \\ 0 \end{Bmatrix} \qquad (5.3.26)$$

You'll note that, just when it seemed we'd left the exciting field of finite-dimensional eigenvalue problems, they're back, since (5.3.26) is of the form $[A]X = 0$. And from what we've already learned, we know that there'll be a solution only if $\|[A]\| = 0$. Taking this determinant gives us

$$\left(\cosh(\beta l) - \cos(\beta l)\right)\left(\cosh(\beta l) - \cos(\beta l)\right) - \left(\sin(\beta l) + \sinh(\beta l)\right)\left(\sinh(\beta l) - \sin(\beta l)\right) = 0 \qquad (5.3.27)$$

Expanding this out and using the trigonometric identities $\sin^2 + \cos^2 = 1$, along with $\cosh^2 - \sinh^2 = 1$, we'll finally obtain

$$\cos(\beta L)\cosh(\beta L) = 1 \qquad (5.3.28)$$

Thus the particular values of $\beta$, those that will determine the mode shapes and associated oscillation frequencies from our beam, are found from this single equation. We'll denote these particular solutions as $\beta_n$. Figure 5.13 shows a plot of both $\cos(\beta l)\cosh(\beta l)$ and 1. The points at which these two curves intersect are where $\cos(\beta l)\cosh(\beta l) = 1$ and therefore indicate what the associated values of $\beta_n l$ are.

You can see from Figure 5.13 that the $\beta$ values no longer increase in a regular manner, as they did for the bar problem (where $\beta_n = \frac{n\pi}{l}$). The first three nonzero values of $\beta_n l$ are 4.7300, 7.8532, and 10.9956. Although these don't differ from each other by a constant amount, the difference approaches a constant as we go to higher and higher modes. This is actually predictable if we think

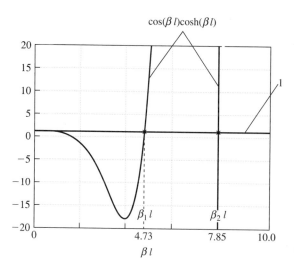

**Figure 5.13**  $\beta_n l$ values for a free-free beam

about what's happening with the cos and cosh terms. The cosh is getting big very quickly as the mode number increases, thus cos has to be very, very small for its product to equal 1. Therefore the roots approach those $\beta$'s at which $\cos(\beta l) = 0$, namely $\dfrac{(2n - 1)\pi}{2}$. It's only when the cosh isn't really huge (as in the lower modes) that the roots move substantially away from the $\cos(\beta l) = 0$ positions.

Now that we have the solution to (5.3.28), we can find the exact mode shape associated with the $\beta_n l$ values.

Using (5.3.25) to solve for $b_1$ in terms of $b_2$, we find that

$$b_2 = b_1 \frac{\sin(\beta_n l) + \sinh(\beta_n l)}{\cos(\beta_n l) - \cosh(\beta_n l)} \tag{5.3.29}$$

Since $b_1 = b_3$ and $b_2 = b_4$, (5.3.16) becomes

$$\mathbf{x} = b_1 \left(\cos(\beta_n x) + \cosh(\beta_n x)\right) + b_2 \left(\sin(\beta_n x) + \sinh(\beta_n x)\right) \tag{5.3.30}$$

Making use of (5.3.29) then gives us

$$\mathbf{x} = b_1 \left(\cos(\beta_n x) + \cosh(\beta_n x) + \frac{\sin(\beta_n l) + \sinh(\beta_n l)}{\cos(\beta_n l) - \cosh(\beta_n l)} \left(\sin(\beta_n x) + \sinh(\beta_n x)\right)\right) \tag{5.3.31}$$

Now (5.3.31) looks pretty complicated because it *is* complicated. But the important thing to remember is that it is fundamentally no different than the simple sine and cosine solutions we found for the bar problem. It's simply the mathematical representation of the actual mode shapes for our problem. It only seems complicated because we're not used to dealing with combinations of sines and hyperbolic sines and such. And if you look at Figures 5.14a–5.14c, you'll see that the system's modes look like quite reasonable mode shapes, just the sort you might expect in a freely vibrating beam. Furthermore, the complexity doesn't really matter unless you're planning on evaluating them by hand. But if you ever plan to really manipulate these modes, you'll undoubtedly use a computer (which won't be at all upset that there's more going on here than just sines and cosines).

Actually, now that we're on the subject, it's a good idea to consider just what would happen if (5.3.31) were entered into a computer program. The reason for wanting to look at this is that there *are* pitfalls associated with beam solutions, pitfalls that have caused hours of frustration to many students who didn't think carefully enough about potential problems before running their programs.

(a)

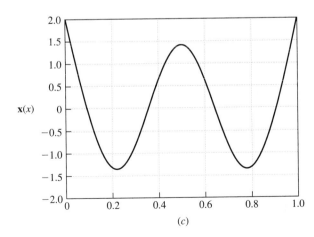

(b)

(c)

**Figure 5.14**  (a) First mode of a free-free beam; (b) second mode of a free-free beam; and (c) third mode of a free-free beam

The main problem lies in the behavior of cosh and sinh. Recall that $\cosh(x) = \dfrac{e^x + e^{-x}}{2}$ and $\sinh(x) = \dfrac{e^x - e^{-x}}{2}$. Both these functions grow very large, very quickly. Since the argument of our mode shape depends upon $\beta x$, it's clear that for large $\beta$'s we'll have large cosh's and sinh's. Furthermore, as their arguments get big, the difference between a cosh and a sinh goes to zero. A glance at the exponential components of $\sinh(\beta x)$ and $\cosh(\beta x)$ shows that the difference between them is simply $e^{-\beta x}$, something that goes to zero very quickly as $x$ increases.

Look carefully at the mode shape given by (5.3.31). If we make the approximation that $\sin(\beta_n l) + \sinh(\beta_n l)$ is basically equal to $\sinh(\beta_n l)$ for large values of $\beta_n$ and $\cos(\beta l) - \cosh(\beta l)$ is basically equal to $-\cosh(\beta_n l)$, then their ratio is just about equal to 1. So we're left with $\mathbf{x} = b_1[\cos(\beta x) + \cosh(\beta x) - (\sin(\beta x) + \sinh(\beta x))]$. Since $\sinh(\beta_n x)$ and $\cosh(\beta_n x)$ are basically equal for large $\beta_n$, $\mathbf{x}$ is approximately equal to $b_1[\cos(\beta_n x) - \sin(\beta_n x)]$. Although this isn't actually a good enough approximation, it is sufficient to illustrate some points. The first thing to realize is that a computer is going to have a much easier time computing $\cos(\beta_n x) - \sin(\beta_n x)$ than it will with the more exact expression. Thus we'd save computation time by using the approximate solution. Second, the computer will actually be unable to calculate the exact expression if too high a mode is being considered. For instance, $\cosh(30)$ is equal to around $5.34 \times 10^{12}$. When you're dealing with numbers this large, even the best computers will start to have difficulties. Why? Well, consider that you're asking the computer to compute $\cos(\beta_n l) - \cosh(\beta_n l)$. If $\beta_n l = 30$, then you're asking the computer to accurately add together $-5.34 \times 10^{12}$ plus something that's around 1. There's no way you're going to do that and still retain accuracy. And take the ratio itself (which we've been saying is around 1). Well, we must be pretty darn sure what it is precisely equal to since we're multiplying by $\sinh(\beta_n x)$, which is going to be huge for values of $x$ away from 0.

Finally, we must be concerned about the fact that some of our arguments involve $\beta_n l$ while others involve $\beta_n x$. Since $x$ goes from 0 to $l$, it's clear that $\cosh(\beta_n x)$ will vary between zero and very large values. Thus it's not true everywhere on the beam that $\cosh(\beta_n x)$ is approximately equal to $\sinh(\beta_n x)$ and both are much greater than $\cos(\beta_n x)$ and $\sin(\beta_n x)$, it's only true well away from $x = 0$.

In a case such as this we must carefully construct an approximation to the actual mode shape that takes into account everything we've been discussing. Luckily for us, most computers will be fine for the first half dozen modes or so. Thus it isn't really necessary to get into this any further unless you're really interested in getting an excellent grasp of modal analysis or are doing detailed analyses. If you have an interest in reading further, [2] goes into more detail on how to construct appropriate approximations.

Before leaving this section, it's time to make a mention of rigid-body modes. All the elastic modes we've looked at so far were associated with nonzero values of $\beta l$. However, we can see from Figure 5.15 that $\beta l = 0$ is also a valid solution. To see what this implies in terms of modal responses, we must go back to (5.3.15). If $\beta l = 0$ and $l \neq 0$, then the unforced equation governing the spatial deflections becomes

$$\mathbf{x}_{xxxx} = 0 \tag{5.3.32}$$

This is straightforward to integrate, and doing so gives us

$$\mathbf{x} = c_1 x^3 + c_2 x^2 + c_3 x + c_4 \tag{5.3.33}$$

The boundary conditions $\mathbf{x}_{xx} = \mathbf{x}_{xxx} = 0$ at $x = 0$ mean that $c_1 = c_2 = 0$. This leaves us with $\mathbf{x} = c_3 x + c_4$. This also satisfies the boundary conditions at $x = l$. Since we have two undetermined constants, we can define two mode shapes. A reasonable choice is to break up the possible motion into a rigid-body translational mode and rigid-body rotational mode. Thus we can set the first mode equal to a constant $d_1$ and the second equal to $d_2(x - \frac{l}{2})$. In this way the first mode is symmetrical and the second is antisymmetrical with respect to the center of the beam. These modes are very important when we are actually analyzing freely supported systems, such as airliners or spacecraft, since they represent motion that occurs at zero frequency. The zero frequency implies that, once the object is pushed into one of these modes, it will continue moving in that way. Thus, if a satellite is given an initial translational velocity, it will keep translating. Since this is often undesirable, we must keep track of these motions (through our rigid-body modes) so that we can be sure to leave them unexcited and/or control them.

With the knowledge you now have, you can address a wide range of problems. For instance, imagine that you're working for a tennis racket manufacturer and you want to improve your racket line. Your first dynamics course may have covered how you can find the center of percussion, that is, the point of a rigid bar at which impact forces will not impart reaction forces to a supporting pivot (which in this case would be your hand). But this analysis completely ignores the fact that tennis rackets are flexible objects. You'll want to consider what the modes are for a tennis racket. You could start your analysis by assuming that the racket is well modeled by a beam and could use free-free boundary conditions (on the assumption that the inertial forces of the ball-racket interaction outweigh the forces of your hand). Then you can go through an analysis similar to the one we've just gone through. Finally, knowing the modes of the racket, you can look at them and determine where they're the greatest and where they're equal to zero. Presumably, you'd like the point at which you grasp the racket to be where a node (point of zero deflection) is located, as this would probably be the most comfortable for the player. Of course, if you want to be really accurate, you'd want to use something more involved than a uniform beam model, but you'd get an easy first estimate from using the simple beam approximation.

If we denote the mode shapes of our problem by $\gamma_n(x)$, then the general solution is given by

$$u(x, t) = \sum_{n=1}^{\infty} a_n \gamma_n(x) \cos(\omega_n t + \phi_n) \tag{5.3.34}$$

Equation (5.3.34) is exactly the same form as (5.2.30), and thus we'd use exactly the same approach in matching any initial conditions. Actually, all solvable continuous vibration problems will have free responses of this form. The general response will be a summation of the system's eigenmodes, each oscillating at its particular natural frequency. Only the form of the modes will change. For a clamped-clamped beam we'd still end up with (5.3.34), except in this case the $\gamma$'s would represent the clamped-clamped modes of the beam, not the free-free ones. If we were looking at plate vibrations, then we'd have to solve a plate equation, which is definitely more complex than the beam equation. But we'd still end up with an infinite number of eigenfunctions, each with its own vibration frequency. For this case we'd also have a $\gamma_{m,n}(x, y)$ because the mode shape would have both an $x$ and $y$ dependence, due to the structure being a two-dimensional plate rather than a one-dimensional beam.

## 5.4 CONTINUOUS SYSTEMS—FORCED VIBRATION

Although we've pretty well covered how to calculate the free vibrations of a continuous system, we haven't yet touched on what we'd do in the face of an external forcing. Just as in the finite-dimensional systems we've already examined, the response of our systems to an applied forcing will very much depend upon the system's free-response characteristics. Thus, having figured out how to get the system's natural modes and frequencies, it's time to turn to the forced problem.

You'll recall that in the finite-dimensional case, applying a force to one of the system's discrete masses excited (in general) all the system modes. The only way that a particular eigenmode wasn't excited was when it had a zero in its modal response that was associated with the mass being excited. Otherwise, all modes would be excited and the amplitude of each contributing mode would depend upon the amplitude of the forcing and on how close the forcing frequency was to that mode's natural frequency. Well, the same thing occurs with continuous systems. It's only a bit more involved if we want to excite *only* one particular mode. For the finite-dimensional case we could apply a force to *every* mass and force the system at the correct frequency and amplitudes to excite just a particular mode. Doing this isn't inconceivable if the number of masses isn't too high. But for a continuous system, we'd need an infinite number of actuators, since the mode is a distributed one. Clearly this isn't possible. However, it's still possible to get pretty close to exciting just a single mode, if appropriate care is taken.

Let's consider the forced-beam problem (all others being similar in approach). We'll rewrite (5.3.7) below:

$$\rho(x)\ddot{u} = -(EI(x)u_{xx})_{xx} + f(x, t) \tag{5.4.1}$$

We're generally concerned with sinusoidal forcing and, since we know how to break down any periodic signal into a summation of sinusoids and apply them individually (from Section 3.2), we'll just concern ourselves with a single sinusoidal force,

$$\rho(x)\ddot{u} = -(EI(x)u_{xx})_{xx} + \bar{f}(x)\cos(\omega t) \tag{5.4.2}$$

Now we're going to exploit some of what we know about eigenfunctions. We know from our finite-dimensional work that any possible vector can be represented by an appropriate summation of the system's eigenvectors. We called this spanning the configuration space. The same is true for continuous problems. Thus we shall represent the force distribution over the beam by an infinite summation of the systems's eigensolutions. Therefore, instead of $\bar{f}(x)$, we'll have the following form:

$$\rho(x)\ddot{u} = -(EI(x)u_{xx})_{xx} + \left(\sum_{n=1}^{\infty} \bar{f}_n \phi_n(x)\right)\cos(\omega t) \tag{5.4.3}$$

In this expression, $\bar{f}_n$ represents the amount of mode $\phi_n(x)$ that's present in $\bar{f}$. Thus by adding up all these terms, we'll obtain $\bar{f}$.

Just as we did in the finite dimensional case, we can represent the response as a summation of all the eigenmodes of the system. We'll represent $u(x, t)$ by

$$u(x, t) = \sum_{n=1}^{\infty} a_n(t)\phi_n(x) \tag{5.4.4}$$

The $a_n(t)$ take the role that the generalized coordinates, $\eta$, took in our finite dimensional studies (Section 4.9). What we're saying here is that each eigenmode will contribute to the overall response, and the amount of that contribution will be controlled by each $a_n(t)$.

Substituting (5.4.4) into (5.4.3), we obtain

$$\rho(x) \sum_{n=1}^{\infty} \ddot{a}_n(t)\phi_n(x) = -\left(EI(x) \sum_{n=1}^{\infty} a_n(t)\phi_n(x)_{xx}\right)_{xx} + \left(\sum_{n=1}^{\infty} \bar{f}_n\phi_n(x)\right)\cos(\omega t) \tag{5.4.5}$$

Very messy indeed. We have a single equation that involves *all* the eigenmodes of the system, and it doesn't look all that solvable. What we need to do is get a set of equations, rather than the summation of terms we're currently looking at. To do this, we'll use the trick we've seen already in Section 4.8, that is, exploit orthogonality. All the eigenmodes are orthogonal (just as the finite-dimension ones were). We'll show this mathematically in Section 5.5, but for now you can just believe it works. To apply orthogonality to this problem, we'll multiply (5.4.5) by one of the eigenmodes ($\phi_m$) and integrate the equation from $x = 0$ to $x = l$,

$$\int_0^l \rho(x)\phi_m(x) \sum_{n=1}^{\infty} \ddot{a}_n(t)\phi_n(x)dx = -\int_0^l \left(EI(x)\phi_m(x) \sum_{n=1}^{\infty} a_n(t)\phi_n(x)_{xx}\right)_{xx} dx$$

$$+ \int_0^l \phi_m(x) \left(\sum_{n=1}^{\infty} \bar{f}_n\phi_n(x)\right)\cos(\omega t)dx \tag{5.4.6}$$

Whoa! This looks even worse than before. But is it really? If a system's eigenmodes are all orthogonal, then we should get some simplifications. Again, if you want the details you must wait until the next section. However, if you're content to believe me, then you can be amazed when I tell you that almost every term in the integrals above goes to zero. Orthogonality for a beam problem means that

$$\int_0^l \rho(x)\phi_m(x)\phi_n(x)dx = 0 \tag{5.4.7}$$

and

$$\int_0^l (EI(x)\phi_m(x)\phi_n(x)_{xx})_{xx} dx = 0 \tag{5.4.8}$$

whenever $m$ and $n$ are different. Therefore, only one term out of each integral will remain, and that is the one associated with $\phi_m(x)$. Thus we're left with

$$\ddot{a}_m(t) \int_0^l \rho(x)\phi_m^2(x)dx = -a_m(t) \int_0^l (EI(x)\phi_m^2(x)_{xx})_{xx}dx + \bar{f}_m \cos(\omega t) \int_0^l \phi_m^2(x)dx$$

$$m = 1, \ldots, \infty \tag{5.4.9}$$

For now we won't pursue the further simplifications that will be introduced in Section 5.5 and will content ourselves with a few observations. First, you'll note that we have an uncoupled set of equations to solve. Furthermore, it's now eminently solvable. The integral terms may look complicated, but they're just equal to scalar constants. Once you've specified what $l$, $EI$, $\rho$, and the $\phi_m(x)$'s are, you can integrate and get some result. Thus what we really have is

$$m_m \ddot{a}_m(t) = -k_m a_m(t) + \bar{f}_m \cos(\omega t) c_m \qquad m = 1, \ldots, \infty \qquad (5.4.10)$$

$c_m$, $m_m$ and $k_m$ are the constants that are the result of numerically integrating the spatial integrals. Bringing the second term to the left gives us

$$m_m \ddot{a}_m(t) + k_m a_m(t) = c_m \bar{f}_m \cos(\omega t) \qquad m = 1, \ldots, \infty \qquad (5.4.11)$$

We're finally down to something reasonable. All we're left with is the old, familiar, forced oscillator problem, and the dependent variable is $a_m(t)$. We have an equivalent mass ($m_m$) and an equivalent spring constant ($k_m$) along with a forcing magnitude ($c_m \bar{f}_m$) and a forcing frequency of $\omega$. So just as in the finite-dimensional case, all we must do is determine the response magnitude of each eigenmode due to the applied forcing, multiply by the associated eigensolution, and then sum the solutions up (as shown in (5.4.4)).

## 5.5 ORTHOGONALITY OF EIGENFUNCTIONS

As we saw in the previous section, the concept of orthogonality exists for continuous systems just as it does for finite-dimensional systems. Having seen that, it is now appropriate to delve a bit more deeply into the concept. In order to clearly see what's going on with continuous systems, let's review what we've already learned about orthogonality for discrete systems. Back in Section 4.8 we saw that the eigenvectors of a discrete vibration problem aren't usually orthogonal to each other (in a geometric sense), but rather are orthogonal with respect to each other *and* the stiffness or mass matrices, that is,

$$X_i^T [M] X_j = 0 \qquad (5.5.1)$$

and

$$X_i^T [K] X_j = 0 \qquad (5.5.2)$$

where $[M]$ and $[K]$ are the mass and stiffness matrices, respectively, and the $X_i$'s are eigenvectors for the system. Only in the special case for which $[M]$ or $[K]$ are proportional to the identity matrix did we have

$$X_i^T X_j = 0 \qquad (5.5.3)$$

Since the mass and stiffness distributions were closely tied up in the concept of orthogonality, it makes sense that they'll still be involved in the continuous case. The only difference is that this time they'll be continuous distributions, rather than lumped quantities.

One of the advantages we had in the finite dimensional problems was that they all were expressed in the same way,

$$[M]\ddot{X} + [K]X = 0 \qquad (5.5.4)$$

where $[M]$ and $[K]$ were $n \times n$ matrices. Therefore once we demonstrated orthogonality for one $[M]$ and $[K]$ we were done; all other combinations were conceptually the same. Not so, however, for the continuous case. When we deal with continuous systems we're dealing with partial differential equations, and there are many qualitatively different partial differential equations that apply to vibratory systems. Rather than go to a correct but abstract operator formulation, we'll look at a few basic continuous problems and indicate how any other particular problem can be handled.

We'll start by examining the bar equation,

$$m(x)\ddot{\xi} = (EA(x)\xi_x)_x \qquad (5.5.5)$$

If we go through the separation of variables procedure, we can find the natural frequencies $\omega_i$, and eigenfunctions $\bar{\xi}_i$, and, substituting them into (5.5.5), obtain

$$-m(x)\omega_i^2\bar{\xi}_i = \frac{\partial}{\partial x}\left(EA(x)\frac{\partial\bar{\xi}_i}{\partial x}\right) \qquad (5.5.6)$$

This equation also holds for $\bar{\xi}_j$, the $j$th eigenfunction,

$$-m(x)\omega_j^2\bar{\xi}_j = \frac{\partial}{\partial x}\left(EA(x)\frac{\partial\bar{\xi}_j}{\partial x}\right) \qquad (5.5.7)$$

We'll now multiply (5.5.6) by $\bar{\xi}_j$ and integrate along the beam, while also multiplying (5.5.7) by $\bar{\xi}_i$ and integrating along the beam as well, giving us

$$-\int_0^l m(x)\omega_i^2\bar{\xi}_j\bar{\xi}_i dx = \int_0^l \bar{\xi}_j\frac{\partial}{\partial x}\left(EA(x)\frac{\partial\bar{\xi}_i}{\partial x}\right)dx \qquad (5.5.8)$$

and

$$-\int_0^l m(x)\omega_j^2\bar{\xi}_i\bar{\xi}_j dx = \int_0^l \bar{\xi}_i\frac{\partial}{\partial x}\left(EA(x)\frac{\partial\bar{\xi}_j}{\partial x}\right)dx \qquad (5.5.9)$$

Next, we want to subtract (5.5.9) from (5.5.8),

$$(\omega_j^2 - \omega_i^2)\int_0^l m(x)\bar{\xi}_i\bar{\xi}_j dx = \int_0^l \bar{\xi}_j\frac{\partial}{\partial x}\left(EA(x)\frac{\partial\bar{\xi}_i}{\partial x}\right)dx$$
$$-\int_0^l \bar{\xi}_i\frac{\partial}{\partial x}\left(EA(x)\frac{\partial\bar{\xi}_j}{\partial x}\right)dx \qquad (5.5.10)$$

Just to keep a grip on where we're going, refer back to the finite-dimensional work. We did a similar thing back there when we premultiplied the equations by $X_j^T$ and $X_i^T$ and then subtracted them. We ended up with a quadratic that involved the $[M]$ matrix, which was multiplied by the difference of two natural frequencies ($\omega_2^2 - \omega_1^2$ for the specific example we were examining). It then turned out that the right-hand side (in that case the difference between two quadratic expressions involving $[K]$) was equal to zero. Thus we found the orthogonality relation that $X_i^T[M]X_j = 0$ for $i \neq j$.

We're going to find a similar result here. The left-hand side is a quadratic relation between $\xi_i$, $\xi_j$, and the mass distribution $m(x)$. It's multiplied by the difference between two natural frequencies. All that's left is to show that the right-hand side equals zero. To do this, we'll use a mathematical operation that we'll be using a lot in Chapter 6: integration by parts. When we integrate by parts, we use the fact that

$$\int_0^l u\,dv = uv\Big|_0^l - \int_0^l v\,du \tag{5.5.11}$$

Our right-hand sides are just begging to be integrated by parts. For instance, if we look at the first term after the equal sign in (5.5.10) we see that identifying $\bar{\xi}_j$ as $u$ and $\left(EA(x)\dfrac{\partial \bar{\xi}_i}{\partial x}\right)_x dx$ as $dv$ allows us to rewrite it as

$$\int_0^l \bar{\xi}_j \frac{\partial}{\partial x}\left(EA(x)\frac{\partial \bar{\xi}_i}{\partial x}\right)dx = \bar{\xi}_j\left(EA(x)\frac{\partial \bar{\xi}_i}{\partial x}\right)\Big|_0^l - \int_0^l EA(x)\frac{\partial \bar{\xi}_i}{\partial x}\frac{\partial \bar{\xi}_j}{\partial x}dx \tag{5.5.12}$$

In the same manner, the last term on the right-hand side of (5.5.10) can be written as

$$\int_0^l \bar{\xi}_i \frac{\partial}{\partial x}\left(EA(x)\frac{\partial \bar{\xi}_j}{\partial x}\right)dx = \bar{\xi}_i\left(EA(x)\frac{\partial \bar{\xi}_j}{\partial x}\right)\Big|_0^l - \int_0^l EA(x)\frac{\partial \bar{\xi}_i}{\partial x}\frac{\partial \bar{\xi}_j}{\partial x}dx \tag{5.5.13}$$

If we substitute both these results into (5.5.10), we'll obtain

$$(\omega_j^2 - \omega_i^2)\int_0^l m(x)\bar{\xi}_i\bar{\xi}_j dx = \bar{\xi}_j\left(EA(x)\frac{\partial \bar{\xi}_i}{\partial x}\right)\Big|_0^l - \bar{\xi}_i\left(EA(x)\frac{\partial \bar{\xi}_j}{\partial x}\right)\Big|_0^l \tag{5.5.14}$$

The integrals' terms are identical, thus they canceled each other out, leaving us with just the boundary-condition terms. Well, what if the beam is fixed at both ends? Then $\bar{\xi}_i$ and $\bar{\xi}_j$ are both zero, thus making the right-hand side zero. And what if it is fixed at one end and free at the other? Then the displacement is zero at one end, the slope is zero at the other, and the boundary terms again go to zero. In fact, you can put in whatever combination of boundary terms you'd like. When you evaluate the right-hand side, you'll see that both terms are either zero or that they exactly cancel each other. Therefore we're left with what we were hoping for, namely

$$(\omega_j^2 - \omega_i^2)\int_0^l m(x)\bar{\xi}_i\bar{\xi}_j dx = 0 \tag{5.5.15}$$

As in the discrete case, we realize that if all the natural frequencies are different, then $\omega_i^2 - \omega_j^2 \neq 0$ and therefore

$$\int_0^l m(x)\bar{\xi}_i\bar{\xi}_j dx = 0, \qquad i \neq j \tag{5.5.16}$$

This is our continuous bar analogue to $X_i^T[M]X_j = 0$. This orthogonality condition doesn't just involve the eigenfunctions, but also includes the mass distribution, just as it did in the discrete case. And whereas the discrete case involved vector/matrix multiplications, here we have multiplications of the functions followed by integration over the length of the bar. Notice that if $m(x)$ is constant, then it factors out of the integral and we're left with

$$\int_0^l \bar{\xi}_i \bar{\xi}_j \, dx = 0, \qquad i \neq j \tag{5.5.17}$$

In this case the functions are actually orthogonal to each other, just as the discrete vectors $X_i$ are for mass matrices that are just constant multiples of the identity matrix.

Finally, (5.5.16) and (5.5.8) give us our other orthogonality relationship:

$$\int_o^l \bar{\xi}_j \frac{\partial}{\partial x} \left( EA(x) \frac{\partial \bar{\xi}_i}{\partial x} \right) dx = 0 \qquad i \neq j \tag{5.5.18}$$

## 5.6 HOMEWORK PROBLEMS

Note: Assume constant material properties for all the following problems.

### Section 5.2

**1.** Determine the boundary conditions for the illustrated system. $T$ is the tension in the string and $\rho$ is the linear mass density.

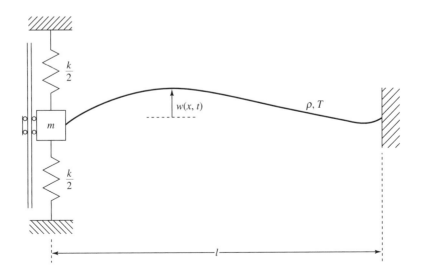

2. Find the natural frequencies and eigenfunctions for the illustrated, spring-restrained string. Let $k_1 = 10$ N/m, $k_2 = 20$ N/m, $T = 120$ N, and $\rho = .1$ kg/m. Show that the frequencies go to $\dfrac{n\pi}{l}\sqrt{\dfrac{T}{\rho}}$ for $k_1 = k_2 = 0$ or $k_1 = k_2 = \infty$ and to $\left(\dfrac{2n-1}{2}\right)\dfrac{\pi}{l}\sqrt{\dfrac{T}{\rho}}$ for $k_1 = 0$, $k_2 = \infty$ or $k_1 = \infty$, $k_2 = 0$. $l = 1$ m.

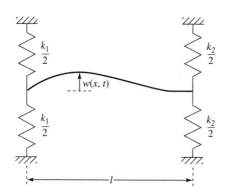

3. Assume that you wish to observe the odd modes of oscillation of a string by using a noncontacting sensor. The string has uniform mass and tension, is pinned at both ends, and is of length 1 m. Would it be better to place the sensor near $x = .5$ m or near $x = \dfrac{\sqrt{2}}{2}$ m ? Why?

4. The figures below show two snapshots in time of a tensioned string undergoing free vibration. Explain how such a situation could arise if the system modes were proportionally damped. Explain how it could also occur under nonproportional damping.

$t = 0$                                                  $t = t_1$

5. When discussing lumped-spring responses in Chapter 1, it was mentioned that the exact solution for the first natural frequency for the illustrated system was .9836 rad/s. Verify this fact and show that for high frequencies the natural frequencies tend toward $n\pi\sqrt{\dfrac{EA}{m}}$. $m = 1$ kg/m, $EA = 10$ N, $m_1 = 10$ kg, $l = 1$ m.

**6.** By how much will the pitch of a note change if the density of the correponding guitar string is doubled?

**7.** If you increase the tension of a guitar string by 10 percent, how much does the pitch alter?

**8.** It is often difficult to get a really good clamped boundary condition experimentally. Assume that, due to insufficient damping conditions, the effective beam length goes to $l+dl$, rather than the $l$ you planned for. By how much will your natural frequencies be altered? Can you use this shift as a tool for determining whether or not your beam is clamped properly?

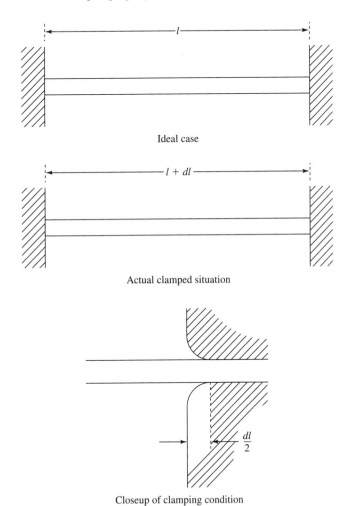

Ideal case

Actual clamped situation

Closeup of clamping condition

**9.** Solve for the eigenmodes and eigenfrequencies of a uniformly tensioned string for which the boundary conditions are $w_x(0) = w_x(l) = 0$, that is, zero slope at each end.

**10.** Two sets of plots are shown that display the natural frequencies of the illustrated system versus a system parameter. What are the possible parameters that might reasonably appear on the $x$ axis of each plot? Why?

(a)

System parameter

(b)

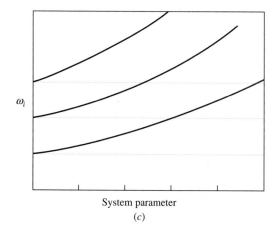

System parameter

(c)

**11.** Determine the correct boundary conditions for the system shown in Problem 10 and indicate graphically where the $\beta$ solutions lie.

**12.** Determine the correct boundary conditions for the system shown in Problem 5 and indicate graphically where the $\beta$ solutions lie.

**13.** Let's look more closely at Problem 11. Let $l = 1$ m, $T = 2$ N, and $\rho = .05$ kg/m. Show that the $\beta$ solutions move from those of a restrained-free string $\left(\left(\frac{2n-1}{2}\right)\pi\right)$ to those of a restrained-restrained string $(n\pi)$, as $k$ is altered from 0 toward $\infty$.

**14.** We'll look more closely at Problem 12 in this exercise. Show that as $m_1$ varies from 0 toward $\infty$, the $\beta_i l$ values move from $\left(\frac{2n-1}{2}\right)\pi$ to $n\pi$, that is, from the fixed-free to the fixed-fixed solutions. Let $m = .2$ kg/m, $l = 2$ m, and $EA = 1000$ N.

**15.** Show that the $\beta$ values for an unrestrained tensioned string are the same as those for a fully restrained tensioned string. Are the eigenfunctions also identical?

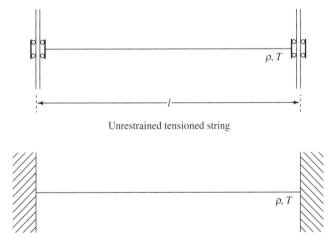

Unrestrained tensioned string

Fully restrained tensioned string

**16.** Show how to calculate the natural frequencies of a tensioned string for which the linear mass density is constant at $\rho_0$ from $x = 0$ to $x = l_1$ and constant at $\rho = \rho_1$ from $x = l_1$ to $x = l$.

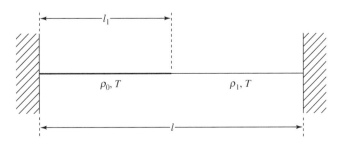

17. Show that the eigenfunctions of an unrestrained, tensioned string are orthogonal to each other.

18. Show that the eigenfunctions of a fully restrained-unrestrained tensioned string are mutually orthogonal.

19. What is the response of a fixed-fixed tensioned string to the initial conditions

$$y(x) = 0 \qquad \text{at } t = 0$$

$$\dot{y}(x) = \sin \frac{\pi x}{l} \qquad \text{at } t = 0$$

20. What is the free response of a fixed-free bar given initial conditions

$$\xi(x, 0) = ax \qquad \text{at } t = 0$$

$$\dot{\xi}(x, 0) = 0 \qquad \text{at } t = 0$$

21. Consider Problem 16. Show that if $\rho_0 = \rho_1$, you obtain the expected uniform tensioned string results.

22. Consider Problem 16. Show that if $\rho_0$ is much greater than $\rho_1$, you'll obtain natural frequencies corresponding to vibrations of a string having tension $T$, density $\rho_1$, and length $l - l_1$. Explain why such a result occurs. Show also that natural frequencies

$$\omega_n = \sqrt{\frac{T}{\rho_0}} \frac{(2n - 1)\pi}{2l_1}$$

are obtained and explain what modes they correspond to. Verify your results numerically.

23. Determine the $\beta$ equation for the system shown below (i.e., the equation that lets you find the $\beta_n$ for the eigenfunctions).

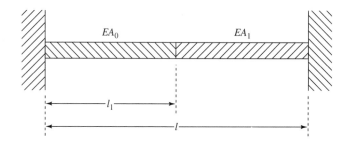

24. Determine the response of a fixed-fixed bar to the initial conditions

$$\xi(x, 0) = a \sin \left( \frac{\pi x}{l} \right)$$

$$\dot{\xi}(x, 0) = b \sin \left( \frac{3\pi x}{l} \right)$$

25. Calculate and plot the response at $x = .5l$ and $.75l$ for a pinned-pinned string having the initial conditions

$$y(x, 0) = 0$$

$$\dot{y}(x, 0) = \sin \frac{\pi x}{l} - \sin \frac{2\pi x}{l}$$

that is, solve for the particular values of the general response

$$y(x,t) = \sum_{n-1}^{N} a_n \sin \beta_n \, \cos(\omega_n t - \phi_n)$$

Plot the responses for $N = 1$ and $N = 2$, that is, a one- and two-term approximation. Comment on any differences between the two cases.

26. If the fundamental frequency of a fixed-fixed tensioned string should equal 440 Hz (a musical A), the string is 1 m long and has a mass of .003 kg, what must the tension be equal to?

27. How do the natural frequencies differ for a 1-m-long bar made of steel, aluminum, and copper? Assume fixed-free boundary conditions.

28. A bar is thrown into a rigid wall as shown below. It contacts the wall with a speed of $v$ m/s at $t = 0$. Determine the response of the bar for $t > 0$. Assume that the right end of the bar remains attached to the wall. $l = 1$ m.

29. A tensioned string is struck in an impulsive manner at $t = 0$ and $x = .5l$. The initial conditions resulting from the impact are given by

$$y(x) = 0 \qquad\qquad \text{at } t = 0$$
$$\dot{y}(x) = a\delta(x - .5l) \qquad \text{at } t = 0$$

Determine the string's response for $t > 0$.

30. Determine the first four natural frequencies for the illustrated system. $\rho = 8 \times 10^3$ kg/m$^3$, $G = 8 \times 10^{10}$ N/m$^2$, $J = 1$ m$^4$, $l = 2$ m.

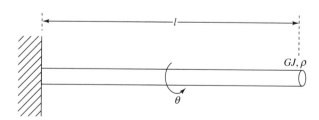

31. Determine the percentage change in the first three natural frequencies for torsional motion if a circular disk is attached to the end of the circular rod shown below. $I_0 = .1$ kg·m$^2$, $l = 1$ m, $G = 7 \times 10^{11}$ N/m$^2$, $\rho = 8 \times 10^3$ kg/m$^3$, $J = 4 \times 10^{-7}$ m$^4$.

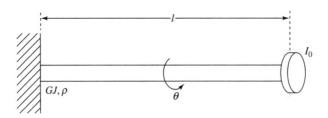

**32.** The figure below shows part of a machine in which a torsional spring restrained, circular shaft is moved by a sinusoidal varying torque applied at $x_1$. At what frequency of excitation will the system stop being well approximated as a rigid shaft in a spring restraint? $l = 1$ m, $G = 6 \times 10^{11}$ N/m$^2$, $\rho = 7 \times 10^3$ kg/m$^3$, $J = 2 \times 10^{-7}$ m$^4$, $k_\theta = 1 \times 10^{-2}$ N·m/rad.

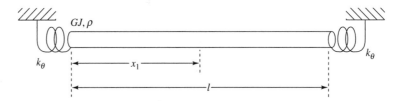

**33.** In Problem 32 we looked at a spring-restrained circular shaft. Numerically determine the lowest-frequency oscillation mode and show that it matches the prediction from a lumped analysis.

**34.** Consider a fixed-free circular rod made of steel. What is lower, the first torsional or the first longitudinal natural frequency? $G = 7.9 \times 10^{10}$ N/m$^2$, $E = 2.1 \times 10^{11}$ N/m$^2$, $\rho = 7.8 \times 10^3$ kg/m$^3$.

**35.** Bicycle designers worry about lightness and stiffness (among other concerns) when designing their frames. Determine which is torsionally stiffer, a smaller-diameter tube or a larger one. Use a constant mass of metal each time. The first tube will have an inside diameter of 3 cm and a wall thickness of .5 mm. The second tube will have an inside diameter of 4 cm. We'll determine stiffness by calculating the first natural frequency of a fixed-free tube upon which we've attached a lumped inertia $I_0$. $G = 7.9 \times 10^{10}$ N/m$^2$, $\rho = 7.8 \times 10^3$ kg/m$^3$, $l = 1$ m, $I_0 = 5 \times 10^{-3}$ kg·m$^2$.

**36.** Is there any difference in the torsional natural frequencies of a solid circular rod and a hollow circular shaft, assuming that they have the same length and mass? Consider the case of fixed-free boundary conditions.

**37.** Consider the system of Problem 32. Let the torsional spring on the left end be infinitely stiff. How large must $k_\theta$ be for the first torsional natural frequency of this system to be within 5 percent of the first torsional natural frequency for fixed-fixed conditions? $l = 1$ m, $G = 5 \times 10^{10}$ N/m$^2$, $J = 2 \times 10^{-8}$ m$^4$, $\rho = 7 \times 10^3$ kg/m$^3$.

**38.** Putting a lumped rotational inertia on the end of a fixed-free, hollow, circular shaft will lower the natural frequencies from those of the original system. How large a rotational inertia can be used before the first frequency shifts by 10 percent? The inner radius of the shaft is 1 cm, the wall thickness is 1 mm, the length is 1 m, $G = 5 \times 10^{10}$ N/m$^2$, and $\rho = 7 \times 10^3$ kg/m$^3$.

**39.** Consider the system of Problem 10. How large a spring constant $k$ should be used to bring the first natural frequency to 185 rad/s from its value when $k = 0$? $l = 2$ m, $T = 1000$ N, $\rho = .02$ kg/m.

**40.** The system below has natural frequencies that are essentially equal to the fixed-fixed ones, as well as a low frequency, rigid-body mode. How small can $m_1$ get before the second natural frequency has increased to 10 percent above its original value? $l = 3$m, $\rho = .02$ kg/m, $T = 3000$ N, $m_1 = 1000$ kg.

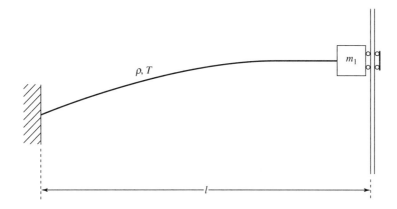

**41.** The illustrated system has natural frequencies of $\omega_1 = 1000$ rad/s, $\omega_2 = 3000$ rad/s, $\omega_3 = 5000$ rad/s ... for $m_1 = 0$. As $m_1$ is increased, the frequencies drop. Can a value of $m_1$ be found such that $\omega_i = 2100$ rad/s and $\omega_{i+1} = 2700$ rad/s, for some $i$? Why?

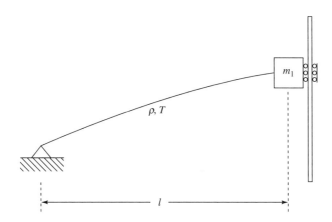

42. What are the first three natural frequencies for the illustrated system (longitudinal motion)? $EA = 2 \times 10^7$ N/m, $\rho = 3 \times 10^3$ kg/m$^3$, $l = 1$ m, $A = 1 \times 10^{-4}$ m$^2$, $k = 5 \times 10^5$ N/m.

43. Treating the illustrated system as two identical bars joined by a lumped spring, determine the first two $\beta$ values for the system. $EA = 2 \times 10^7$ N/m, $\rho = 3 \times 10^3$ kg/m$^3$, $l = 1$ m, $A = 1 \times 10^{-4}$ m$^2$, k $= 5 \times 10^5$ N/m.

44. Show analytically that the odd harmonics (1st, 3d, 5th, etc.) of the lowest natural frequency for a fixed-fixed string of the length $l$ are exactly equal to the natural frequencies of a fixed-unrestrained string of length $l/2$. Explain why this is the case.

45. Why can't the illustrated modal response exist as a free-vibration eigenfunction of a fixed-spring, re-strained, tensioned string?

Mode shape

(a)

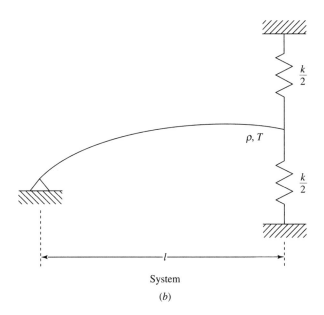

System

(b)

**46.** Can you predict any of the natural frequencies for System *A*, given the natural frequencies of System *B*?

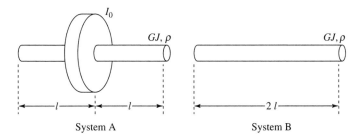

System A                          System B

**47.** What is the largest change in the natural frequencies that you'd expect if you added a spring restraint to the original tensioned string?

Original system

(a)

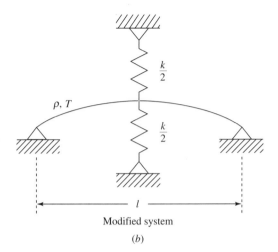

Modified system

(b)

**48.** Solve for the eigenfunctions of a fixed-free bar. Use this information to find the strain (first spatial derivative with respect to $x$). Where is the strain a maximum for the eigenfunction? Where is it a minimum? How would you use this information if you were using strain gauges as sensors?

## Section 5.3

**49.** Show that the $\beta$ values (and hence the natural frequencies) are the same for both a free-free beam and a clamped-clamped beam.

**50.** Determine the first four $\beta l$ values for a beam that is pinned on one end and both pinned and torsionally spring-restrained at the other end.

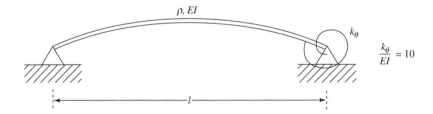

**51.** What is the equation of motion for free vibration of a beam with nonconstant stiffness and nonconstant mass distribution?

**52.** The eigenfunctions for a fixed-fixed bar were given by $\sin(n\pi x)$ and the natural frequencies were evenly spaced: $\omega_n = \beta_n \sqrt{\frac{EA}{m}}$ where $\beta_n = n\pi$. Show that the eigenfunctions for a pinned-pinned beam are also given by $\sin(n\pi x)$. Are the natural frequencies also evenly spaced?

**53.** What sort of beam deflection would you have if the displacement, slope, moment, and shear were zero at $x = 0$, with no conditions specified at $x = l$?

**54.** Derive another (equivalent) form for the deflection of the beam in Example 5.1.

## Section 5.4

**55.** Calculate the response of a fixed-free bar if the applied forcing is of the form

$$f(x, t) = \bar{f} \sin\left(\frac{\pi x}{2l}\right) \cos(\omega t)$$

for which $\omega$ is not equal to any of the bar's natural frequencies.

**56.** Determine the forced response of a tensioned string to an applied forcing of the form shown below. $\omega$ isn't equal to any of the string's natural frequencies. The boundary conditions of the string are given by $w(o, t) = w(l, t) = 0$.

$$f(x, t) = \bar{f}\delta(x - x_0) \cos(\omega t)$$

**57.** What is the forced response of a fixed-free circular rod subjected to a time varying torque $\tau$ at $x = x_1$?

$$\tau = \bar{\tau}\partial(x - x_1) \cos(\omega t)$$

**58.** Determine the general response of a pinned-pinned beam when subjected to a distributed force of the form

$$f(x) = 3 \sin(\pi x) \cos(4.1t) + 4 \sin(3\pi x) \cos(10t)$$

Assume constant beam properties.

**59.** What will the response be of a free-free beam to a transverse-distributed force of 11 N/m? Note that the force is constant, both in time and with respect to $x$.

## Section 5.5

**60.** Numerically solve for two different eigenfunctions for the system shown below and demonstrate numerically that they satisfy orthogonality. $l = 1.0$ m, $EA = 2.0 \times 10^7$ N/m, $m = 0.8$ kg, $k = 5.0 \times 10^5$ N/m, $m_1 = .3$ kg.

**61.** Numerically calculate the first two eigenfunctions of a fixed-fixed bar and verify that they satisfy orthogonality. $EA = 8.0 \times 10^6$ N, $m = 2.0$ kg/m, $l = 1.0$ m.

**62.** Consider the system of Problem 60. Set $m_1$ equal to zero. Numerically calculate the first two eigenmodes and demonstrate that they satisfy orthogonality. $EA = 5.0 \times 10^6$ N, $k = 7.787 \times 10^6$ N/m, $l = 1.0$ m, $m = 1.0$ kg/m.

**63.** Consider the system shown in Problem 5. Show that the first two numerically calculated modes satisfy orthogonality (i.e., generate the modes numerically [vector representation] and numerically evaluate $\int_0^l m\phi_i\phi_j\,dx + m_1\phi_i(1)\phi_j(1)$). $m = 1$ kg/m, $l = 1$ m, $EA = 1.3 \times 10^6$ N, $m_1 = 1.5$ kg.

**64.** Show that the first two modes of the illustrated system satisfy orthogonality. (Note: The first mode is a rigid-body mode). Numerically calculate the $\beta$'s and the eigenfunctions and numerically verify orthogonality. $G = 7 \times 10^9$ N/m$^2$, $J = 2 \times 10^{-3}$ m$^4$, $\rho = 3 \times 10^3$ kg/m$^3$, $l = 1$ m, $I_1 = 12$ kg$\cdot$m$^2$, $I_2 = 12$ kg$\cdot$m$^2$.

**65.** Consider the system of Problem 64. Set $I_2$ equal to zero. Show that orthogonality holds by numerically generating the first two eigenfunctions for the system and then numerically evaluate

$$\int_0^l \rho J\phi_i(x)\phi_j(x)\,dx + I_1\phi_i(0)\phi_j(0)$$

$G = 7 \times 10^9$ N/m$^2$, $J = 2 \times 10^{-3}$ m$^4$, $\rho = 3 \times 10^3$ kg/m$^3$, $l = 1$ m, $I_1 = 12$ kg $\cdot$ m$^2$.

# 6

## Approximate Solutions Methods

## 6.1 INTRODUCTION

Although we've learned a lot in the preceding pages, it should also be clear that there exist a great number of problems that are still beyond our analysis abilities. Anything that is more complex than a group of lumped spring-mass systems or a simple continuous system, such as a string, is currently beyond our skills. The simple addition of a nonuniform mass to a bar or a spatially varying stiffness in a beam makes the problem too difficult for closed-form analysis. These are the problems that this chapter addresses. Our solutions will not be exact, but merely approximate. Luckily, since we'll be able to create these approximate solutions to whatever degree of accuracy we desire, they'll meet all our needs. With the techniques to be introduced in the following pages you'll be able to analyze general systems and be able to do so in an efficient and accurate manner.

## 6.2 LUMPED APPROXIMATIONS

This section is probably the shortest in the book. The main focus of this chapter is on so-called modal approximations, that is, methods that presume the existence of distributed modes for a structure. The opposite view would be one that pays no attention to whether or not distributed modes might exist, but instead breaks the problem into little chunks. The finite element method is a well-known approach that takes this point of view. In this approach the actual continuous body is broken into simpler units (elements) for which displacement/stress relations can be easily calculated. These elements are then joined together in a mesh so as to approximate the original system. The technique is very widely used and is usually taught as a separate course at the first-year graduate level. Many books are available that are devoted solely to the subject.

A different approach is to forget continuous elements entirely and try to approximate the continuous system by a collection of springs and masses (and perhaps dampers). In this way the system would look like a MDOF lumped system, something we already know how to handle. This approach will be the focus of this section.

Here's how you might proceed for a particular example. Let's assume you're examining a uniform bar. You might decide to approximate the distributed mass by $n$ equal lumped masses located along the length of the bar. You could find the springs that connect the masses by determining the static deflection of the bar under a fixed load at the free end (the other end being restrained) and then by determining what combination of discrete springs would yield this overall stiffness. As the number of masses and springs are increased, the quality of the approximation would improve, assuming you've been reasonable in assigning the mass and spring values.

The advantage of this approach is that it can be used to analyze quite general systems. The complexity of the continuous system isn't a big issue except in that it will generate a large number of lumped masses (which, of course, impacts the ease of solution).

**Example 6.1**

**Problem**  Determine the first two natural frequencies for a uniform tensioned string (fixed-fixed boundary conditions) using a lumped approach. Compare the results to the exact analytical predictions. Let the length $l$ of the string be .9 m, the total mass $m_t$ be .05 kg, and the tension $T$ be 1000 N/m.

**Solution**  We'll first divide the mass of the string into $n$ equal portions. Thus, instead of having a string of length .9 m, we'll have $n$ sections, each $\Delta = \frac{.9}{n}$ m long. These sections will only contribute a stiffness to the problem; they'll have no mass of their own. Between each string section we'll put a lumped mass. If we imagine each mass as a little round ball, then we could view each string section as having half a ball on each end. If this is the case, then we'd have to put a half ball of mass at each of the string endpoints to maintain consistency. Of course, these lumped masses won't move since they're fixed in place due to the boundary conditions. Nonetheless it's important to include them, since ignoring them would move more mass onto the interior of the string than there should be.

Counting the two half masses at the endpoints, we have a total of $n+1$ complete "balls," each having a mass $m = \frac{m_t}{n+1}$ kg. Since the string sections are massless, they'll always be line segments, and we can therefore determine the forces acting on each mass element in a straightforward way. If we concentrate our attention on the $i$th mass element, we can see that the equations of motion are given by

$$m\ddot{y}_i = \left(\frac{y_{i+1} - y_i}{\Delta}\right)T - \left(\frac{y_i - y_{i-1}}{\Delta}\right)T \tag{6.2.1}$$

We've used the approximation that the component of vertical force due to the tension is proportional to the angular deflection (small angle approximation) and that the angle is approximated by the difference between the vertical deflections of two neighboring masses divided by the distance between them.

Rearranging (6.2.1) gives us

$$m\ddot{y}_i - \frac{T}{\Delta}y_{i-1} + \frac{2T}{\Delta}y_i - \frac{T}{\Delta}y_{i+1} = 0 \tag{6.2.2}$$

This can be expressed in matrix form as

$$\begin{bmatrix} m & 0 & 0 & \cdots \\ 0 & m & 0 & \cdots \\ 0 & 0 & m & \cdots \\ \vdots & \vdots & \vdots & \ddots \end{bmatrix} \begin{Bmatrix} \ddot{y}_1 \\ \ddot{y}_2 \\ \ddot{y}_3 \\ \vdots \end{Bmatrix} + \begin{bmatrix} \frac{2Tn}{l} & -\frac{Tn}{l} & 0 & \cdots \\ -\frac{Tn}{l} & \frac{2Tn}{l} & -\frac{Tn}{l} & \cdots \\ 0 & -\frac{Tn}{l} & \frac{2Tn}{l} & \cdots \\ \vdots & \vdots & \vdots & \ddots \end{bmatrix} \begin{Bmatrix} y_1 \\ y_2 \\ y_3 \\ \vdots \end{Bmatrix} = \begin{Bmatrix} 0 \\ 0 \\ 0 \\ \vdots \end{Bmatrix} \tag{6.2.3}$$

Dividing all the equations by $m$ and reexpressing $\Delta$ in terms of $l$ and $n$ gives us

$$\begin{bmatrix} 1 & 0 & 0 & \cdots \\ 0 & 1 & 0 & \cdots \\ 0 & 0 & 1 & \cdots \\ \vdots & \vdots & \vdots & \ddots \end{bmatrix} \begin{Bmatrix} \ddot{y}_1 \\ \ddot{y}_2 \\ \ddot{y}_3 \\ \vdots \end{Bmatrix} + \frac{(n+1)^2 T}{l m_t} \begin{bmatrix} 2 & -1 & 0 & \cdots \\ -1 & 2 & -1 & \cdots \\ 0 & -1 & 2 & \cdots \\ \vdots & \vdots & \vdots & \ddots \end{bmatrix} \begin{Bmatrix} y_1 \\ y_2 \\ y_3 \\ \vdots \end{Bmatrix} = \begin{Bmatrix} 0 \\ 0 \\ 0 \\ \vdots \end{Bmatrix} \tag{6.2.4}$$

If we perform an eigenanalysis on this set of equations using $n = 10$, we'll obtain a first and second natural-frequency estimate of $\tilde{\omega}_1 = 466.7$ rad/s and $\tilde{\omega}_2 = 924.0$ rad/s. The actual natural frequencies of the continuous system are given by

$$\omega_n = n\pi\sqrt{\frac{T}{m_t l}} \tag{6.2.5}$$

which yields results of $\omega_1 = 468.3$ rad/s and $\omega_2 = 936.6$ rad/s. As you can easily observe, the lumped approach did a good job of approximating the continuous system's natural frequencies.

As you might surmise, this lumped approach can be used in many ways. There is nothing to stop you from using nonuniform lumped masses, for instance. Thus, you might want each mass element to be spaced uniformly along a structure, but you might vary the amount of mass at each location to accurately reflect the mass distribution in the original structure.

## 6.3 RAYLEIGH'S QUOTIENT

One of the most amazing people in the field of vibrations was Lord Rayleigh. He was born wealthy, but decided to devote his life to science (a pretty rare occurrence). At age 35, he published the *Theory of Sound* [10] in which he laid out the fundamental principles of both acoustics and vibrations. Of course, much of this material was already known, but Lord Rayleigh added a great deal of his own and was the first to put the information on paper in an easily readable form. Once you've finished the book you're reading now, you might find a copy of the *Theory of Sound* and read Volume One, the volume concerned with vibrations. You'll find a good bit that is very similar to what you're learning now.

Although we'll run into Rayleigh again, later in the chapter, the reason for mentioning him now is because we're going to learn about what's called Rayleigh's quotient. The idea behind Rayleigh's quotient is relatively straightforward. We'd like to be able to find a system's natural frequencies without having to solve the associated eigenvalue problem. Now you might think this isn't possible; it would be like getting something for nothing. However, if we already know the eigensolutions to a problem that's close to the one we're currently interested in, we can exploit this knowledge to our advantage. We'll consider discrete systems first and then move on to continuous systems in Section 6.5. In this way we'll get a good insight into the method since we'll still be using a familiar framework. Once we understand what's going on with respect to discrete systems, making the jump to continuous systems will be relatively easy.

If we're dealing with an undamped, lumped mass system, we know that the equations of motion are of the form

$$[M]\ddot{X} + [K]X = O \tag{6.3.1}$$

We can recognize that this is an eigenvalue problem, and thus we know the solutions are given by

$$-\omega_i^2[M]X_i + [K]X_i = O \tag{6.3.2}$$

where $X_i$ is the $i$th eigenvector and $\omega_i$ is the $i$th natural frequency of oscillation. Next we'll premultiply both sides of this equation by $\frac{1}{2}X_i^T$,

$$-\frac{1}{2}\omega_i^2 X_i^T[M]X_i + \frac{1}{2}X_i^T[K]X_i = O \tag{6.3.3}$$

It's worth discussing why we might want to do this. For ease, let's look at a 3-DOF system (Figure 6.1). A 2-DOF system would be even easier, but we'll be looking at some characteristics that require at least three degrees of freedom in a few pages, so we'll stick with 3-DOF.

We know that the potential and kinetic energies of the system can be written as

$$p_e = \frac{1}{2}k_1 x_1^2 + \frac{1}{2}k_2(x_2 - x_1)^2 + \frac{1}{2}k_3(x_3 - x_2)^2 + \frac{1}{2}k_4 x_3^2 \tag{6.3.4}$$

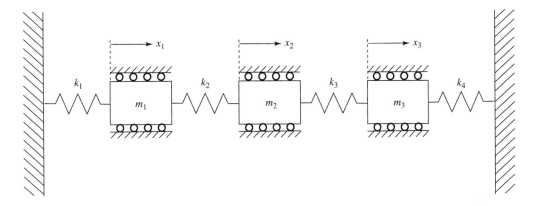

**Figure 6.1**   3-DOF spring-mass system

and

$$k_e = \tfrac{1}{2}m_1\dot{x}_1^2 + \tfrac{1}{2}m_2\dot{x}_2^2 + \tfrac{1}{2}m_3\dot{x}_3^2 \tag{6.3.5}$$

These are all quadratic terms, that is, everything is raised to the second power. Thus we can write them in matrix form as

$$p_e = \tfrac{1}{2}\{x_1\ \ x_2\ \ x_3\} \begin{bmatrix} k_1 + k_2 & -k_2 & 0 \\ -k_2 & k_2 + k_3 & -k_3 \\ 0 & -k_3 & k_3 + k_4 \end{bmatrix} \begin{Bmatrix} x_1 \\ x_2 \\ x_3 \end{Bmatrix} \tag{6.3.6}$$

and

$$k_e = \tfrac{1}{2}\{\dot{x}_1\ \ \dot{x}_2\ \ \dot{x}_3\} \begin{bmatrix} m_1 & 0 & 0 \\ 0 & m_2 & 0 \\ 0 & 0 & m_3 \end{bmatrix} \begin{Bmatrix} \dot{x}_1 \\ \dot{x}_2 \\ \dot{x}_3 \end{Bmatrix} \tag{6.3.7}$$

More compactly, this is just

$$p_e = \tfrac{1}{2}X^T[K]X \tag{6.3.8}$$

and

$$k_e = \tfrac{1}{2}\dot{X}^T[M]\dot{X} \tag{6.3.9}$$

where $[M]$ and $[K]$ are our familiar mass and stiffness matrices. If we go one step further and ask what the energies would be if the system were oscillating at one of its natural frequencies, say $\omega_i$, then (6.3.8) and (6.3.9) would become

$$p_e = \tfrac{1}{2}X_i^T[K]X_i \cos^2(\omega_i t) \tag{6.3.10}$$

$$k_e = \tfrac{1}{2}\omega_i^2 X_i^T[M]X_i \sin^2(\omega_i t) \tag{6.3.11}$$

where $X_i$ is the $i$th eigenvector and $X = X_i \cos(\omega_i t)$ describes the motion.

Both these energies are time-varying functions that oscillate between zero and a maximum value. As a glance at (6.3.10) and (6.3.11) reveals, the maximum value of the potential energy is equal to $\frac{1}{2}X_i^T[K]X_i$, while the maximum kinetic energy is equal to $\frac{1}{2}\omega_i^2 X_i^T[M]X_i$. Using an overbar to represent the maximum, we therefore have

$$\bar{p}_e = \tfrac{1}{2}X_i^T[K]X_i \tag{6.3.12}$$

and

$$\bar{k}_e = \tfrac{1}{2}\omega_i^2 X_i^T[M]X_i \tag{6.3.13}$$

We're going to need another term soon, so we'll introduce it now. If we factor the $\omega^2$ out of (6.3.13), we'll obtain what we'll call the *zero-frequency kinetic energy*, which we'll indicate by a zero superscript,

$$\bar{k}_e^0 = \tfrac{1}{2}X_i^T[M]X_i \tag{6.3.14}$$

Now compare (6.3.12) and (6.3.13) with (6.3.3). We can see that the first term of (6.3.3) is just the maximum kinetic energy of the system and the second is the maximum potential energy. So what we've done by premultiplying by $X_i^T$ is to change force terms into energy terms. If we now divide (6.3.3) by $X_i^T[M]X_i$, we'll have

$$\omega_i^2 = \frac{\frac{1}{2}X_i^T[K]X_i}{\frac{1}{2}X_i^T[M]X_i} \tag{6.3.15}$$

This tells us that the square of a system's natural frequency is *always* equal to the ratio of the system's maximum potential energy to the system's zero-frequency kinetic energy. Since we're concerned with the ratio of the two energies, the $\frac{1}{2}$ factors aren't a problem; we can just as well factor them out,

$$\omega_i^2 = \frac{X_i^T[K]X_i}{X_i^T[M]X_i} \tag{6.3.16}$$

The more general form of (6.3.16), one that simply depends on an arbitrary input vector, is given by

$$R_Q(X) = \frac{X^T[K]X}{X^T[M]X} \tag{6.3.17}$$

$R_Q$ stands for, not surprisingly, Rayleigh's quotient. Any given problem has an infinite number of values for the Rayleigh's quotient, since changing $X$ will change $R_Q(X)$. Notice that simply scaling the input vector doesn't alter Rayleigh's quotient, that is, $R_Q(X) = R_Q(nX)$, where $n$ is any constant. This is in keeping with our notions of eigensolutions and the fact that the eigenvectors can be determined only up to a multiplicative constant. It is important to remember that Rayleigh's quotient will produce a continuum of values, depending upon $X$. For the special cases when the input $X$ is equal to an eigenvector, then Rayleigh's quotient will equal the square of the natural frequency associated with that eigenvector.

Equation (6.3.17) is very significant, not just for our discussion of Rayleigh's quotient, but for our later work in approximation methods. That is a way of getting the eigenvalues for our problem without solving an eigenvalue problem. If (and that's a big if right now) someone hands us the actual eigenvectors of our problem, then simply substituting them into (6.3.17) will give us the associated eigenvalue (or equivalently, the square of the natural frequency).

As it now stands, this may not seem like a big deal. First, you have to wonder how often someone will conveniently hand you the system eigenvectors. And if you don't have the exact eigenvectors to work with, but only have an approximation of them, then who's to say that (6.3.17) won't give us an inaccurate estimation of the natural frequencies? Well, the good news is that there's actually a bit more to (6.3.17) than there initially appears to be. In the first place, design invariably involves an iterative process in which a system is *tweaked* in various ways in order to improve its performance. Generally, you know a good deal about the system before the tweaking starts. In a similar vein, improvements of an existing product often consist of small modifications to the basic unit, modifications that are, again, meant to improve performance in some way. Here also, the basic system is well understood before modifications begin. Thus there is a very good chance that the system eigenmodes and natural frequencies have already been catalogued and are ready for use in later analyses. Since the changes to the existing system are generally not too large, it seems reasonable that the eigenmodes of the system will not change drastically. Therefore the old eigenmodes make very reasonable starting estimates for the modified system's eigenmodes.

Let's say that someone gives you (or you estimate for yourself) an approximation to the $i$th eigenvector of the modified system. You then take this approximation and use it in (6.3.17). You obviously won't get $\omega_i^2$, because the only way to get that would be if you had used the exact $i$th eigenvector. But what you will get is an extremely good estimate of $\omega_i^2$. In fact, the error in the estimate of $\omega_i^2$ will be far less than the error in your approximation of the $i$th eigenvector. That's the power of (6.3.17); it gives a good quality estimate for the natural frequency even if your input eigenvector estimate isn't very good.

## Example 6.2

**Problem**   Let's look at the system shown in Figure 6.2 to illustrate how well Rayleigh's quotient can work. Figure 6.2*a* represents our original system. We then, due to some design considerations, make changes in the system (resulting in Figure 6.2*b*) and want to determine how the natural frequencies have changed by using Rayleigh's quotient.

**Solution**   The original system has the following mass and stiffness matrices

$$[M] = \begin{bmatrix} 10 & 0 & 0 \\ 0 & 10 & 0 \\ 0 & 0 & 10 \end{bmatrix}, \quad [K] = \begin{bmatrix} 200 & -100 & 0 \\ -100 & 200 & -100 \\ 0 & -100 & 200 \end{bmatrix} \quad (6.3.18)$$

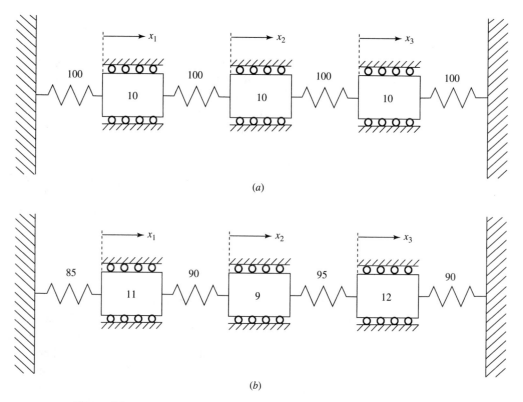

**Figure 6.2**   Two 3-DOF spring-mass systems; (*a*) unperturbed, and (*b*) perturbed

which support the eigenvectors

$$X_1 = \begin{Bmatrix} .5000 \\ .7071 \\ .5000 \end{Bmatrix}, \qquad X_2 = \begin{Bmatrix} .7071 \\ .0000 \\ -.7071 \end{Bmatrix}, \qquad X_3 = \begin{Bmatrix} .5000 \\ -.7071 \\ .5000 \end{Bmatrix} \qquad (6.3.19)$$

with $\omega_1^2 = 5.858$ rad/s, $\omega_2^2 = 20.00$ rad/s, and $\omega_3^2 = 34.14$ rad/s.

The modified system (indicated by a hat) has mass and stiffness matrices equal to

$$[\hat{M}] = \begin{bmatrix} 11 & 0 & 0 \\ 0 & 9 & 0 \\ 0 & 0 & 12 \end{bmatrix}, \quad [\hat{K}] = \begin{bmatrix} 175 & -90 & 0 \\ -90 & 185 & -95 \\ 0 & -95 & 185 \end{bmatrix} \qquad (6.3.20)$$

which support the eigenvectors

$$\hat{X}_1 = \begin{Bmatrix} .5130 \\ .6831 \\ .5198 \end{Bmatrix}, \qquad \hat{X}_2 = \begin{Bmatrix} .7308 \\ .0216 \\ -.6823 \end{Bmatrix}, \qquad \hat{X}_3 = \begin{Bmatrix} .4315 \\ -.8064 \\ .4045 \end{Bmatrix} \qquad (6.3.21)$$

with $\hat{\omega}_1^2 = 5.014$ rad/s, $\hat{\omega}_2^2 = 15.67$ rad/s and $\hat{\omega}_3^2 = 31.20$ rad/s.

Keep in mind that this problem was designed to be simple, and so there's no real reason to use Rayleigh's quotient; it isn't a big deal to solve the problem analytically. That's the reason that both problems have been solved exactly: we can see how well Rayleigh's quotient works. Notice that the eigenvalues of the two problems (the square of the natural frequencies) have changed substantially, as have the eigenvectors.

What we'll do now is use the *original* eigenvectors as *estimates* of the new eigenvectors and use Rayleigh's quotient to determine the new natural frequencies. Thus, for the first natural frequency estimate (which we'll denote by a tilde), we have

$$\tilde{\omega}_1^2 = R_Q(X_1) = \frac{X_1^T[\hat{K}]X_1}{X_1^T[\hat{M}]X_1} \tag{6.3.22}$$

Thus we're trying to estimate the new natural frequency by using a good guess for the eigenvector and using our knowledge of what the new mass and stiffness matrices are actually equal to. Using the values given above, what we'll find is

$$\tilde{\omega}_1^2 = 5.043 \text{ rad/s} \tag{6.3.23}$$

If we go ahead and determine the estimates of the second and third natural frequencies, using $X_2$ and $X_3$ respectively, we'll obtain

$$\tilde{\omega}_2^2 = 15.65 \text{ rad/s}, \qquad \tilde{\omega}_3^2 = 30.57 \text{ rad/s} \tag{6.3.24}$$

Compare these results with the squares of the actual natural frequencies (5.014, 15.67, and 31.20 rad/s). The results are fantastic! 5.043 rad/s is off only by .5 percent from 5.014 rad/s, 15.65 rad/s is off by .1 percent, and 30.57 rad/s is off by 2 percent. Rayleigh's quotient gave us excellent estimates of the modified natural frequencies and did it using eigenvectors that were noticeably off from the actual eigensolutions.

Having seen that Rayleigh's quotient does this nifty job of giving us better answers than the data we feed in, it's time to show why this happens. The answer has a lot to do with what's happening in a geometric sense with the system's eigenstructure. This is a bit complicated to visualize, so we'll break the problem down and look at bits of it individually before putting it all together.

First, let's see how Rayleigh's quotient varies as we use different amounts of a particular eigenvector. Figure 6.3 shows Rayleigh's quotient for our original system (6.3.18). In these plots we're using the actual eigenvectors of the system and simply changing how much is used. As you can see, the plots are flat; it makes no difference whether we use a lot or a little. This is completely in line with our notion of eigenvectors. All that matters are the relative displacements, not the actual magnitudes.

Now we'll go on to something more interesting. In Figure 6.4, we're again plotting Rayleigh's quotient. However, this time our estimated vector consists of one eigenvector ($X_1$) plus a varying amount of another. More specifically, in Figure 6.4a we're letting our estimated vector equal $X_1 + a_2 X_2$, while keeping $a_3$ (the amount of the third eigenvector) equal to zero. When $a_2 = 0$, Figure 6.4a shows that $R_Q$ gives us the correct value for $\omega_1^2$, namely 5.858. We expect the graph to exactly equal the square of the first natural frequency when $a_2 = 0$, and it does. The real question is what happens if $a_2 \neq 0$? As we add more and more of a contamination vector (by letting $a_2$ grow in Figure 6.4a) the $R_Q$ estimate increases. Near the origin, the plot looks quadratic and it

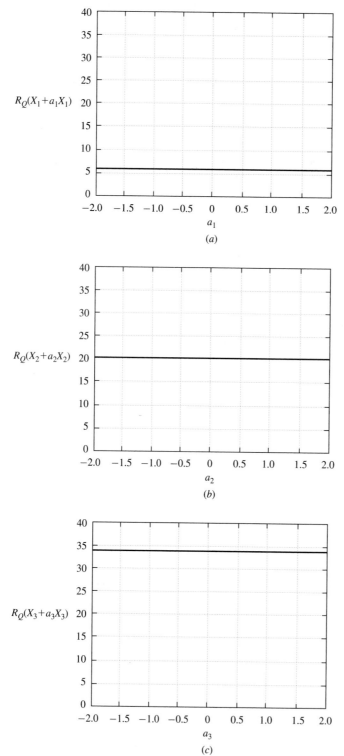

**Figure 6.3** Behavior of Rayleigh's quotient about each eigenvector when perturbed by that eigenvector

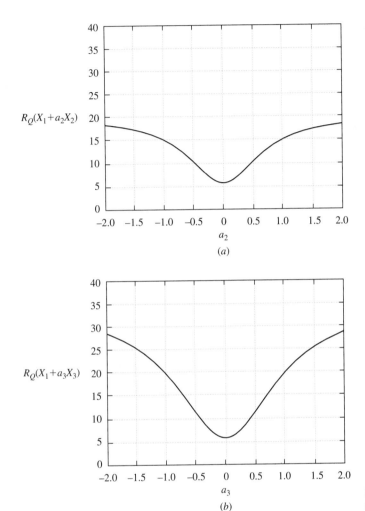

**Figure 6.4** Behavior of Rayleigh's quotient about the first eigenvector

levels off for high values of $a_2$. The same qualitative behavior is seen in Figure 6.4b, where we're contaminating $X_1$ with $X_3$.

This behavior is always present: the lowest value that Rayleigh's quotient can produce is the square of the first natural frequency. No input vector can produce a lower output. Therefore one extremely inefficient way to determine the modified system's first natural frequency would be to try every conceivable trial vector. The one giving you the lowest value for Rayleigh's quotient will be the first eigenvector. Of course, nobody would ever use such an approach, but at least we know we could if all else failed.

The next set of plots (Figure 6.5) shows how Rayleigh's quotient varies when we let the estimated vector equal $X_3 + a_1 X_1$ (Figure 6.5a) and then let it equal $X_3 + a_2 X_2$ (Figure 6.5b). We

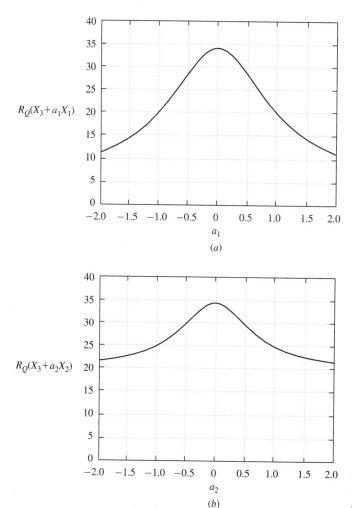

**Figure 6.5** Behavior of Rayleigh's quotient about the third eigenvector

see that the plots are maximums when the estimated vector is simply equal to $X_3$, that is, the value decreases as we add components of $X_1$ or $X_2$ to it. Thus we have another nugget of information, namely that the highest possible value for Rayleigh's quotient is equal to the square of the largest natural frequency (in this case, the third natural frequency).

Now it's time for the weirder result. The first two cases were nicely behaved. Contaminating the first eigenvector (the one associated with the lowest natural frequency) with any other eigenvector simply made Rayleigh's quotient increase, while adding some other eigenvector to the last eigenvector (the one associated with the highest natural frequency) caused Rayleigh's quotient to decrease. These are the two simplest cases: using either the first or the last eigenvector as the one that's being contaminated. The behavior is more complex for any other eigenvec-

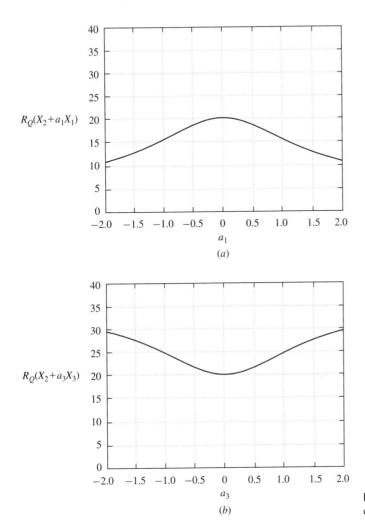

**Figure 6.6** Behavior of Rayleigh's quotient about the second eigenvector

tor. Of course, in this example there is only one other eigenvector to consider since it's only a three-mass system. But for more complex systems, all eigenvectors other than those corresponding to the lowest or highest natural frequency will display the kind of behavior we'll see next.

Figure 6.6 shows how Rayleigh's quotient changes as we contaminate the second eigenvector with either the first eigenvector ($X_2 + a_1 X_1$, Figure 6.6a) or with the third eigenvector ($X_2 + a_3 X_3$, Figure 6.6b). We can see from these plots that Rayleigh's quotient decreases when we add some of the first eigenvector ($X_1$), but increases if we add some of the third eigenvector ($X_3$). This actually makes sense if you think about what Rayleigh's quotient does. As we've seen, Rayleigh's quotient spits out a scalar that corresponds to the system's natural frequencies if the input is an eigenvector. We know that the second natural frequency is less than the third and more than the first. By adding some of $X_3$ to our estimated vector, we're reorienting the estimated vector toward the third eigenvector. If we add a lot of $X_3$, then the second becomes trivial in comparison, thus making the estimated vector approximate $X_3$. And Rayleigh's quotient for the third eigenvector is bigger than for the second. So we should expect Rayleigh's quotient to increase as we add more and more of $X_3$, otherwise how would the result smoothly increase to the higher value?

The same logic holds when we add some of $X_1$. In this case Rayleigh's quotient will end up smaller as we get more and more of the first eigenvector involved, and so we'd expect it to decrease steadily toward this lower value.

Of course, it's more complicated if we add a little of both remaining eigenvectors to an initial eigenvector. When the initial eigenvector is the first or the last, then we still get a minimum or maximum (Figure 6.7) since both lead in the same direction (toward a lower or higher value depending on whether the initial eigenvector is the highest or lowest one). But for the case of the second eigenvector we have two competing trends. Adding some $X_1$ will cause Rayleigh's quotient to decrease, but adding some $X_3$ will cause it to increase. Therefore the overall result will depend upon how much of each is added. As an example, Figure 6.8 illustrates how Rayleigh's quotient will vary as an equal amount of $X_1$ and $X_3$ are added to $X_2$. Note that the result in this case is *no* variation, the effect of $X_1$ is exactly canceled by that of $X_3$, resulting in no change to Rayleigh's quotient.

This behavior is the same as we'd see in higher-dimensional systems for all eigenvectors, except those associated with the lowest and highest natural frequencies. Assuming that the natural frequencies have been ordered from lowest to highest, perturbing off $X_i$ with some of $X_j$ will cause the Rayleigh's quotient to increase if $j$ is greater than $i$ and decrease if $j$ is less than $i$.

It's time for the final observation. You may recall from your multivariable calculus class that when a function has a minimum, the derivative at that minimum with respect to the independent variable is equal to zero. The same holds true at a maximum and also at inflection points. We can see this clearly in Figure 6.9. The slope of Rayleigh's quotient versus $a_1$, $a_2$, etc., is always zero at the minimum. For the case of $X_2$ (Figure 6.6) we still have a slope of zero with respect to the $a_i$'s; the only thing different is that we have both a maximum and a minimum, depending upon which independent variable we're looking at. Even when we're looking at combinations of the $a_i$'s (Figure 6.7), we still have a zero slope when the $a_i$'s equal zero. Let's take a closer look at a particular case. Figure 6.10 is an enlargement of Figure 6.4b, and shows how Rayleigh's quotient varies when $a_2 = 0$ and $X_1 + a_3X_3$ is the input. If we're near $a_3 = 0$, then, because the slope is zero, small changes in $a_3$ don't appreciably alter Rayleigh's quotient. The point labeled A on the graph shows the change in Rayleigh's quotient when $a_3 = .10$. As you can see, the value doesn't change much at all from what it was when $a_3 = 0$. This is a fundamental property of minima, maxima, or inflection points. These kinds of points are called stationary points because the dependent function is essentially stationary about them; it doesn't change (much) as the dependent variable is altered.

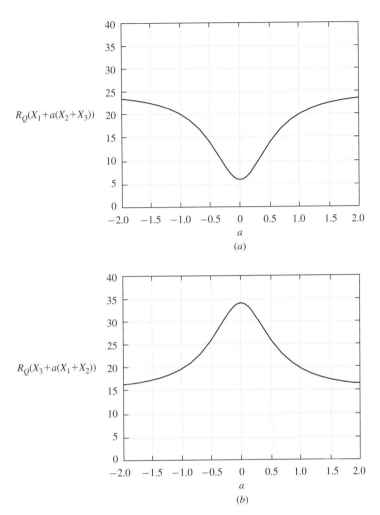

$R_Q(X_1 + a(X_2 + X_3))$

$(a)$

$R_Q(X_3 + a(X_1 + X_2))$

$(b)$

**Figure 6.7**    Behavior of Rayleigh's quotient about the first and third eigenvectors when perturbed by a combination of both remaining eigenvectors

This is the result we've been looking for! If we have a halfway decent guess for a modified
system's eigenvector, then our estimated vector has a lot of the actual eigenvector in it, along with
some contamination from the other eigenvectors. What we've now found out is that this contamination doesn't really alter the result very much, as long as there isn't too much contamination.
Therefore pretty good guesses as to the eigenvectors will give us very good estimates of the natural
frequencies.

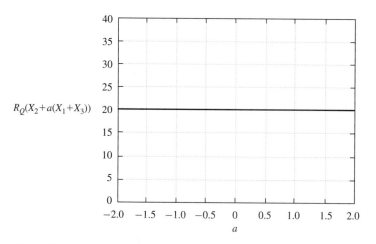

**Figure 6.8** Behavior of Rayleigh's quotient about the second eigenvector when perturbed by a combination of both remaining eigenvectors

Having numerically demonstrated all of this, it's time to look at the mathematics to see why Rayleigh's quotient behaves in the way that it does. We'll start by examining (6.3.17) more closely. We already know that the eigenvectors of a system span the system's configuration space. Thus, any vector that we come up with can be represented by an appropriate combination of the system's eigenvectors. We also know from our recent work that Rayleigh's quotient is stationary about the system's eigenvectors. We'll assume that our estimated vector is given by

$$\hat{X} = a_1 X_1 + a_2 X_2 + \cdots + a_{j-1} X_{j-1} + X_j + a_{j+1} X_{j+1} + \cdots + a_n X_n \qquad (6.3.25)$$

that is,

$$\hat{X} = \sum_{i=1}^{n} a_i X_i \qquad (6.3.26)$$

where $a_j = 1$. We'll further assume that the remaining $a_i$'s are small ($a_i \ll 1$). Thus $\hat{X}$ is mainly equal to the $j$th eigenvector $X_j$, with only small contamination from the other eigenvectors. This is the definition of a *good* estimate; it's close to, but not exactly equal to, the actual eigenvector. We don't have to pin down the $a_i$'s exactly, it's enough to know that they're small.

Our next step is to see what Rayleigh's quotient looks like in terms of our estimate $\hat{X}$. Using (6.3.26) in (6.3.17) gives us

$$R_Q(\hat{X}) = \frac{\left(\sum_{i=1}^{n} a_i X_i^T\right) [K] \left(\sum_{i=1}^{n} a_i X_i\right)}{\left(\sum_{i=1}^{n} a_i X_i^T\right) [M] \left(\sum_{i=1}^{n} a_i X_i\right)} \qquad (6.3.27)$$

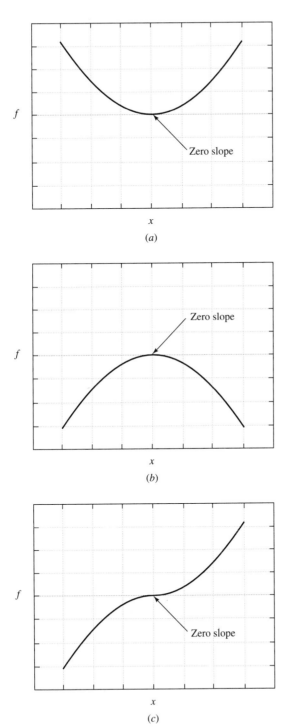

**Figure 6.9**  Zero-slope behavior when a function is evaluated at a minumum, maximum, and inflection point

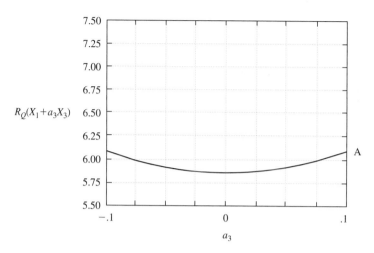

**Figure 6.10** Behavior of Rayleigh's quotient about the first eigenvector (enlarged)

For clarity, we can explicitly expand the summation out

$$R_Q(\hat{X}) = \frac{(a_1 X_1^T + a_2 X_2^T + \cdots)[K](a_1 X_1 + a_2 X_2 + \cdots)}{(a_1 X_1^T + a_2 X_2^T + \cdots)[M](a_1 X_1 + a_2 X_2 + \cdots)} \tag{6.3.28}$$

We now need to recall that the eigenvectors of a system are orthogonal with respect to the $[K]$ and $[M]$ matrices. Therefore many of the multiplications in (6.3.28) will simply yield zero. Let's also decide that our eigenvectors have been normalized so that

$$X_i^T[M]X_i = 1 \tag{6.3.29}$$

This step isn't actually necessary, it just makes the mathematics a bit clearer. If the vectors have been normalized in this way, then we know from Chapter 4 that

$$X_i^T[K]X_i = \omega_i^2 \tag{6.3.30}$$

If we now use the orthogonality properties on (6.3.28) we'll be left with

$$R_Q(\hat{X}) = \frac{a_1^2 \omega_1^2 + a_2^2 \omega_2^2 + \cdots + \omega_j^2 + \cdots a_n^2 \omega_n^2}{a_1^2 + a_2^2 + \cdots + 1.0 + \cdots + a_n^2} \tag{6.3.31}$$

The denominator can be rearranged slightly,

$$R_Q(\hat{X}) = \frac{a_1^2 \omega_1^2 + a_2^2 \omega_2^2 + \cdots + \omega_j^2 + \cdots a_n^2 \omega_n^2}{1 + a_1^2 + a_2^2 + \cdots} \tag{6.3.32}$$

Since all the $a_i$'s are small, the denominator is in the form

$$1 + \epsilon \tag{6.3.33}$$

where $\epsilon$ is small. You may recall that the inverse of $1 + \epsilon$ is given by

$$\frac{1}{1 + \epsilon} = 1 - \epsilon + \epsilon^2 + \cdots \tag{6.3.34}$$

If we use this approximation in (6.3.32), we can rewrite Rayleigh's quotient as

$$R_Q(\hat{X}) = \left(a_1^2 \omega_1^2 + a_2^2 \omega_2^2 + \cdots + \omega_j^2 + \cdots a_n^2 \omega_n^2\right) \left(1 - (a_1^2 + a_2^2 + \cdots)\right.$$
$$+ (a_1^2 + a_2^2 + \cdots)^2 - \cdots) \tag{6.3.35}$$

All we'll need to be interested in are the terms up to second order (terms like $a_1^2$), and so when we expand out (6.3.35) we will neglect all higher-order terms. Therefore

$$R_Q(\hat{X}) \approx \omega_j^2 + a_1^2(\omega_1^2 - \omega_j^2) + a_2^2(\omega_2^2 - \omega_j^2) + \cdots + a_{j-1}^2(\omega_{j-1}^2 - \omega_j^2)$$
$$+ a_{j+1}^2(\omega_{j+1}^2 - \omega_j^2) + \cdots + a_n^2(\omega_n^2 - \omega_j^2) \tag{6.3.36}$$

Let's take a second to think about what we have. We see that our Rayleigh's quotient is equal to the square of the $j$th natural frequency plus error terms. Furthermore, all of the error terms are second-order terms! Even though our input vector had first-order errors (terms such as $a_1 X_1$), no equivalent errors are showing up in our Rayleigh's quotient approximation. If they had, we'd be seeing terms such as $a_1 \omega_1^2$, $a_2 \omega_j^2$, etc. This is what gives us our great accuracy. The errors in the input depend on the $a_i$'s, which are small. Well, if the $a_i$'s are small, then the $a_i^2$ are absolutely tiny. Equation (6.3.36) shows that our approximate value is equal to the actual eigenvalue we're seeking (the $\omega_j^2$ term), plus some tiny errors that are proportional to the $a_i^2$'s. A decent guess leads to an excellent approximation, just as we've already seen.

Finally, can (6.3.36) also predict the precise kind of behavior we saw in Figures 6.4–6.8? Absolutely. Let's first consider what we get if we're looking for the first natural frequency. In this case, $j = 1$ and (6.3.36) becomes

$$R_Q(\hat{X}) = \omega_1^2 + a_2^2(\omega_2^2 - \omega_1^2) + a_3^2(\omega_3^2 - \omega_1^2) + \cdots \tag{6.3.37}$$

Since $\omega_1$ is the lowest natural frequency, all of the error terms are positive (the $a_i^2$ are always positive and the $\omega_i^2 - \omega_1^2$ terms are also positive since $\omega_i > \omega_1$). Since all the error terms are positive, we see that $R_Q$ is a minimum around the first eigenvector; all contaminating vectors can serve only to increase its value.

If we go to the other end of the spectrum and look at the highest natural frequency, then

$$R_Q(\hat{X}) = \omega_n^2 + a_1^2(\omega_1^2 - \omega_n^2) + a_2^2(\omega_2^2 - \omega_n^2) + \cdots \tag{6.3.38}$$

For this case $R_Q$ is a maximum, since any contamination will reduce its value (the $a_i^2$ terms are still positive and all the differences between natural frequencies are negative).

For any other natural frequency, we'll have some positive and some negative terms. This can be seen in (6.3.36). Whenever $i > j$, the $a_i^2(\omega_i^2 - \omega_j^2)$ terms will be positive (causing positive errors), and whenever $i < j$, the $a_i^2(\omega_i^2 - \omega_j^2)$ terms will be negative (causing negative errors).

All these characteristics can be seen in Figures 6.4–6.8. The only thing to keep in mind when looking at these figures and our mathematical results is that our analysis of Rayleigh's quotient has all been based on input vectors that are close to the actual eigenvectors. The results are, therefore, only valid for small $a_i$, as previously stated. The plots in Figures 6.4–6.8 hold for both small and large values of $a_i$. Thus the analysis and figures will match for small values of $a_i$ but diverge as the $a_i$ get too large. For instance, Figure 6.4$b$ shows how Rayleigh's quotient behaves for the first eigenvector, with some contamination from the third. Our mathematical analysis says that the plot should look like

$$R_Q(\hat{X}) = \omega_1^2 + a_3(\omega_3^2 - \omega_1^2) \tag{6.3.39}$$

which it does, for small values of $a_3$. However, as $a_3$ increases, the plot levels off, approaching $\omega_3^2$, not a surprising result since $X_3$ becomes the dominant component of $\hat{X}$ as $a_3$ increases. Figure 6.11 superimposes both these solutions, showing how the two systems behave similarly for $a_3$ around zero and diverge as the magnitude of $a_3$ grows.

One additional observation comes directly from looking at Rayleigh's quotient. Since Rayleigh's quotient is equal to $\dfrac{X^T[K]X}{X^T[M]X}$ we can see that increasing $[K]$, while holding $[M]$ fixed, will increase the resulting frequency estimates. Thus increasing $[K]$, which means increasing the

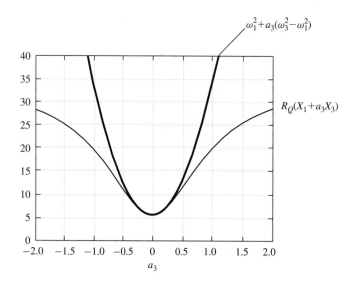

**Figure 6.11** Comparison of Rayleigh's quotient and a second-order polynomial approximation

stiffness in the system, results in higher natural frequencies. Similarly, increasing the mass of the system (leading to larger values of $[M]$) serves to decrease the natural frequencies, since Rayleigh's quotient is proportional to $\frac{1}{X^T[M]X}$. This is in line with our intuition and with results we've already seen.

## 6.4 RAYLEIGH-RITZ METHOD: DISCRETE SYSTEMS

Rayleigh's quotient is a powerful tool for estimating system natural frequencies, but we can go quite a bit further to improve our results. The first person to realize this was Walter Ritz, the reason that the method is called the Rayleigh-Ritz method. The real power of the Rayleigh-Ritz method is how it allows us to analyze complicated continuous problems. However, in order to illustrate how the method works, we're first going to confine our attention to finite-dimensional systems. This will make it much easier to grasp what the mathematics is doing for us. Only after we've mastered a finite-dimensional application of the method will we move on to the more physically meaningful continuous problems.

Okay, back to the method. Ritz looked at results that were the same as those he just finished deriving in the last section and asked himself if there was any way in which he could derive even more information. His logic went something like this:

> Rayleigh's quotient gives us a good way of estimating a system's fundamental frequency. If I use a pretty good estimate for the first eigenvector, I'll get an excellent approximation. But I'd like to know more than just the fundamental frequency; I'd like several natural frequencies and I'd like to know their associated eigenvectors. Is there a way that I could use more than one estimated vector, and mix them around in an appropriate way, so that I end up with estimates for several natural frequencies and several eigenvectors? And can I work it so that, as I use more and more estimated vectors, my results get better and better?

Well, what Ritz decided, after a bit of rumination, was that the fundamental reality of Rayleigh's quotient was the fact that it was stationary when perturbed around any of the system's actual eigenvectors. (We studied this fact in detail in the previous section.) So what Ritz decided to do was to generalize the concept of an estimated vector. Rather than use just a single vector as an estimated vector, Ritz used several. Thus, he first decided on a *set* of trial vectors, $\hat{X}_1$, $\hat{X}_2$, .... Keep in mind that these trial vectors are not the eigenvectors (unless you're real lucky in choosing your trial vectors). Each $\hat{X}_i$ is simply your best estimate of the system's eigenvectors. They might be good estimates or they might be lousy. What we'll see is that, either way, you'll get good results from the Rayleigh-Ritz method as long as you use enough trial vectors. However, the better your initial estimates, the quicker and more easily you'll get a good final result. Also, it's important to make sure that the trial vectors are independent from each other. In the work to follow, we'll be adding the vectors together to get new vectors, something that wouldn't work well if we had dependent vectors floating around.

Once he'd decided on his trial vectors, Ritz decided to work *as if* these were really the system eigenvectors (which, again, they aren't). But *IF* they actually were the eigenvectors, then Rayleigh's quotient would be stationary with respect to perturbations around them. This is the key to Ritz's breakthrough. He chose to treat his trial vectors as if they were the eigenvectors, and thus he proposed forcing Rayleigh's quotient to be stationary with respect to perturbations about them. Again, this is a crucial point. Rayleigh's quotient is not generally stationary with respect to perturbations about some arbitrary set of trial vectors; it's stationary with respect to the actual eigenvectors. Ritz's conceptual leap was to realize that if he forced Rayleigh's quotient to be stationary with respect to his trial vectors, it would cause Rayleigh's quotient to put these trial vectors together in such a way as to end up being as close as possible to the actual eigenvectors.

Since the trial vectors are independent, it is certainly possible to add them together in order to get new vectors. And it's also true that, if we knew what the actual eigenvectors looked like, we could add them up in such a way as to best approximate the eigenvectors. The Rayleigh-Ritz method will figure out how to do this for us, without actually ever knowing what the real eigenvectors are. Pretty good trick, no?

Here's how it's done. We know that after we've appropriately added together our trial vectors in order to approximate an eigenvector, the resultant vector will look like

$$\text{Est} = \sum_{i=1}^{n} b_i \hat{X}_i \tag{6.4.1}$$

I'm really sorry about all the subscripts and superscripts, the hats and tildes and whatnot. They're necessary, but I know they're a real pain to keep track of. That's why I won't introduce yet another one to stand for "the best estimate of the eigenvector that's constructed from the trial vectors." I'll simply call this Est, for "Estimated eigenvector." I hope this makes it a little easier to follow. The $\hat{X}_i$ are our trial vectors and the $b_i$ are the coefficients that cause the overall summation to best approximate the eigenvector.

The only tricky part, therefore, is to determine exactly what these $b_i$ are equal to. If we know that, then we're done, since we'd just use (6.4.1) to find Est. As mentioned, it's the process of forcing Rayleigh's quotient to be stationary that will give us these $b_i$.

If we write out Rayleigh's quotient in terms of our estimated eigenvector, we'll have

$$R_Q(\text{Est}) = \frac{\text{Est}^T[K]\text{Est}}{\text{Est}^T[M]\text{Est}} \tag{6.4.2}$$

However, if we look closely at what Est is, we see that the real unknowns are the $b_i$. So, since the only real unknowns in Rayleigh's quotient are the $b_i$, we can can indicate this by writing (6.4.2) as

$$R_Q(\text{Est}) = R_Q(b_i) \tag{6.4.3}$$

Equation (6.4.3) states that the unknown quantities are the $b_i$, nothing more. Having said this, we'll refrain from writing the $b_i$ dependency explicitly, since it will make the following derivations even more cluttered than they already are. Just keep in mind that when you see $R_Q$, this means the

Rayleigh's quotient that depends on all the unknown $b_i$, as well as the particular $\hat{X}_i$ we've chosen (which are already known).

It's now time to make Rayleigh's quotient stationary. Recall that the definition of stationarity is that the first partial derivatives with respect to each of the independent variables must be zero. This is exactly what we saw in Figures 6.4–6.8. We looked at Rayleigh's quotient around each of the eigenvectors and saw that as we varied our trial vector in the direction of another eigenvector, holding the others fixed, the value of Rayleigh's quotient didn't change. This is precisely what is meant by the first partials being equal to zero. We therefore require

$$\frac{\partial R_Q}{\partial b_i}\bigg|_{\text{all } b\text{'s}=0} = 0, \quad i = 1, 2, \dots, n \tag{6.4.4}$$

for every $b_i$.

Since both the numerator and denominator of Rayleigh's quotient depend on the $b_i$, we'll have to use the chain rule to perform this differentiation. If we write Rayleigh's quotient as

$$R_Q = \frac{\text{Num}}{\text{Den}} \tag{6.4.5}$$

where $\text{Num} = \text{Est}^T[K]\text{Est}$ and $\text{Den} = \text{Est}^T[M]\text{Est}$, then

$$\frac{\partial R_Q}{\partial b_i} = \frac{\partial \left[\text{Num}\,(\text{Den})^{-1}\right]}{\partial b_i} = \frac{-\text{Num}\frac{\partial \text{Den}}{\partial b_i} + \text{Den}\frac{\partial \text{Num}}{\partial b_i}}{\text{Den}^2} = 0 \tag{6.4.6}$$

Den is always positive since it involves the positive masses of the system and quadratic (positive) combinations of the trial vectors. Consequently, the only way we can satisfy (6.4.6) is if

$$-\text{Num}\frac{\partial \text{Den}}{\partial b_i} + \text{Den}\frac{\partial \text{Num}}{\partial b_i} = 0 \tag{6.4.7}$$

or, after dividing by Den,

$$\frac{\partial \text{Num}}{\partial b_i} - R_Q\frac{\partial \text{Den}}{\partial b_i} = 0 \tag{6.4.8}$$

What we have to do now is figure out what partial differentiation of the numerator and denominator gives us. If we focus on the first term, what we have is

$$\frac{\partial \text{Num}}{\partial b_i} = \frac{\text{Est}^T[K]\text{Est}}{\partial b_i} = \frac{\partial}{\partial b_i}\left[(b_1\hat{X}_1^T + b_2\hat{X}_2^T + \cdots)[K](b_1\hat{X}_1 + b_2\hat{X}_2 + \cdots)\right]$$

$$= \frac{\partial}{\partial b_i}\big[b_1^2\hat{X}_1^T[K]\hat{X}_1 + b_2^2\hat{X}_2^T[K]\hat{X}_2 + \cdots + b_1b_2\hat{X}_1^T[K]\hat{X}_2 + b_2b_1\hat{X}_2^T[K]\hat{X}_1$$

$$+ b_1b_3\hat{X}_1^T[K]\hat{X}_3 + b_3b_1\hat{X}_3^T[K]\hat{X}_1 + b_2b_3\hat{X}_2^T[K]\hat{X}_3 + b_3b_2\hat{X}_3^T[K]\hat{X}_2 + \cdots\big] \tag{6.4.9}$$

What we see in (6.4.9) is a series of terms that involves only one particular eigenvector (the $\hat{X}_i^T[K]\hat{X}_i$ terms) and another series that involves all the cross-terms (terms such as $\hat{X}_i^T[K]\hat{X}_j$). Since we're differentiating only with respect to one of the $b_i$ at a time, most of the terms will drop out, leaving us with

$$\frac{\partial \mathrm{Num}}{\partial b_i} = 2b_i \hat{X}_i^T[K]\hat{X}_i + b_1 \hat{X}_1^T[K]\hat{X}_i + b_1 \hat{X}_i^T[K]\hat{X}_1 + b_2 \hat{X}_2^T[K]\hat{X}_i$$

$$+ b_2 \hat{X}_i^T[K]\hat{X}_2 + b_3 \hat{X}_3^T[K]\hat{X}_i + b_3 \hat{X}_i^T[K]\hat{X}_3 + \cdots \tag{6.4.10}$$

This can be made a bit more compact by remembering that $[K] = [K]^T$ and $[M] = [M]^T$. Since the terms $\hat{X}_i^T[K]\hat{X}_j$ are just scalar quantities, they are equal to their own transpose. Thus $\left(\hat{X}_i^T[K]\hat{X}_j\right)^T = \hat{X}_j^T[K]^T \hat{X}_i = \hat{X}_j^T[K]\hat{X}_i$. So we see that each member of the crossproduct pairs in (6.4.10) are equal to the same value, leading to

$$\frac{\partial \mathrm{Num}}{\partial b_i} = 2b_i \hat{X}_i^T[K]\hat{X}_i + 2b_1 \hat{X}_1^T[K]\hat{X}_i + 2b_2 \hat{X}_2^T[K]\hat{X}_i + 2b_3 \hat{X}_3^T[K]\hat{X}_i + \cdots \tag{6.4.11}$$

You'll note that there's a pretty clear pattern shaping up in these results. To see this, let's first make a couple of new definitions. Rather than writing out all these $\hat{X}_i^T[K]\hat{X}_j$ terms, we'll say that

$$\mathcal{K}_{ij} \equiv \hat{X}_i^T[K]\hat{X}_j \tag{6.4.12}$$

and

$$\mathcal{M}_{ij} \equiv \hat{X}_i^T[M]\hat{X}_j \tag{6.4.13}$$

Combining (6.4.1) and (6.4.5) and making use of (6.4.12) and (6.4.13) lets us write out the denominator and numerator of Rayleigh's quotient in the following form:

$$\mathrm{Num} = \sum_{i=1}^{n}\sum_{j=1}^{n} b_i b_j \mathcal{K}_{ij} \tag{6.4.14}$$

and

$$\mathrm{Den} = \sum_{i=1}^{n}\sum_{j=1}^{n} b_i b_j \mathcal{M}_{ij} \tag{6.4.15}$$

Written like this, it is clear that Rayleigh's quotient depends only upon the $b$, all the $\mathcal{K}_{ij}$'s and $\mathcal{M}_{ij}$ being known constants.

Using this new $\mathcal{K}_{ij}$ and $\mathcal{M}_{ij}$ notation, we can rewrite (6.4.11) as

$$\frac{\partial \mathrm{Num}}{\partial b_i} = 2(b_1\mathcal{K}_{i1} + b_2\mathcal{K}_{i2} + b_3\mathcal{K}_{i3} + \cdots) \tag{6.4.16}$$

which can be written as

$$\frac{\partial \mathrm{Num}}{\partial b_i} = 2 \begin{bmatrix} \mathcal{K}_{i1} & \mathcal{K}_{i2} & \mathcal{K}_{i3} & \cdots \end{bmatrix} \begin{Bmatrix} b_1 \\ b_2 \\ b_3 \\ \vdots \end{Bmatrix} \tag{6.4.17}$$

We can now follow exactly the same procedure for $\frac{\partial \mathrm{Den}}{\partial b_i}$. Since the form of the denominator and numerator are identical, the only difference being the use of $[M]$ instead of $[K]$, we can immediately realize that

$$\frac{\partial\,Den}{\partial b_i} = 2\,[\mathcal{M}_{i1}\ \ \mathcal{M}_{i2}\ \ \mathcal{M}_{i3}\ \ \cdots]\begin{Bmatrix} b_1 \\ b_2 \\ b_3 \\ \vdots \end{Bmatrix} \tag{6.4.18}$$

So, using (6.4.16) and (6.4.18), we can write the $i$th equation (6.4.8) as

$$[\mathcal{K}_{i1}\ \ \mathcal{K}_{i2}\ \ \mathcal{K}_{i3}\ \ \cdots]\begin{Bmatrix} b_1 \\ b_2 \\ b_3 \\ \vdots \end{Bmatrix} - R_Q\,[\mathcal{M}_{i1}\ \ \mathcal{M}_{i2}\ \ \mathcal{M}_{i3}\ \ \cdots]\begin{Bmatrix} b_1 \\ b_2 \\ b_3 \\ \vdots \end{Bmatrix} = 0 \tag{6.4.19}$$

This is obviously true for any $i$, and so we can write out all $n$ equations by filling in all the other rows of $\mathcal{M}$ and $\mathcal{K}$,

$$\begin{bmatrix} \mathcal{K}_{11} & \mathcal{K}_{12} & \vdots \\ \mathcal{K}_{12} & \mathcal{K}_{22} & \vdots \\ \cdots & \cdots & \mathcal{K}_{nn} \end{bmatrix}\begin{Bmatrix} b_1 \\ b_2 \\ \vdots \\ b_n \end{Bmatrix} - R_Q\begin{bmatrix} \mathcal{M}_{11} & \mathcal{M}_{12} & \vdots \\ \mathcal{M}_{12} & \mathcal{M}_{22} & \vdots \\ \cdots & \cdots & \mathcal{M}_{nn} \end{bmatrix}\begin{Bmatrix} b_1 \\ b_2 \\ \vdots \\ b_n \end{Bmatrix} = \begin{Bmatrix} 0 \\ 0 \\ \vdots \\ 0 \end{Bmatrix} \tag{6.4.20}$$

or, more compactly,

$$[\mathcal{K}]B - R_Q[\mathcal{M}]B = 0 \tag{6.4.21}$$

What does this look like? An eigenvalue problem! (What else?) If we realize that $R_Q$ is an unknown scalar, and the elements of $B$ are also unknown, then we see we have $n+1$ unknowns and $n$ equations, just like all our previous eigenvalue problems. If we further realize that $R_Q$ is our approximation to the square of the system's natural frequencies, then (6.4.21) looks exactly like our old friend $[M]\ddot{X} + [K]X = 0$, since the eigenvalue form of this problem is given by $[K]X_i - \omega_i^2[M]X_i = 0$.

Now, let's think a bit about what (6.4.21) is telling us. If we use only one trial vector, then (6.4.21) becomes a scalar equation. In fact, it becomes exactly equivalent to the Rayleigh's quotient approach. If we add another trial vector, then the matrices in (6.4.21) are $2 \times 2$. Three trial vectors makes the problem three dimensional. And so on. What we have is an eigenvalue problem that has dimensions equal to the number of trial vectors we're using. Let's assume that we want to use three trial vectors. Then, when we solve the associated eigenvalue problem, we'll find three values for the system natural frequencies (corresponding to the three roots of the characteristic equation for the problem). For each of these solutions we'll have an associated eigenvector. Note that this is *not* a system eigenvector. Rather, it tells us what the $b_i$ should be to best approximate a system eigenvector using our trial functions. The following examples will illustrate this point.

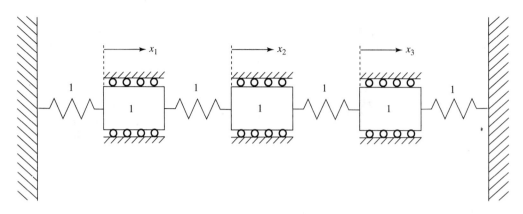

**Figure 6.12**   3-DOF spring-mass system, spatially symmetric masses and springs

**Example 6.3**

**Problem**   We'll start by looking at a simple example, one that's very easy to analyze analytically and that will therefore allow us to compare the Rayleigh-Ritz approach to known results. The system we'll consider is shown in Figure 6.12 and is simply that well-used example, a serial chain of masses and springs. For this example we'll let all the spring constants be equal to 1 N/m and all the masses equal 1 kg.

**Solution**   We know that the eigenvalue problem is given by

$$\left[[K] - \omega^2[M]\right] X = \{0\} \tag{6.4.22}$$

and has solutions $\omega_1^2 = .5858$ (rad/s)$^2$, $\omega_2^2 = 2.000$ (rad/s)$^2$, and $\omega_3^2 = 3.4142$ (rad/s)$^2$, with associated eigenvectors

$$X_1 = \left\{ \begin{array}{c} 1 \\ \sqrt{2} \\ 1 \end{array} \right\}, \quad X_2 = \left\{ \begin{array}{c} 1 \\ 0 \\ -1 \end{array} \right\}, \quad X_3 = \left\{ \begin{array}{c} 1 \\ -\sqrt{2} \\ 1 \end{array} \right\} \tag{6.4.23}$$

Notice that these eigenvectors are somewhat special due to the system's configuration. Since the system has identical masses and identical springs, it will produce eigenvectors that are symmetric with respect to the midpoint of the vector or antisymmetric. Thus $X_2$, an antisymmetric eigenvector, is fundamentally different from $X_1$ and $X_3$, which are symmetric. We'll see that this will affect our Rayleigh-Ritz results.

To begin, we'll use

$$\hat{X}_1 = \left\{ \begin{array}{c} 1 \\ 2 \\ 1 \end{array} \right\} \tag{6.4.24}$$

as our initial trial vector. If we calculate $\mathcal{M}_{11}$ and $\mathcal{K}_{11}$ as suggested in (6.4.12) and (6.4.13), we'll find that

$$\mathcal{M}_{11} = 6 \tag{6.4.25}$$

and

$$\mathcal{K}_{11} = 4 \tag{6.4.26}$$

The Rayleigh-Ritz equation is then

$$4 - 6R_Q = 0 \tag{6.4.27}$$

or $R_Q = .667$ (rad/s)$^2$. Since our estimated trial function was quite a bit off the actual eigenvector, we obtained only a reasonably accurate eigenvalue estimate for the actual value of .5858 (rad/s)$^2$. We'll now move on to a two-trial function analysis.

**Example 6.4**

**Problem**   Use two trial vectors to analyze the problem of Example 6.3.

**Solution**   This time, we'll say that

$$\hat{X}_1 = \left\{ \begin{array}{c} 1 \\ 2 \\ 1 \end{array} \right\}, \qquad \hat{X}_2 = \left\{ \begin{array}{c} 1 \\ 0 \\ -1 \end{array} \right\} \tag{6.4.28}$$

This case is special in two ways. For one thing, we've added an antisymmetric vector to our trial set, which previously consisted of only a symmetric vector. Since the original system is symmetric, it can be broken into two simpler parts, the symmetric parts being independent of the antisymmetric ones; and it would seem that adding this vector won't do us any good. Since the actual first eigenvector is symmetric, adding an antisymmetric correction by means of the second trial vector shouldn't improve the estimate.

In addition, this second vector happens to be equal to the second eigenvector. Therefore there is no reason for the Rayleigh-Ritz method to add anything to it; it's already as close as it can get to the correct vector.

Having said all that, let's see what we actually get. Our $\mathcal{K}$ and $\mathcal{M}$ matrices are

$$\mathcal{M} = \begin{bmatrix} 6 & 0 \\ 0 & 2 \end{bmatrix}, \qquad \mathcal{K} = \begin{bmatrix} 4 & 0 \\ 0 & 4 \end{bmatrix} \tag{6.4.29}$$

and so our eigenvalue problem becomes

$$\begin{bmatrix} 4 & 0 \\ 0 & 4 \end{bmatrix} \left\{ \begin{array}{c} b_1 \\ b_2 \end{array} \right\} - R_Q \begin{bmatrix} 6 & 0 \\ 0 & 2 \end{bmatrix} \left\{ \begin{array}{c} b_1 \\ b_2 \end{array} \right\} = \{0\} \tag{6.4.30}$$

Notice that there are no off-diagonal terms in this set of equations. As we saw in Chapter 4, this means that the equations are completely uncoupled. Thus we have two individual equations, neither of which influences the other. Solving them gives us

$$R_{Q_1} = .\bar{6} \text{ (rad/s)}^2, \quad B_1 = \left\{ \begin{array}{c} 1 \\ 0 \end{array} \right\}$$

$$R_{Q_2} = 2.0 \text{ (rad/s)}^2, \quad B_2 = \left\{ \begin{array}{c} 0 \\ 1 \end{array} \right\} \tag{6.4.31}$$

Note that we are using $B_i$ to denote the vector of $b_i$'s associated with $R_{Q_i}$ and that the $B_i$'s components are used to calculate our estimate of the system's eigenvectors. As we expected, the addition of $\hat{X}_2$ didn't alter Est$_1$ in the slightest, as indicated by the fact that the second component is equal to zero. Thus

$$\text{Est}_1 = b_1\hat{X}_1 + b_2\hat{X}_2 = 1\begin{Bmatrix} 1 \\ 2 \\ 1 \end{Bmatrix} + 0\begin{Bmatrix} 1 \\ 0 \\ -1 \end{Bmatrix} = \begin{Bmatrix} 1 \\ 2 \\ 1 \end{Bmatrix} \tag{6.4.32}$$

Similarly, the estimate for the second natural frequency and eigenvector is unaffected by $\hat{X}_1$.

## Example 6.5

**Problem**   Repeat Example 6.4 with different trial vectors.

**Solution**   Having seen that the choice of trial vectors in the preceding example didn't induce any improvement in the eigenvalue estimate, let's see how a different choice fares. For this example we'll use two dissimilar vectors that are neither symmetric nor antisymmetric. Hopefully the Rayleigh-Ritz procedure will use them to create approximations to the actual eigenvalues and eigenvectors.

The two new trial vectors are

$$\hat{X}_1 = \begin{Bmatrix} 1 \\ 3 \\ 2 \end{Bmatrix}, \qquad \hat{X}_2 = \begin{Bmatrix} 2 \\ 1 \\ -1 \end{Bmatrix} \tag{6.4.33}$$

Using these new values for our trial vectors, we can calculate $\mathcal{K}$ and $\mathcal{M}$ and obtain

$$\left( \begin{bmatrix} 10 & 0 \\ 0 & 10 \end{bmatrix} \begin{Bmatrix} b_1 \\ b_2 \end{Bmatrix} - R_Q \begin{bmatrix} 14 & 3 \\ 3 & 6 \end{bmatrix} \begin{Bmatrix} b_1 \\ b_2 \end{Bmatrix} \right) = \{0\} \tag{6.4.34}$$

If we divide $\mathcal{K}_{11}$ by $\mathcal{M}_{11}$ we get .7143 (rad/s)$^2$ as our single-vector estimate for the square of the first natural frequency. You'll note that this is worse than our previous single-vector estimate (which was $.\bar{6}$ (rad/s)$^2$). This makes sense since this trial vector is even further from the first eigenvector than the original trial vector was. If we calculate the solution to (6.4.34), we'll obtain

$$R_{Q_1} = .\bar{6} \text{ (rad/s)}^2, \quad B_1 = \begin{Bmatrix} 3 \\ 1 \end{Bmatrix}$$

$$R_{Q_2} = 2.0 \text{ (rad/s)}^2, \quad B_2 = \begin{Bmatrix} -1 \\ 3 \end{Bmatrix} \tag{6.4.35}$$

Interesting. We've obtained the exact same estimates for the natural frequencies as we did last time. However, the associated eigenvectors have changed. Let's take a look at what our new estimated eigenvectors are. The first is

$$\text{Est}_1 = b_1\hat{X}_1 + b_2\hat{X}_2 = 3\begin{Bmatrix} 1 \\ 3 \\ 2 \end{Bmatrix} + 1\begin{Bmatrix} 2 \\ 1 \\ -1 \end{Bmatrix} = \begin{Bmatrix} 5 \\ 10 \\ 5 \end{Bmatrix} \tag{6.4.36}$$

This is simply five times our old first trial vector! And since constant multiples make no difference to eigenvalues, we can just as well say it *is* our old one. The second is given by

$$\text{Est}_2 = -1 \begin{Bmatrix} 1 \\ 3 \\ 2 \end{Bmatrix} + 3 \begin{Bmatrix} 2 \\ 1 \\ -1 \end{Bmatrix} = \begin{Bmatrix} 5 \\ 0 \\ -5 \end{Bmatrix} \tag{6.4.37}$$

which is also five times our old second trial vector. So the new set of two trial vectors must actually be just a linear combination of the old set. This is worth remembering. If you have a set of trial vectors and another set that are combinations of the first, then your new frequency estimates will be the same as the old ones. Along with this goes the fact that your estimated eigenvectors will also remain unchanged. The $b_i$'s will change, of course, since the trial vectors have been altered. But when you examine the final estimated eigenvectors that are indicated by the $b_i$'s, you'll see that you get the same answers.

We can gain some more insight into this by examining the $\mathcal{K}$ and $\mathcal{M}$ matrices and recalling our knowledge of normal coordinates. We know from Chapter 4 that

- the normal coordinates are those that decouple our equations of motion.
- the decoupled mass and stiffness matrices are diagonal.
- the normal coordinates are always linear combinations of our original coordinates.

Well, we can view our new set of $\hat{X}_i$'s as our original coordinate set. The Rayleigh-Ritz method takes these vectors and combines them so that Rayleigh's quotient is stationary about them. There's only one set of vectors that will satisfy this property of stationarity. We'll know that we have them because the $\mathcal{K}$ and $\mathcal{M}$ matrices associated with them will be diagonal. Thus the resultant eigenvectors $B_i$ will be unitary (ones on the diagonal with everything else equal to zero), which will mean that we can't improve the trial vectors by combining them; they're optimal as they are.

With this in mind, take a look at the $\mathcal{K}$ and $\mathcal{M}$ matrices of our first trial set. They are already in diagonal form. Thus any linear combination of them, when run through the Rayleigh-Ritz method, will simply reproduce them.

It's important to note that, even though these matrices are diagonal, they aren't necessarily associated with the best possible results. It would be best if the vectors were exactly equal to the actual eigenvectors. In this case the $\mathcal{K}$ and $\mathcal{M}$ matrices would be diagonal and the eigenvalue estimates would be exact (which they aren't for our examples). These are the two ways by which we can obtain completely uncoupled $\mathcal{K}$ and $\mathcal{M}$ matrices: by using the actual eigenvectors or by choosing (or calculating) the combination of any original trial set that renders Rayleigh quotient stationary.

Having seen what happened with our first two trial sets, its time to try a third.

**Example 6.6**

**Problem**   Repeat Example 6.4 with different trial vectors.

**Solution**   This time, we'll use

$$
\hat{X}_1 = \begin{Bmatrix} 4 \\ 3 \\ 0 \end{Bmatrix}, \qquad \hat{X}_2 = \begin{Bmatrix} 0 \\ 2 \\ 4 \end{Bmatrix} \tag{6.4.38}
$$

The associated $\mathcal{K}$ and $\mathcal{M}$ matrices are

$$
\mathcal{K} = \begin{bmatrix} 26 & -8 \\ -8 & 24 \end{bmatrix}, \qquad \mathcal{M} = \begin{bmatrix} 25 & 6 \\ 6 & 20 \end{bmatrix} \tag{6.4.39}
$$

Just using the first trial vector gives us

$$
R_Q = \frac{\mathcal{K}_{11}}{\mathcal{M}_{11}} = 1.04 \; (\text{rad/s})^2 \tag{6.4.40}
$$

This is not terribly close to the actual value of .5858 rad/s² (having an error of 78 percent). Let's see how adding the second trial vector helps. If we go ahead and solve the $2 \times 2$ eigenvalue problem, we'll find

$$
R_{Q_1} = \; .5961 \; (\text{rad/s})^2, \quad B_1 = \begin{Bmatrix} .2500 \\ .2396 \end{Bmatrix}
$$

$$
R_{Q_2} = 2.0246 \; (\text{rad/s})^2, \quad B_2 = \begin{Bmatrix} .2500 \\ -.3054 \end{Bmatrix} \tag{6.4.41}
$$

These are more typical results. Our original $\mathcal{K}$ and $\mathcal{M}$ matrices were not decoupled, thus our trial vectors were improved by the Rayleigh-Ritz procedure. Unlike last time, neither of the frequency estimates was exact, both were simply approximations to the actual eigenvalues. At .5961 $(\text{rad/s})^2$, our estimate for $\omega_1^2$ is greatly improved from the results we had using our first trial vector. Our original 78 percent error has been reduced to 2 percent. To find the associated eigenvector estimate, we must add .2500 times the first trial vector to .2396 times the second, that is, $B_1(1, 1) * \hat{X}_1 + B_1(2, 1) * \hat{X}_2$. If we do this, we'll obtain

$$
\text{Est}_1 = \begin{Bmatrix} 1.000 \\ 1.229 \\ .959 \end{Bmatrix} \tag{6.4.42}
$$

Clearly, this is a lot closer to the actual eigenvector (shown in (6.4.23)) than was the initial trial vector. Our first trial vector was $\{4 \;\; 3 \;\; 0\}^T$, a crummy guess. By appropriately adding a judicious amount of the second trial vector ($\{0 \;\; 2 \;\; 4\}^T$), we came up with $\{1.000 \;\; 1.229 \;\; .959\}^T$, a really good estimate. Similarly, neither initial vector is close to the system's second eigenvector, but by subtracting .3054 times the second trial vector from .2500 times the first, we get $\{1.000 \;\; .139 \;\; -1.222\}^T$, which is quite close to the actual eigenvector. Furthermore, the error in the estimate of $\omega_2^2$ is only 1 percent.

We've seen in Example 6.6 that the Rayleigh-Ritz approach has done an excellent job. We fed in some trial vectors and got out good estimates of the actual system solutions. Of course, the next step is to ask what happens if we add a third trial vector. Well, if we have three independent, $3 \times 1$ vectors to work with, we know that we can create any $3 \times 1$ vector we might wish to; any three independent trial vectors span the system's configuration space. Therefore, if you give the Rayleigh-Ritz formulation three independent vectors and ask it for the combination that best approximates the actual solution, it will give you the actual solution. That's only fair, since at this point you're doing just as much work as solving the original problem would have required. Remember, the actual point of the Rayleigh-Ritz approach is to save time and get good estimates of the first few natural frequencies and eigenfunctions of a continuous system, one that has an infinite number of modes. In this section we restricted the problem to a finite-dimensional one in order to better understand the method. If the original problem has order $n$, then to find the natural frequencies and eigenvectors we must solve an $n$th-order eigenvalue problem. By going ahead and using $n$ trial functions, we'll again get an $n$th-order eigenvalue problem to solve, that is, no savings. But that's okay because we'd never use the method in this way. The whole point is to use it when you want only a few of the modes, not all of them.

**Example 6.7**

**Problem**    Add another trial function to Example 6.6.

**Solution**    Just to demonstrate that we do indeed get the precise answer back if we expand our trial function set, we'll again do a Rayleigh-Ritz analysis, this time using

$$\hat{X}_1 = \begin{Bmatrix} 4 \\ 3 \\ 0 \end{Bmatrix}, \qquad \hat{X}_2 = \begin{Bmatrix} 0 \\ 2 \\ 4 \end{Bmatrix}, \qquad \hat{X}_3 = \begin{Bmatrix} 1 \\ 1 \\ 1 \end{Bmatrix} \qquad (6.4.43)$$

For these three trial vectors we'll find that

$$\mathcal{K} = \begin{bmatrix} 26 & -8 & 4 \\ -8 & 24 & 4 \\ 4 & 4 & 2 \end{bmatrix}, \qquad \mathcal{M} = \begin{bmatrix} 25 & 6 & 7 \\ 6 & 20 & 6 \\ 7 & 6 & 3 \end{bmatrix} \qquad (6.4.44)$$

The full solutions to the Rayleigh-Ritz problem using these matrices are

$$\begin{aligned} R_{Q_1} &= .5858 \ (\text{rad/s})^2 \\ R_{Q_2} &= 2.0000 \ (\text{rad/s})^2 \\ R_{Q_3} &= 3.4142 \ (\text{rad/s})^2 \end{aligned} \qquad (6.4.45)$$

with the associated $B_i$ vectors

$$B_1 = \begin{Bmatrix} .4142 \\ .4142 \\ -.6569 \end{Bmatrix}, \qquad B_2 = \begin{Bmatrix} .0 \\ -.5 \\ 1.0 \end{Bmatrix}, \qquad B_3 = \begin{Bmatrix} -2.4142 \\ -2.4142 \\ 10.6569 \end{Bmatrix} \qquad (6.4.46)$$

As predicted, these natural frequency estimates are exact. Furthermore, if you construct the first estimated eigenvector, you'll see it's

$$\text{Est}_1 = .4142 \begin{Bmatrix} 4 \\ 3 \\ 0 \end{Bmatrix} + .4142 \begin{Bmatrix} 0 \\ 2 \\ 4 \end{Bmatrix} + -.6569 \begin{Bmatrix} 1 \\ 1 \\ 1 \end{Bmatrix} = \begin{Bmatrix} 1 \\ \sqrt{2} \\ 1 \end{Bmatrix} \tag{6.4.47}$$

the actual first eigenvector. Similarly, $\text{Est}_2$ and $\text{Est}_3$ also give us the exact second and third eigenvector.

**Example 6.8**

**Problem**   Example 6.7 gave us a good insight into how the Rayleigh-Ritz method functions. What we'll do now is try a more complicated problem, the solutions of which will more closely resemble real-world results. The system to be analyzed is shown in Figure 6.13. You'll note that it is again a chain of springs and masses; however, this time it is not symmetrical. All the springs are the same, but the first mass is one-half the mass of the others.

**Solution**   The $[K]$ and $[M]$ matrices are given by

$$[K] = \begin{bmatrix} 2 & -1 & 0 & 0 & 0 \\ -1 & 2 & -1 & 0 & 0 \\ 0 & -1 & 2 & -1 & 0 \\ 0 & 0 & -1 & 2 & -1 \\ 0 & 0 & 0 & -1 & 2 \end{bmatrix} \tag{6.4.48}$$

and

$$[M] = \begin{bmatrix} 1 & 0 & 0 & 0 & 0 \\ 0 & 2 & 0 & 0 & 0 \\ 0 & 0 & 2 & 0 & 0 \\ 0 & 0 & 0 & 2 & 0 \\ 0 & 0 & 0 & 0 & 2 \end{bmatrix} \tag{6.4.49}$$

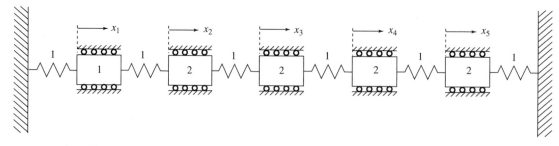

**Figure 6.13**   5-DOF spring-mass system, spatially nonsymmetric masses and springs

Our trial functions will be

$$\hat{X}_1 = \begin{Bmatrix} 2 \\ 3 \\ 4 \\ 3 \\ 1 \end{Bmatrix}, \quad \hat{X}_2 = \begin{Bmatrix} 2 \\ 2 \\ 0 \\ -1 \\ -1 \end{Bmatrix}, \quad \hat{X}_3 = \begin{Bmatrix} 2 \\ 0 \\ -1 \\ 0 \\ 1 \end{Bmatrix}, \quad \hat{X}_4 = \begin{Bmatrix} 2 \\ -1 \\ 0 \\ 1 \\ -1 \end{Bmatrix}$$

and

$$\hat{X}_5 = \begin{Bmatrix} .5 \\ -.4 \\ 1. \\ -.3 \\ .4 \end{Bmatrix} \tag{6.4.50}$$

Based on these trial vectors and the given $[M]$ and $[K]$ matrices, the $[\mathcal{K}]$ and $[\mathcal{M}]$ matrices are found to be

$$[\mathcal{K}] = \begin{bmatrix} 12 & 2 & -1 & 4 & 1.8 \\ 2 & 10 & 4 & 2 & -.9 \\ -1 & 4 & 12 & 7 & 1.2 \\ 4 & 2 & 7 & 20 & 2 \\ 1.8 & -.9 & 1.2 & 2 & 5.36 \end{bmatrix} \tag{6.4.51}$$

and

$$[\mathcal{M}] = \begin{bmatrix} 74 & 8 & -2 & 2 & 5.6 \\ 8 & 16 & 2 & 0 & -.8 \\ -2 & 2 & 8 & 2 & -.2 \\ 2 & 0 & 2 & 10 & .4 \\ 5.6 & -.8 & -.2 & .4 & 3.07 \end{bmatrix} \tag{6.4.52}$$

If we now go through and solve all the possible Rayleigh-Ritz problems (one trial vector, two trial vectors, etc.), we'll finally obtain the estimates shown in Table 6.4.

**TABLE 6.1** $\omega_i^2$ ESTIMATES FOR 1–5 TRIAL VECTORS (rad/s)$^2$

|  | Number of trial vectors | | | | |
|---|---|---|---|---|---|
|  | 1 | 2 | 3 | 4 | 5 |
| $R_{Q1}$ | .1622 | .1612 | .1546 | .1424 | .1393 |
| $R_{Q2}$ |  | .6423 | .5705 | .5688 | .5558 |
| $R_{Q3}$ |  |  | 1.511 | 1.176 | 1.158 |
| $R_{Q4}$ |  |  |  | 2.153 | 1.733 |
| $R_{Q5}$ |  |  |  |  | 2.414 |

Here we see that every time a new vector is added to our trial set, the estimates get better. We start with the Rayleigh's quotient answer of .1622. By adding a second vector, this estimate goes to .1612 and we also get an estimate of the square of the second natural frequency (.6423). Note that when a third trial vector is added, the first estimate drops by quite a bit more than it did when we went from one vector to two (.1612 down to .1546, as opposed to .1622 to .1612). The reason for this is that the second trial vector didn't really do much for us in terms of improving how well we could approximate the first eigenvector. However, the first trial vector plus the third allow us to get a much better approximation, thus the bigger improvement in our estimate. The answers for five trial vectors are exact since the original system only involves five eigenvectors, just as we saw in the previous example.

It's time for a brief recap before moving on to continuous systems. Let's say, for the moment, that the $[M]$ and $[K]$ matrices for our actual system are $10 \times 10$. The easiest attempt at finding the fundamental frequency involves a single equation. This is the Rayleigh's quotient approach. If we decide to use more than one trial vector, we must use the Rayleigh-Ritz approach. As we add more and more trial vectors, our estimates for the system's natural frequencies and eigenvectors become better and better. Each time we add a trial vector we're solving an eigenvalue problem whose size is equal to the number of trial vectors. If we go all the way up to 10 independent trial vectors, we're solving a problem that's every bit as hard as the actual problem we were trying to approximate.

We can also add a few more observations. First, you'll probably have noticed by now that our Rayleigh-Ritz estimates always went down or remained unchanged as we added trial vectors. This isn't a coincidence. If you think about the problem in physical terms, what the Rayleigh-Ritz method does is force the system to behave as if it could move only in certain vector directions. For instance, the actual system might involve five masses, as in the previous example. But by using only three trial vectors, we're constraining the system. We're saying to it, "We don't care how the system wants to move, it can *only* move in the ways allowed by these three vectors." Constraints invariably stiffen systems, and stiffer systems have higher natural frequencies. Thus our estimates are generally greater than the actual values. If, by chance, we happen to pick the actual eigenvector, or if our trial set can be combined to give us an eigenvector, then the natural frequency estimate will equal the actual value. But the estimated value won't ever be lower than the true result. That's one of the nice things about the Rayleigh-Ritz approach: it always gives results that are greater than or equal to the actual values. Of course, if we use $n$ independent trial vectors for an $n$-mass system, then we're still telling the system it has to move in the manner allowed by these vectors. However, any vector (including the eigenvectors) can be formed from the five independent vectors, and thus our results will be precisely equal to the actual values, because our constraints will then have gone to zero.

Another item to be concerned with is how we decide how many trial vectors are enough. Each time we added a new trial vector our estimates improved. However, we don't want to use too many or the method isn't buying us anything in reduced computation. The first thing to do is make sure that our trial vectors are as good as we can make them. The convergence rate is certainly going to be faster if the trial vectors are close to the actual eigenvectors. In fact, if they're equal to the eigenvectors, then the convergence is immediate. Once we have our trial vector set established, then we're still left with the problem of deciding when our results are good enough. Although there's no absolute way to determine how many are enough, a rough rule of thumb is to use twice as many vectors as the number of eigenfrequencies we wish to estimate. Thus, if we want good estimates of the first three eigenfrequencies, we should start with six trial vectors. To be really confident that we have a handle on the actual eigenfrequencies, we would then solve the problem again, this time with even more trial vectors. We would then compare the results for the first three eigenfrequencies. If they don't change too much, we can be confident that we're close to the actual answer. Of course, now you're going to ask how much is too much. Again, the answer will vary depending upon your own application, but generally if the result doesn't change by more than 2 percent, you'd be safe in assuming you're close to the right result.

Keep in mind that this problem of convergence is usually encountered when dealing with continuous problems, in which there are literally an infinite number of eigenfrequencies (at least mathematically). In this case you know the eigenfrequency estimates will continue to change as long as you add new trial functions, and therefore it becomes important to know when to stop.

## 6.5 RAYLEIGH-RITZ METHOD: CONTINUOUS PROBLEMS

Having read the last section, you're now an expert on how the Rayleigh-Ritz method works for finite-dimensional systems. Unfortunately, you'll rarely be able to apply this method to actual finite-dimensional systems since most engineering structures are continuous in nature. Fortunately, the differences between using it for continuous systems and finite-dimensional ones are minimal. As we've already seen, there are several qualitatively different types of continous problems we can examine: one-dimensional string-type problems, two-dimensional membrane problems, fourth-order beam problems, etc. Whereas we easily showed what the Rayleigh-Ritz method was like for all finite dimensional problems (they were all matrix problems), it's not as straightforward for the continuous case. We'll demonstrate by example how the method works and then indicate how you can apply it to any other type of problem.

The first example we'll look at is the bar-vibration problem from Chapter 5. You'll recall from (5.2.6) that the governing equation for free vibrations of a bar is given by

$$m(x)\ddot{\xi} = (EA(x)\xi_x)_x \tag{6.5.1}$$

We'll write $m$ and $EA$ from now on, instead of $m(x)$ and $EA(x)$, in order to keep the notation clean, but don't forget that we're allowing both of these to depend upon $x$. In Chapter 5 we were able to deal only with the case of constant mass and stiffness distributions, but much of the power of the Rayleigh-Ritz method lies in the fact that it can handle nonconstant distributions of mass and stiffness just as easily as it handles the constant cases. If we assume sinusoidal motion at a frequency $\omega$ (i.e., $\xi = \bar{\xi} \cos(\omega t)$), then (6.5.1) becomes

$$\omega^2 m\bar{\xi} = -(EA\bar{\xi}_x)_x \tag{6.5.2}$$

and dividing by $m\bar{\xi}$ gives us

$$\omega^2 = \frac{-(EA\bar{\xi}_x)_x}{m\bar{\xi}} \tag{6.5.3}$$

At this point we'll introduce what looks like a trick. You can consider it one if you wish, or simply view it as another mathematical tool in your black bag of mathematical techniques. It consists of multiplying the numerator and denominator of (6.5.3) by the dependent variable ($\bar{\xi}$), and then integrating each of them over the length of the bar. Since we're performing the same operation to both the numerator and denominator, the overall ratio is unaffected, and we obtain

$$\omega^2 = \frac{-\displaystyle\int_0^l \bar{\xi}\left(EA\bar{\xi}_x\right)_x dx}{\displaystyle\int_0^l m\bar{\xi}^2 dx} \tag{6.5.4}$$

What we have here is an expression for $\omega^2$. If we use an eigenfunction for $\bar{\xi}$, then we come up with

$$\omega_i^2 = \frac{-\displaystyle\int_0^l \bar{\xi}_i\left(EA(\bar{\xi}_i)_x\right)_x dx}{\displaystyle\int_0^l m\bar{\xi}_i^2 dx} \tag{6.5.5}$$

This is a very close parallel to (6.3.16). We might even go so far as to think that we could use (6.5.4) to define a continuous form of Rayleigh's quotient:

$$R_Q(\bar{\xi}) \equiv \frac{-\displaystyle\int_0^l \bar{\xi}(EA\bar{\xi}_x)_x dx}{\displaystyle\int_0^l m\bar{\xi}^2 dx} \tag{6.5.6}$$

This would give us a scalar that equals $\omega_i^2$ when the input is the $i$th eigenfunction and gives us some other answer if the input isn't exactly equal to an eigenfunction. And if we did think this, we'd be right. This is a perfectly acceptable form of Rayleigh's quotient for continuous bar problems. It's also true in this case (as it was for the finite-dimensional case) that the minimum value for Rayleigh's quotient is equal to $\omega_1^2$ and that Rayleigh's quotient is stationary about any eigenfunction. The only difference is that for the finite-dimensional case we knew that the highest obtainable value of Rayleigh's quotient was equal to $\omega_n^2$, if the problem had $n$ natural frequencies. Since the number of natural frequencies goes to infinity for a continuous problem, Rayleigh's quotient is unbounded for these systems.

Just as in the finite-dimensional case, we can use (6.5.6) to get a quick estimate of the first natural frequency of our system by using a trial function that we feel is close to the actual eigenfunction. And just as before, if the guess is pretty good, the estimate will be extremely good. In fact, the only other point in which our continuous Rayleigh's quotient seems to differ from that of the finite-dimensional case is in the form of the numerator. If you first compare the denominators of both (6.5.6) and (6.3.17), you'll see that they're both quadratic in form and both represent the zero-frequency kinetic energy, $\bar{k}_e^0$. This can be clearly shown for our continuous case. For a bar, the kinetic energy of a differential element is simply

$$dk_e = \tfrac{1}{2} m \dot{\xi}^2 dx \qquad (6.5.7)$$

The total kinetic energy is found by integrating over the length of the bar,

$$k_e = \tfrac{1}{2} \int_0^l m \dot{\xi}^2 dx \qquad (6.5.8)$$

Finally, using $\xi = \bar{\xi} \cos(\omega t)$ leads to

$$k_e = \left( \tfrac{1}{2} \omega^2 \int_0^l m \bar{\xi}^2 dx \right) \sin^2(\omega t) \qquad (6.5.9)$$

Since the maximum value of $\sin^2(\omega t)$ is 1.0, the maximum value of (6.5.9) is just $\tfrac{1}{2} \omega^2 \int_0^l m \bar{\xi}^2 dx$, and the zero-frequency kinetic energy is found by factoring out the $\omega^2$ term. Again, just as in the finite-dimensional case, we're factoring out the common factor of $\tfrac{1}{2}$ since it will appear in both the numerator and denominator.

Since the denominator of (6.5.6) seems to be the zero-frequency kinetic energy, it seems reasonable that the numerator should represent the potential energy of the system. And yet it certainly doesn't look like a quadratic form. It is, however, in a form that's just made for integration by parts. Recall that integration by parts is given by

$$\int_0^l u\,dv = uv \Big|_0^l - \int_0^l v\,du \qquad (6.5.10)$$

We can identify $(EA\bar{\xi}_x)_x dx$ as $dv$ and $\bar{\xi}$ as $u$ to find

$$-\int_0^l \bar{\xi}(EA\bar{\xi}_x)_x dx = -EA\bar{\xi}\bar{\xi}_x \Big|_0^l + \int_0^l EA\bar{\xi}_x^2 dx \qquad (6.5.11)$$

The $EA\bar{\xi}_x \big|_0^l$ terms are boundary terms. Notice that, for either a fixed or free boundary, they're equal to zero. Thus we'll pay no attention to them for the moment. Now look at the last term of the equation. It's quadratic. Not only that, it's (minus the $\tfrac{1}{2}$) exactly equal to the potential energy of the bar. This isn't a course on deformable bodies, but if it were you'd probably be asked to find the internal energy of a bar and it would turn out to be equal to this. So by integrating by parts we see that the numerator really is proportional to the system's potential energy, just as in the finite-dimensional case. Therefore, we can just as easily write Rayleigh's quotient as

$$R_Q(\bar{\xi}) = \frac{-EA\bar{\xi}\bar{\xi}_x\Big|_0^l + \int_0^l EA\bar{\xi}_x^2\, dx}{\int_0^l m\bar{\xi}^2\, dx} \tag{6.5.12}$$

We have two acceptable forms of Rayleigh's quotient, (6.5.6) and (6.5.12). All our continuous problems will support qualitatively similar forms, one in which part of the differential equation itself appears in the numerator and one in which the potential energy and boundary conditions appear. To distinguish them, we'll refer to the the first type (illustrated here by [6.5.6]) as a *force formulation*, since the numerator contains a force term (in this case $\left(EA\bar{\xi}_x\right)_x$). We'll refer to the second type (6.5.12) as an *energy formulation*, since the primary term in the numerator is the system's potential energy.

Because we have two acceptable forms for the Rayleigh's quotient, it follows that we'll have two versions of the Rayleigh-Ritz equations. We'll derive one version in detail and simply present the other, as its derivation exactly parallels the first.

Just as in the case of finite-dimensional systems, we'll need to pick a set of trial functions and then make Rayleigh's quotient stationary with respect to them. The resulting equations, which will be in exactly the same form as the ones for the discrete case, will tell us how much of each trial function to use to approximate the eigenfunctions, and, of course, we'll also get the best approximation to the system's natural frequencies. There's actually a bit more to choosing trial functions than you might think. However, now is not the time to explain it. We'll assume for the present that any set of independent functions is okay to use and then we'll go into more depth later in the chapter.

If we denote our trial functions by $\phi_i(x)$, then

$$\text{Est}(x) = \sum_{i=1}^{n} b_i \phi_i(x) \tag{6.5.13}$$

Note the similarity between (6.5.13) and (6.4.1). In both cases we're approximating our displacement function (or vector) by a finite summation of trial functions (or vectors). Therefore, the only difference between the finite and the continuous formulation at this point (beyond the fact that the dependent variable in the first case is a vector, whereas in the second it is a function) is the fact that one involves matrix operations while the other requires integration. Once again, we're using Est to indicate that this is going to be our estimate of the eigenfunction. The $(x)$ is included to make it doubly clear that we're dealing with a function, not a vector.

Substituting (6.5.13) into (6.5.12) yields

$$R_Q\left(\text{Est}(x)\right) = \frac{-EA\left(\text{Est}(x)\right)\left(\text{Est}(x)\right)_x\Big|_0^l + \int_0^l EA\left(\frac{\partial \text{Est}(x)}{\partial x}\right)^2 dx}{\int_0^l m\left(\text{Est}(x)\right)^2 dx} \tag{6.5.14}$$

or, in terms of our trial functions,

$R_Q(\text{Est}(x)) =$

$$\frac{-EA\left(\sum_{i=1}^{n}b_i\phi_i(x)\right)\left(\sum_{j=1}^{n}(b_j\phi_j(x))_x\right)\Big|_0^l + \int_0^l EA\left(\sum_{i=1}^{n}b_i\frac{\partial\phi_i(x)}{\partial x}\right)\left(\sum_{j=1}^{n}b_j\frac{\partial\phi_j(x)}{\partial x}\right)dx}{\int_0^l m\left(\sum_{i=1}^{n}b_i\phi_i(x)\right)\left(\sum_{j=1}^{n}b_j\phi_j(x)\right)dx}$$

$$(6.5.15)$$

Notice that the terms such as $(\text{Est}(x))^2$ were expanded out as the product of two summations. This is always necessary when dealing with products of summations and subsequent differentiations (as we'll be doing).

What we can now do is recognize that, as before, Rayleigh's quotient can be written as

$$R_Q = \frac{\text{Num}}{\text{Den}} \qquad (6.5.16)$$

where

$$\text{Num} = -EA\left(\sum_{i=1}^{n}b_i\phi_i(x)\right)\left(\sum_{j=1}^{n}(b_j\phi_j(x))_x\right)\Big|_0^l$$

$$+ \int_0^l EA\left(\sum_{i=1}^{n}b_i\frac{\partial\phi_i(x)}{\partial x}\right)\left(\sum_{j=1}^{n}b_j\frac{\partial\phi_j(x)}{\partial x}\right)dx \qquad (6.5.17)$$

and

$$\text{Den} = \int_0^l m\left(\sum_{i=1}^{n}b_i\phi_i(x)\right)\left(\sum_{j=1}^{n}b_j\phi_j(x)\right)dx \qquad (6.5.18)$$

At this point we can start to tidy things up a bit. For instance, the summations can be pulled out of the various terms as follows.

$$\text{Num} = -\sum_{i=1}^{n}\sum_{j=1}^{n}b_ib_j EA\phi_i(x)\frac{\partial}{\partial x}\phi_j(x)\Big|_0^l + \sum_{i=1}^{n}\sum_{j=1}^{n}b_ib_j\int_0^l EA\phi_i(x)\phi_j(x)dx \qquad (6.5.19)$$

and

$$\text{Den} = \sum_{i=1}^{n}\sum_{j=1}^{n}b_ib_j\int_0^l m\phi_i(x)\phi_j(x)dx \qquad (6.5.20)$$

Our expression for Num can be tightened up further by combining the two terms, giving us

$$\text{Num} = \sum_{i=1}^{n}\sum_{j=1}^{n}b_ib_j\left(-EA\phi_i(x)\frac{\partial\phi_j(x)}{\partial x}\Big|_0^l + \int_0^l EA\frac{\partial\phi_i(x)}{\partial x}\frac{\partial\phi_j(x)}{\partial x}dx\right) \qquad (6.5.21)$$

Notice that everything within the parentheses is known and, once evaluated, will simply leave us with a constant. You've probably also figured out that this is shaping up to be the same form we had in the finite-dimensional case. By differentiating with respect to the $b$'s and grouping terms, we're going to end up with equations that look just like (6.4.14) and (6.4.15). The only difference is the particular components of the $\mathcal{K}$ and $\mathcal{M}$ matrices. For the bar, after carrying out the appropriate differentiations, we'll obtain

$$\mathcal{K}_{ij} \equiv -\frac{1}{2}EA\left(\phi_i(x)\frac{\partial\phi_j(x)}{\partial x} + \phi_j(x)\frac{\partial\phi_i(x)}{\partial x}\right)\Big|_0^l + \int_0^l EA\frac{\partial\phi_i(x)}{\partial x}\frac{\partial\phi_j(x)}{\partial x}dx \qquad (6.5.22)$$

and

$$\mathcal{M}_{ij} \equiv \int_0^l m\phi_i(x)\phi_j(x)dx \qquad (6.5.23)$$

In other words, the denominator and numerator of Rayleigh's quotient can be written as

$$\text{Num} = \sum_{i=1}^n \sum_{j=1}^n b_i b_j \mathcal{K}_{ij} \qquad (6.5.24)$$

and

$$\text{Den} = \sum_{i=1}^n \sum_{j=1}^n b_i b_j \mathcal{M}_{ij} \qquad (6.5.25)$$

Because the form of Rayleigh's quotient is the same, at this point, for both the finite- and infinite-dimensional cases, there is no need to perform any more derivations; we've already done them in the previous section. We immediately can see that the final form of the Rayleigh-Ritz equations will be

$$[\mathcal{K}]B - R_\varrho[\mathcal{M}]B = 0 \qquad (6.5.26)$$

just as it was in the previous section.

If we wanted to use the force formulation (6.5.6), then our definition of $\mathcal{K}$ would change to

$$\mathcal{K}_{ij} \equiv \frac{1}{2}\int_0^l \left(\phi_i(x)\frac{\partial}{\partial x}\left(EA\frac{\partial\phi_j}{\partial x}\right) + \phi_j(x)\frac{\partial}{\partial x}\left(EA\frac{\partial\phi_i}{\partial x}\right)\right)dx \qquad (6.5.27)$$

What we've seen here is that the Rayleigh-Ritz approach for finite-dimensional problems is really just the same as it was for continuous problems. The only difference is that in one case we're adding together trial *vectors*, whereas in the other we're adding trial *functions*. Similarly, the $\mathcal{K}_{ij}$ and $\mathcal{M}_{ij}$ values for the finite-dimensional case are found from matrix operations on vectors, while for the continuous case they're made up of integral functions. However, the end result for each case is the same—a constant scalar quantity.

Having established all this, some of you might still be a trifle unsatisfied. After all, the continuous Rayleigh's quotient did have one little qualitative difference in the numerator, namely those boundary-condition terms that popped in the energy formulation. The exact analogue of the finite-dimensional case's $X^T[K]X$ term is the continuous case's $\int_0^l EA\xi^2 dx$ term. The $EA\xi\xi_x\Big|_0^l$ term would seem to be unique to the continuous case. Well, in a way there really isn't a difference since the boundary conditions for the finite-dimensional case were implicitly included by whether or not the last mass was free or spring restrained. However, boundary conditions are going to pop up frequently in continuous problems, and they'll be more bothersome than in finite-dimensional problems. Thus we will be looking more closely at boundary conditions and how they affect our solution procedures. But first, an example of the Rayleigh-Ritz method for a continuous problem.

### Example 6.9

**Problem**   Apply the Rayleigh-Ritz method to a tensioned string with a spring at one end, shown in Figure 6.14.

**Solution**   Before we even start, we can get a feel for what the natural frequencies and eigenfunctions will be by considering the simpler examples shown in Figures 6.15a and 6.15b. The first case is a fixed-fixed string and the other is a fixed-free string. The free condition simply means that the string is free to move vertically, but not horizontally. These represent limiting cases for the more general problem of a fixed-spring restrained string. If the spring constant goes to infinity, then the boundary condition becomes equivalent to a fixed one (Figure 6.15a) and if the stiffness goes to zero, then we approach the fixed-free configuration (Figure 6.15b). Since we know that adding stiffness to a system will increase (or leave unchanged) the natural frequencies of the system, we also know that our natural frequencies for a finite spring constant will lie between the natural frequencies associated with a spring and those corresponding to the infinite spring constant case. Figure 6.16 shows these limiting values of the natural frequencies and indicates the range within which the actual frequencies must lie for the finite-stiffness case. The values at $k = 0$ correspond to the natural frequencies of a fixed-free system and the limiting values shown at the rightmost end of the plot ($k = 50$) closely correspond to the fixed-fixed solutions.

We've already seen how to obtain these results in Chapter 5. The equation of motion for free vibrations of a string (Appendix A) is given by

$$\rho\ddot{y} = (Ty_x)_x \tag{6.5.28}$$

For sinusoidal motion at a frequency $\omega$, we therefore have

$$\rho\omega^2\bar{y} + (T\bar{y}_x)_x \tag{6.5.29}$$

**Figure 6.14**   Fixed-spring restrained tensioned string

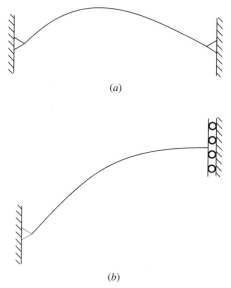

(a)

(b)

**Figure 6.15** (a) Fixed-fixed string; and (b) fixed-free string

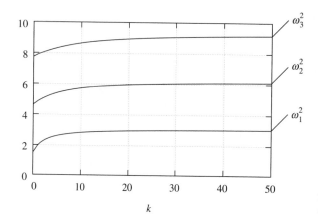

$k$

**Figure 6.16** Variation of natural frequencies as a function of spring constant

If the mass density $\rho$ and the tension $T$ are constant, then we can solve this problem in the same way that we solved the bar problem of Chapter 5. We can rewrite (6.5.29) as

$$\bar{y}_{xx} + \beta^2 \bar{y} = 0 \qquad (6.5.30)$$

where $\beta^2 = \dfrac{\rho \omega^2}{T}$.

Equation (6.5.30) has solutions

$$\bar{y} = a_1 \sin(\beta x) + a_2 \cos(\beta x) \qquad (6.5.31)$$

Applying the fixed boundary condition at $x = 0$ shows us that $a_2$ must equal zero, leaving us with

$$\bar{y} = a_1 \sin(\beta x) \qquad (6.5.32)$$

If we're concerned with the fixed-free case, then $y_x = 0$ at $x = 1$, giving us

$$\frac{\partial \bar{y}(1)}{\partial x} = a_1 \beta \cos(\beta) = 0 \qquad (6.5.33)$$

which means that

$$\beta_n = \frac{(2n - 1)\pi}{2} \qquad (6.5.34)$$

On the other hand, if we look at the fixed-fixed case, then $y = 0$ at $x = 1$, and so

$$\bar{y}(1) = a_1 \sin(\beta) = 0 \qquad (6.5.35)$$

In this case we see that

$$\beta_n = n\pi \qquad (6.5.36)$$

These results are the values indicated in Figure 6.16. Since the finite spring-stiffness case is only a bit more difficult than our limiting case examples, we can go ahead and solve the problem in closed form, giving us an exact answer to compare our approximate results with. Note that for this example we're using $T = 1$ N and $\rho = 1$ kg/m, not because they're particularly physically meaningful, but because they simplify the example. In addition, we'll take $l$ to be equal to 1 m.

An easy way to approach the problem of formulating the correct boundary conditions is to add a small mass to the end-point you're concerned with. You can then apply Newton's law to this mass. For our problem, the appropriate force balance will be

$$m\ddot{y}(1) = -ky(1) - T\frac{\partial y(1)}{\partial x} \qquad (6.5.37)$$

The $ky(1)$ term is the restoring force due to the spring and the $-T\dfrac{\partial y(1)}{\partial x}$ term is the force due to the tension in the string. Since the angle of the string is approximated by $\dfrac{\partial y(1)}{\partial x}$, and when this is negative the tension is oriented upward, we must put in a minus sign so that negative slopes of the string generate upward forces on the mass.

Once we have a force balance, all we need do is set $m$ to zero (since in actuality there really isn't any mass at the end) to obtain

$$ky(1) = -T\frac{\partial y(1)}{\partial x} \qquad (6.5.38)$$

This represents a combination of the two boundary-condition types we just examined, both displacement and slope. Since $\bar{y} = a_1 \sin(\beta x)$, evaluating (6.5.38) at $x = 1$ gives us

$$ka_1 \sin(\beta) = -Ta_1 \beta \cos(\beta) \qquad (6.5.39)$$

which can be rearranged to yield

$$\frac{-k}{T} \tan(\beta) = \beta \qquad (6.5.40)$$

As we saw in Chapter 5, an easy way to visualize the answer to this kind of problem is to plot both the left- and right-hand sides versus $\beta$ and see where they intersect. Such a plot is shown in Figure 6.17 and the solutions for the first six $\beta$'s are $\beta_1 = 2.8044$, $\beta_2 = 5.667$, $\beta_3 = 8.6031$, $\beta_4 = 11.5993$, $\beta_5 = 14.6374$, and $\beta_6 = 17.7032$.

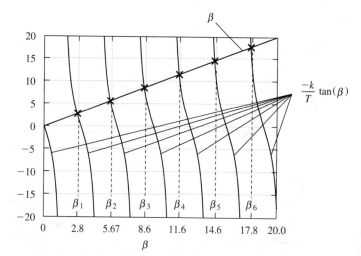

**Figure 6.17** $\beta$ values for a fixed-spring restrained string

For this problem, we'll choose the trial functions

$$\phi_n(x) = \sin\left(\frac{(2n-1)\pi x}{2}\right) \tag{6.5.41}$$

As we recently saw, these include both the eigenfunctions for a fixed-free string, as well as those of a fixed-fixed string. Since the actual boundary condition lies somewhere between these, it seems reasonable that both sets will allow a convergence to the correct solution. Note that the actual solution will have a finite displacement and finite slope at the spring end of the string. Thus none of our trial functions matches the boundary condition. The fixed-free solutions have a finite displacement, but zero slope, and the fixed-fixed solutions exhibit finite slope with zero displacement at the spring end. Combining sets of trial functions in this way can often improve the performance of the Rayleigh-Ritz method, and we'll examine this more closely in a few pages. For now, let's see what sort of results we get when using these trial functions in the Rayleigh-Ritz equations.

Since the functions don't match the boundary conditions, we must use the energy formulation. We can obtain the correct form (6.5.22–6.5.26) simply by replacing the parameters of the bar with the appropriate string parameters. Doing so will leave us with

$$\mathcal{K}_{ij} = -\frac{1}{2}T\left(\phi_i(x)\frac{\partial\phi_j(x)}{\partial x} + \phi_j(x)\frac{\partial\phi_i(x)}{\partial x}\right)\Bigg|_0^1 + \int_0^1 T\frac{\partial\phi_i(x)}{\partial x}\frac{\partial\phi_j(x)}{\partial x}dx \tag{6.5.42}$$

and

$$\mathcal{M}_{ija} = \int_0^1 \rho\phi_i(x)\phi_j(x)dx \tag{6.5.43}$$

Since our trial functions don't satisfy the boundary conditions, we must get the boundary information into the problem. Luckily, the energy formulation lets us do that quite efficiently. The string is pinned at the left and so the boundary terms go to zero. We already know that the right boundary condition is

$$ky(1) = -T\frac{\partial y(1)}{\partial x} \tag{6.5.44}$$

We simply have to apply the same condition for each of our trial functions:

$$k\phi_i(1) = -T\frac{\partial \phi_i(1)}{\partial x} \tag{6.5.45}$$

Now we'll look at (6.5.42) and where we see terms such as $T\dfrac{\partial \phi_i(1)}{\partial x}$ we will simply replace them with $-k\phi_i(1)$. Doing so will change $\mathcal{K}$ to

$$\mathcal{K}_{ij} = k\phi_i(1)\phi_j(1) + \int_0^1 T\frac{\partial \phi_i(x)}{\partial x}\frac{\partial \phi_j(x)}{\partial x}dx \tag{6.5.46}$$

Note that the boundary contribution appears as energy. If we multiplied everything by one-half we'd see that the first term of $\mathcal{K}$ looks like a potential energy of the discrete spring, while the second term looks like the potential energy of the string itself. Similarly, $\mathcal{M}$ will be the kinetic energy of the string (with $\omega^2$ removed).

Since the trial functions are just sines and the tension and mass distribution are constant, we can evaluate $\mathcal{K}$ and $\mathcal{M}$ by hand. In more complex problems you'd obviously want to use the computer. Just so you can check yourself if you want, the upper $2 \times 2$ block of each matrix is equal to

$$[\mathcal{K}] = \begin{bmatrix} 9.2337 & 1.0472 \\ 1.0472 & 4.9348 \end{bmatrix} \tag{6.5.47}$$

and

$$[\mathcal{M}] = \begin{bmatrix} .50000 & .42441 \\ .42441 & .50000 \end{bmatrix} \tag{6.5.48}$$

Using these matrices will give you the estimated natural frequencies of the first two modes. If we record these results, as well as those obtained for 1, 3, 4, 5, and 6 trial functions, we'll obtain Table 6.1.

Here we see behavior similar to that seen in the finite-dimensional example. As more trial functions are added, the approximation becomes better and better. Note that when the second trial function is added, the estimate for $\beta_1$ becomes much better, going from 4.30 to 2.81. This is quite

**TABLE 6.1** $\beta$ ESTIMATES FOR 1–6 TRIAL FUNCTIONS

| | | | Number of trial functions | | | |
|---|---|---|---|---|---|---|
| Exact | 6 | 5 | 4 | 3 | 2 | 1 |
| 2.80443 | 2.80443 | 2.80443 | 2.80443 | 2.80448 | 2.80675 | 4.29737 |
| 5.66687 | 5.66687 | 5.66687 | 5.66695 | 5.69515 | 8.98819 | |
| 8.60307 | 8.60308 | 8.60393 | 8.82614 | 14.9677 | | |
| 11.5993 | 11.62067 | 12.3305 | 22.6011 | | | |
| 14.6374 | 16.3272 | 32.0714 | | | | |
| 17.7032 | 43.4629 | | | | | |

close to the exact solution of 2.80. The reason for this abrupt improvement is that the first trial function correctly has a finite displacement at the spring end, but doesn't have a finite slope. As we've already noted, the exact solution will have both a finite displacement, as well as a finite slope, at the end. We're clearly not matching this boundary condition with just a single $\sin(\frac{\pi x}{2})$ trial function. Adding the second trial function doesn't change the displacement (since $\sin(\pi x)$ vanishes at $x = 1$), but it does contribute a finite slope. Thus, by combining two functions, we're doing a very good job of matching both the boundary conditions, as well as the overall shape of the eigenfunction. For reference, Figures 6.18 and 6.19 show the actual eigenfunction, as well as

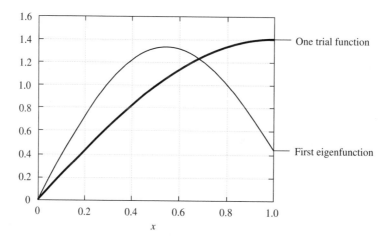

**Figure 6.18**   One trial-function approximation to the first mode of a fixed-spring restrained string

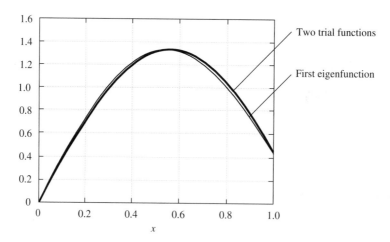

**Figure 6.19**   Two trial-function approximation to the first mode of a fixed-spring restrained string

what the one- and two-mode approximations actually look like. Note also that the rule of thumb is holding. Looking at the six-mode case, we see that the first three $\beta$ estimates are excellent and also that the error increases markedly for higher $\beta$ estimates.

Now that we've seen a simple example of how we can use the Rayleigh-Ritz method, it's appropriate to look at a more complex problem, that of a beam undergoing transverse vibrations. At the same time, we'll get more specific about the ways in which we can formulate Rayleigh-Ritz problems and see how the boundary conditions affect our approach.

We've already seen how uniform beams behave and know that their eigenfunctions can be somewhat complicated, involving sines, cosines, sinhs, and coshs. One of the advantages of the Rayleigh-Ritz approach is that we needn't concern ourselves with how complex the actual eigenfunctions are; we'll be approximating them by means of much simpler functions.

**Example 6.10**

**Problem**   Determine how to apply the Rayleigh-Ritz approach to the problem of a vibrating beam.

**Solution**   You'll recall from Chapter 5 that the equation of motion for transverse beam vibrations is

$$\rho(x)\ddot{y} = -(EI(x)y_{xx})_{xx} + f(x,t) \tag{6.5.49}$$

and setting the forcing to zero gives us

$$\rho(x)\ddot{y} = -(EI(x)y_{xx})_{xx} \tag{6.5.50}$$

We can now do what we did for the bar problem: assume sinusoidal motion ($y(x,t) = \bar{y}e^{i\omega t}$) and divide by $-\rho(x)\bar{y}$ to obtain

$$\omega^2 = \frac{(EI(x)\bar{y}_{xx})_{xx}}{\rho(x)\bar{y}} \tag{6.5.51}$$

As before, we can multiply by the dependent variable and integrate over the domain (the bar's length) in order to define our Rayleigh's quotient

$$R_Q = \frac{\displaystyle\int_0^l \bar{y}(EI(x)\bar{y}_{xx})_{xx}dx}{\displaystyle\int_0^l \rho(x)\bar{y}^2dx} \tag{6.5.52}$$

This is the first way in which we can express Rayleigh's quotient. In line with our bar example, we'll refer to this as the force formulation.

We can also use integration by parts in order to bring the boundary conditions out and put the numerator into an energy form, thus giving us our energy formulation. Our first integration by parts gives us

$$R_Q = \frac{\bar{y}(EI(x)\bar{y}_{xx})_x \Big|_0^l - \displaystyle\int_0^l \bar{y}_x(EI(x)\bar{y}_{xx})_x dx}{\displaystyle\int_0^l \rho(x)\bar{y}^2dx} \tag{6.5.53}$$

and a second brings us to

$$
R_Q = \frac{\bar{y}\left(EI(x)\bar{y}_{xx}\right)_x \Big|_0^l - EI(x)\bar{y}_x\bar{y}_{xx}\Big|_0^l + \int_0^l EI(x)\bar{y}_{xx}^2\,dx}{\int_0^l \rho(x)\bar{y}^2\,dx}
\tag{6.5.54}
$$

What can we say about the differences between the energy formulation and the force formulation? Well, for one thing, the boundary conditions are explicitly included in the energy formulation. We saw in the previous example that this allowed us to include the energy of the boundary condition elements and we didn't have to worry about having our trial functions satisfy all the problem's boundary conditions. Since the boundary conditions are not explicitly present in the force formulation, our trial functions must all match the appropriate boundary conditions. If they don't, then we don't have a hope of getting a good answer since there's no other way to get the boundary conditions into the problem. And it's not enough to match some of them, the trial functions must match all the boundary conditions of the problem if you're planning to use the force formulation approach. If you think about it, it's clearly easier to come up with trial functions that don't meet all the boundary conditions than to try and find ones that do. Thus, using the force formulation implies we'll be doing more work in getting our trial functions.

The energy formulation, on the other hand, already has the boundary terms in the expression for Rayleigh's quotient. This means that our trial functions needn't satisfy all the boundary conditions of the problem. If they don't, the boundary terms in Rayleigh's quotient's numerator will include them for us. This is generally good, since it means we have a wider choice of trial functions (there are more trial functions that don't satisfy all the boundary conditions than there are ones that do).

You should also notice that the dependent variable in the force formulation gets differentiated four times, while in the energy formulation it's only differentiated twice. This means that we can get by with trial functions in the energy formulation that are only half as differentiable as in the force formulation. Although this isn't overly important from a theoretical standpoint (as most of our trial functions will be infinitely differentiable), it is important in a practical sense. We may well be getting our trial functions from a finite-element analysis or from experimental data, and therefore we will not have precise information about the functions at all points and the information we have will contain some amount of noise contamination. Since differentiation amplifies the effects of noise, the less we differentiate, the better off we are. Thus the energy formulation again looks like a good choice.

## Break for Boundary Conditions

We've sidestepped the issue of boundary conditions long enough; it's time to look carefully at them and how they affect our solution procedures. We'll start by considering our trial functions. How can we decide if a given set is a good one or not? To do this, it helps to first ask what the best set would be. If we could find a set of functions that satisfied the differential equation and satisfied all the boundary conditions of the problem, then we'd have an excellent set of trial functions. In fact, we'd have the eigenfunctions themselves, since they're the only functions that do both of these things. The next best set of functions would be functions that don't satisfy the differential equation, but do satisfy all the boundary conditions of the problem. Except for the actual eigenfunctions, these are the best kind of function to have. It's obvious that we can solve the Rayleigh-Ritz problem with these trial functions. After all, we know that the Rayleigh-Ritz method works to add together our trial functions so as to best approximate an eigenfunction. This is analogous to a Fourier series analysis, in which we add together our Fourier series functions to approximate some other function. The only difficulty that can turn up in a Fourier series analysis is if the functions we've chosen don't match the end conditions of the function we're trying to approximate. For example, say we're using sines to approximate one period of a cosine. Unfortunately, all the sines go to zero at the end-points (the boundary conditions), whereas the cosine goes to one. Therefore it's impossible to exactly match the cosine with a finite number of terms. As the number of included sines' functions is increased, we'll come closer to approximating our function everywhere but at the end-points. An unfortunate consequence of using sines, one that is not mitigated by adding more terms, is the fact that the end-points are always going to have zero deflections. Zero deflection at the ends is physically impossible for some situations (such as a free-free beam) and certainly suggests to us that our Rayleigh-Ritz approach will probably encounter some difficulties if we insist on using trial functions that violate physicality in such a way.

Clearly, if our trial functions had been composed of cosines, rather than sines, a single trial function would have been sufficient to exactly match the trial function. By choosing a set of trial functions that didn't match the proper boundary conditions (the sines), we doomed ourselves to needing a tremendous number of trial functions, hardly the thing for an efficient approximation technique. So if you can choose functions that match the system's boundary conditions, it behooves you to do so.

Functions that match all the boundary conditions are special enough to merit their own name, and are called *comparison functions*. The easiest way to remember this is to say to yourself, "These are the ones we *compare* all the others with because they're the best." Comparison functions are the ones we must use if we're going to be applying a force formulation form of the Rayleigh-Ritz equations, since there's no way to add any corrections for boundary condition mismatches. Clearly, the functions we choose must be $2p$ times differentiable, since they'll be differentiated this much in the numerator. (Recall from Chapter 5 that $2p$ refers to the number of differentiations occuring in the stiffness part of the differential equation.) If the trial functions weren't $2p$ times differentiable, then the numerator would be equal to zero, hardly conducive to a good approximation.

If we now consider the energy formulation approach, we see that there is only one differentiation for the bar problem and two for the beam problem. This means that the trial functions need be only half as differentiable as in the force formulation approach. The trial functions also need only satisfy the geometric boundary conditions of the problem. We saw in Chapter 5 that we have two general kinds of boundary conditions: geometric and natural. The geometric boundary conditions involved displacements and/or slopes at the end-points, while the natural boundary conditions involved forces and/or moments. The functions that we can use in the energy formulation are called admissible functions. You can remember this by saying, "We'll *admit* these function into our bag of trial functions." They need be only half as differentiable as comparison functions, and they need only satisfy the geometric boundary conditions.

Now that we've discussed the different types of boundary conditions and the differences between admissible and comparison functions, we'll present an in-depth look at a particular problem and see how the different types of trial functions affect our results.

**Example 6.11**

**Problem**   Apply the Rayleigh-Ritz approach to a pinned-spring restrained beam (Figure 6.20).

**Solution**   We'll take the length of the beam to be 1.0 m and let the mass and stiffness distributions be constant. This example is an excellent one for us because we'll be able to see precisely what happens as we alter the degree to which we're matching the boundary conditions. Our trial functions will be chosen so that they always match the left boundary conditions (zero displacement and moment), but they'll differ in how well they match the conditions at the right end. To start, we'll let $EI = 1$ N·m$^2$ and $\rho = 1$ kg/m. We'll set $k$ (the spring stiffness) equal to 100 N/m.
    Recall from Chapter 5 that the general boundary condition at the right end is given by

$$u_{xx} = 0 \quad \text{at } l = 1 \tag{6.5.55}$$

and

$$EIu_{xxx} - ku = 0 \quad \text{at } l = 1 \tag{6.5.56}$$

We'll use

$$\phi_i(x) = \sin\left(\frac{i\pi x}{2}\right) \tag{6.5.57}$$

as our first set of trial functions and we'll be using an energy formulation for the Rayleigh-Ritz equations. As in the previous example, this choice will highlight the power of the Rayleigh-Ritz method to get good answers from what seem like inappropriate choices for the trial functions. If we examine (6.5.55), we notice that $u_{xx}$ is always zero at the right end of the beam, while (6.5.56) tells us that there is a specific ratio between $u$ and $u_{xxx}$ at the right end: $\dfrac{u_{xxx}}{u} = \dfrac{k}{EI}$.
    Do our trial functions all satisfy these boundary conditions? Certainly not! $\phi_1(1) = 1.0$ at the right end (which is good; the right-end displacements will be nonzero), but $\dfrac{\partial^2 \phi_1(1)}{\partial x^2}$ is also going

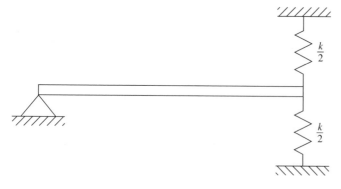

**Figure 6.20**   Pinned-spring restrained beam

to be nonzero, whereas the boundary conditions call for it to be zero. In addition, $\dfrac{\partial^3 \phi_1(1)}{\partial x^3} = 0$, making it impossible to satisfy (6.5.56) when the actual displacement is nonzero. We have similar problems with $\phi_2$. $\phi_2(1) = 0$, which doesn't match reality; the spring will definitely deflect. The first derivative and third derivatives are nonzero, a good match to the physical behavior.

What we see is that we're not matching any of the natural boundary conditions exactly with any one trial function. However, by combining the trial functions, we should be able to get a good approximation. We're further helped by the fact that we're using an energy formulation. Thus the spring energy will be accounted for, relieving us of the constraint of exactly matching $\dfrac{\partial^3 u}{\partial x^3}$ at $x = 1$. Since our trial functions do not all match the boundary conditions, they are admissible functions, not comparison functions.

We've already seen that the terms in the $\mathcal{K}$ and $\mathcal{M}$ matrices are simply the energies of the system, expressed in terms of the trial functions. Thus, since the energies of the beam are given by

$$k_e = \frac{1}{2} \int_0^1 \rho \dot{u}^2 dx \tag{6.5.58}$$

and

$$p_e = \frac{1}{2} \int_0^1 E I u_{xx}^2 dx \tag{6.5.59}$$

we can immediately identify our $\mathcal{K}$ and $\mathcal{M}$ elements as

$$\mathcal{K}_{ij} = \int_0^1 E I \frac{\partial^2 \phi_i(x)}{\partial x^2} \frac{\partial^2 \phi_j(x)}{\partial x^2} dx + k \phi_i(1) \phi_j(1) \tag{6.5.60}$$

and

$$\mathcal{M}_{ij} = \int_0^1 \rho \phi_i(x) \phi_j(x) dx \tag{6.5.61}$$

Just as we saw in the previous section, the discrete energy of the spring is added to the overall energy of the beam in the $\mathcal{K}$ entries.

Of course, it would be convenient if we could compare our approximate results with the exact answer. And luckily, since the problem is a relatively simple one, we can find the exact results ahead of time. From Section 5.3, we recall that our general solution for the eigenfunction of a beam is

$$b_1 \cos(\beta x) + b_2 \sin(\beta x) + b_3 \cosh(\beta x) + b_4 \sinh(\beta x) \tag{6.5.62}$$

Applying the boundary conditions for the left end of the beam gives us

$$b_1 + b_3 = 0 \tag{6.5.63}$$

and

$$\beta^2 \left[ -b_1 + b_3 \right] = 0 \tag{6.5.64}$$

Together, these tell us that $b_1 = b_3 = 0$. Using this knowledge and applying the boundary conditions for the right end gives us

$$\beta^2\big[ - b_2 \sin(\beta) + b_4 \cos(\beta)\big] = 0 \qquad (6.5.65)$$

and

$$EI\beta^3\big[ - b_2 \cos(\beta) + b_4 \cosh(\beta) - k(b_2 \sin(\beta) + b_4 \sinh(\beta))\big] = 0 \qquad (6.5.66)$$

As we saw in Chapter 5, this is simply an eigenvalue problem, which we can write out as

$$\begin{bmatrix} - \sin(\beta) & \sinh(\beta) \\ -EI\beta^3 \cos(\beta) - k \sin(\beta) & EI\beta^3 \cosh(\beta) - k \sinh(\beta) \end{bmatrix} \begin{Bmatrix} b_2 \\ b_4 \end{Bmatrix} = \begin{Bmatrix} 0 \\ 0 \end{Bmatrix} \qquad (6.5.67)$$

The solutions of (6.5.67) give us the $\beta_i$'s for our beam-vibration problem and the associated values of $b_2$ and $b_4$ let us construct the eigenfunctions. The characteristic equation of (6.5.67) is

$$2k \tan(\beta) \tanh(\beta) + EI\beta^3\big[ \tanh(\beta) - \tan(\beta)\big] = 0 \qquad (6.5.68)$$

and the first five values of $\beta$ are equal to 2.9886, 5.1482, 7.3872, 10.310, and 13.395.

Now that we know what our answers should be (if the approximation works), we can move ahead and see how well we do. Using the trial functions given by (6.5.57) and the $\mathcal{K}$ and $\mathcal{M}$ formulations given by (6.5.60) and (6.5.61), we find that our $\mathcal{M}$ and $\mathcal{K}$ matrices for a six-function approximation are equal to

$$\mathcal{M} = \begin{bmatrix} .5 & .42441 & 0.0 & -.16977 & 0.0 & .10913 \\ .42441 & .5 & .25465 & 0.0 & -.06063 & 0.0 \\ 0.0 & .25465 & .5 & .36378 & 0.0 & -.14147 \\ -.16977 & 0.0 & .36378.5 & .28294 & 0.0 & 0.0 \\ 0.0 & -.06063 & 0.0 & .28294 & .5 & .34725 \\ .10913 & 0.0 & -.14147 & 0.0 & .34725 & .5 \end{bmatrix} \qquad (6.5.69)$$

and

$$\mathcal{K} = \begin{bmatrix} 103.04 & 10.335 & -100.00 & -16.537 & 100.00 & 23.919 \\ 10.335 & 48.705 & 55.811 & 0.0 & -36.912 & 0.0 \\ -100.00 & 55.811 & 346.57 & 318.92 & -100.00 & -279.06 \\ -16.537 & 0.0 & 318.92 & 779.27 & 689.03 & 0.0 \\ 100.00 & -36.912 & -100.00 & 689.03 & 2,002.5 & 1,902.7 \\ 23.919 & 0.0 & -279.06 & 0.0 & 1,902.7 & 3,945.1 \end{bmatrix}. \qquad (6.5.70)$$

Solving the Rayleigh-Ritz equations associated with these matrices gives us $\beta$ values of 2.9886, 5.1482, 7.3874, 10.310, 14.345, and 31.125. A comparison of these results with the actual $\beta$ values shows that the first four are right on the money. Only with the fifth approximation do we see some significant error start to creep in.

Having seen how accurate a good set of trial functions were in Example 6.10, we'll now look at the consequences of reducing the size of the trial set.

**Example 6.12**

**Problem**  Revisit Example 6.10 but drop all $\sin(n\pi x)$ trial functions.

**Solution**  Dropping these terms means that our reduced set will include only the $\sin\left(\dfrac{(2n-1)\pi x}{2}\right)$ terms as trial functions. If we restrict our trial function set in this way, then the $\beta$ estimates will be 3.3772, 5.2136, 7.9638, 11.034, 14.155, and 17.289. Notice that the errors of the lower $\beta$ estimates have gone up. This is because we've lost the ability to match the slope boundary conditions, thus hurting our approximation ability. Figure 6.21 shows a comparison between the actual eigenfunction for the first mode and the approximations generated by the full and reduced sets of trial functions.

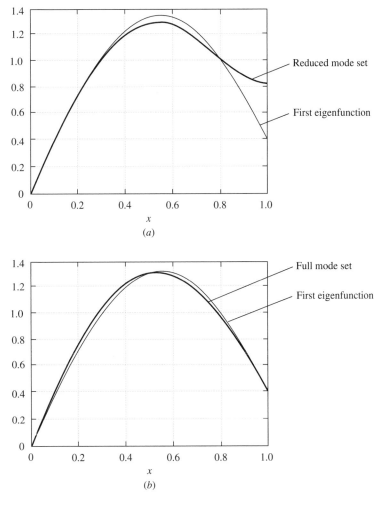

**Figure 6.21**  Comparison of reduced and full mode set approximations for a fixed-spring restrained beam, first eigenfunction

Figure 6.21*a* compares the reduced set's approximation to the actual eigenfunction. Note that the approximation is very good at the left end, but becomes poorer as it approaches the right end. At $x = 1$, the slope of the trial function approximation is zero (unavoidable considering the type of trial functions used), while the actual solution has a finite, negative slope. Because of the mismatch, we generated a poorer natural frequency estimate. Now look at Figure 6.21*b*. The match is now excellent for all points along the beam. As a consequence, our natural frequency estimate is also excellent.

The same type of result can be seen in Figure 6.22, in which the two approximations are compared to the exact result for the second eigenfunction.

(a)

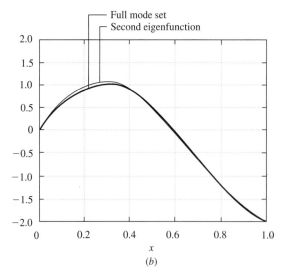

(b)

**Figure 6.22** Comparison of reduced and full mode set approximations for a pinned-spring restrained beam, second eigenfunction

This illustrates the sort of problem that can occur with a boundary-condition mismatch. If you have a complete set of functions (such as those in Example 6.10), then convergence is quick and painless. Failure to match the physical constraints of the system, such as occurred in Example 6.11, will severely compromise the performance of the method. Thus it pays to think carefully about boundary conditions.

## 6.6  ASSUMED MODES METHOD

The Rayleigh-Ritz method has shown itself to be a powerful tool in calculating the modal responses of complicated systems. However, another approach is perhaps even more widespread—the *assumed modes method*. This approach has the advantage of being readily applicable to both linear and nonlinear systems. Although the approach seems different from that of the Rayleigh-Ritz method, it turns out that the final equations are of the same form if the problem is linear and the energy formulation is used in obtaining the equations of motion.

The name is actually very descriptive. What we do is simply *assume* that a given set of trial functions will work for us. Then we find the equations of motion using Lagrange's equations. To see how it works, we'll consider the same example that we examined in Section 6.5. We'll call our assumed modes $\phi_i(x)$. We'll assume that each trial function $\phi_i$ is multiplied by a coordinate that depends on time and that it is denoted by $a_i(t)$. This is analogous to what we know will result from a modal analysis, namely a summation of the system's eigenfunctions, each multiplied by a function of time. In the free-vibration case this function of time is just sinusoidal motion at the system's natural frequency, and for the forced case it's a more involved function. Thus we're saying

$$u(x,t) = \sum_{i=1}^{n} a_i(t)\phi_i(x) \tag{6.6.1}$$

To illustrate the method, we'll examine the same problem that we just discussed, a pinned-spring restrained beam.

The potential energy of the beam is given by

$$p_e = \frac{1}{2}\int_0^1 EI(x)u_{xx}^2 dx \tag{6.6.2}$$

and the kinetic is given by

$$p_e = \frac{1}{2}\int_0^1 \rho(x)\dot{u}^2 dx \tag{6.6.3}$$

The potential energy of the spring is given by $\frac{1}{2}ku(1)^2$. Using (6.6.1–6.6.3) allows us to form the

Lagrangian $(k_e - p_e)$,

$$L = \frac{1}{2}\int_0^1 \rho(x)\left(\sum_{i=1}^n \dot{a}_i(t)\phi_i(x)\right)^2 - \frac{1}{2}\int_0^1 EI(x)\left(\sum_{i=1}^n a_i(t)(\phi_i(x))_{xx}\right)^2$$

$$+\frac{1}{2}k\left(\sum_{i=1}^n a_i(t)\phi_i(1)\right)^2 \tag{6.6.4}$$

Now we must apply Lagrange's equations, which means we must differentiate with respect to the system's generalized coordinates (in this case the $a_i$'s and $\dot{a}_i$'s). The only tricky part is that we have a summation that's squared. This means that we must be careful to catch all the cross terms during the differentiation. For example, consider the simple summation

$$y = \left(\sum_{i=1}^2 a_i\phi_i\right)^2 \tag{6.6.5}$$

Expanding this we get

$$y = a_1^2\phi_1^2 + 2a_1a_2\phi_1\phi_2 + a_2^2\phi_2^2 \tag{6.6.6}$$

Differentiating with respect to $a_1$ gives us

$$\frac{dy}{dx_1} = 2a_1\phi_1^2 + 2a_2\phi_1\phi_2 \tag{6.6.7}$$

You'll note the cross-term $2a_2\phi_1\phi_2$. This is the sort of term that we must make sure to account for correctly.

Applying Lagrange's equations to (6.6.4) (with respect to the $i$th generalized coordinate) results in

$$\int_0^1 \rho(x)\left(\ddot{a}_1\phi_1(x)\phi_i(x) + \ddot{a}_2\phi_2(x)\phi_i(x) + \cdots\right)dx$$

$$+\int_0^1 EI\left(a_1\frac{\partial^2\phi_i(x)}{\partial x^2}\frac{\partial^2\phi_1(x)}{\partial x^2} + \frac{\partial^2\phi_i(x)}{\partial x^2}\frac{\partial^2\phi_2(x)}{\partial x^2} + \cdots\right)dx$$

$$+k\left(a_1\phi_1(1)\phi_i(1) + a_2\phi_2(1)\phi_i(1) + \cdots\right) = 0 \tag{6.6.8}$$

Just as in the Rayleigh-Ritz development, this can be seen as the $i$th row of a the matrix expression

$$[\mathcal{M}]\{\ddot{a}\} + [\mathcal{K}]\{a\} = 0 \tag{6.6.9}$$

where

$$\mathcal{K}_{ij} = \int_0^1 EI\frac{\partial^2\phi_i(x)}{\partial x^2}\frac{\partial^2\phi_j(x)}{\partial x^2}dx + k\phi_i(1)\phi_j(1) \tag{6.6.10}$$

and

$$\mathcal{M}_{ij} = \int_0^1 \rho\phi_i(x)\phi_j(x)dx \tag{6.6.11}$$

Note that these matrices are identical to (6.5.60) and (6.5.61). These equations aren't in precisely the same form as the Rayleigh-Ritz equations, because these are equations of motion while the Rayleigh-Ritz equations are already in eigenvalue form. However, if we're looking for oscillatory solutions, we'd assume that $\{a\} = \bar{A}e^{i\omega t}$, leading to

$$[\mathcal{K}]\bar{A} - \omega^2[\mathcal{M}]\bar{A} = 0 \qquad (6.6.12)$$

This is exactly the form of the Rayleigh-Ritz equations. Thus, either viewpoint, the Lagrangian approach of assumed modes or the stationarity approach of Rayleigh-Ritz, ultimately leads to the same system of equations.

## 6.7 HOMEWORK PROBLEMS

### Section 6.2

**1.** In this problem we'll come up with a lumped discretization model for a uniform bar and compare its performance to the exact analytical results. Start by approximating the bar as a series of lumped spring and mass elements. A first cut might be to make all the springs and masses equal. The difficulty with this approach is that it causes a modeling problem. Putting two of our models together would produce a bar that has a spring nonuniformity in the center, since the two end springs would result in an equivalent spring of only half the stiffness of the other springs. Thus you might want to have uniform interior springs and a spring with twice the interior stiffness at the ends. Compare the results generated from this model to those found from an exact analysis. Consider both free-free and fixed-fixed conditions. Use an eight-mass approximation.

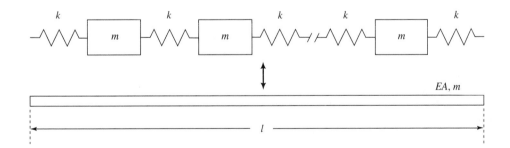

**2.** Repeat Problem 1 but this time use the approximation shown below. In this case the interior stiffness will be uniform, but the end masses will differ from the interior ones.

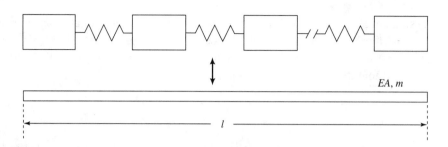

3. In this problem we'll consider the effect of adding a lumped mass to the end of a bar, using the approximation for the bar that was developed in Problem 1. Thus we're considering the system shown in Problem 5, Chapter 5. How accurate are the results of the MDOF approximation, which simply considers the lumped mass and the equivalent stiffness of a beam? How do they compare to the exact result? Use $m_1 = 50$ ml and discretize the beam into eight-mass elements.

4. For this problem we'll again consider the approximation of Problem 1, but include a spring constraint on the right end of the bar. How well does the approximation method do in solving this problem? Compare your approximate results with the exact solution. $k_5 = \frac{20EA}{l}$.

## Section 6.3

5. Consider the system shown in Problem 83, Chapter 4. $f_i = 0$. Calculate the exact eigenvectors and natural frequencies. Mass normalize your eigenvectors and determine how much contamination from the second-mode eigenvector is needed before the natural frequency estimate is off by 5 percent. $m_i = 1$ kg, $k_1 = 1000$ N/m, $k_2 = 2000$ N/m, $k_3 = 4000$ N/m, and $k_4 = 1000$ N/m.

6. Consider the system of Problem 83, Chapter 4. Set $f_i$ and $k_4$ to zero. If $k_1$ were zero, the system would have a rigid-body mode in which all $x_1$ were equal. How much error is there in assuming a rigid-body mode for finite values of $k_1$? How large is $k_1$ when the error hits 5 percent? What does the actual first mode look like at this point? $k_2 = 1000$ N/m, $k_3 = 1000$ N/m, $m_1 = 2$ kg, $m_2 = 2$ kg, and $m_3 = 2$ kg.

7. For the system below we'll ignore rotation of $m_3$. For $m_3$ going to zero we'd have a mode for which all $x_i$ were equal. Use this as your assumed solution for a Rayleigh's quotient approximation for the case of finite $m_3$. How large can $m_3$ get before the error between the approximation and the exact result exceeds 5 percent? What does the actual first mode look like? $k_i = 10$ N/m, $m_1 = 10$ kg, and $m_2 = 10$ kg.

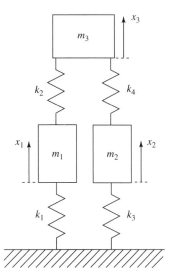

8. The system shown below supports a mode for which $x_1 = a$, $x_2 = -a$, and $x_3 = 0$, for uniform mass and stiffness values. Let $k_1 = 400$ N/m, $k_2 = 400$ N/m, $k_3 = 800$ N/m, $m_3 = 15$ kg, $m_1 = 8 + \alpha$ kg, and $m_2 = 8 - \alpha$ kg. How large can $\alpha$ get before the frequency error between the Rayleigh's quotient estimate found from using the mode described above and the actual natural frequency exceeds 2 percent? What do you notice in your Rayleigh's quotient results as you vary $\alpha$? Do not allow rotation of $m_3$.

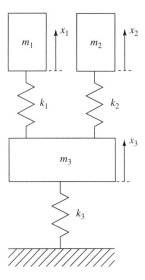

9. Look at the system of Problem 83, Chapter 4. Take an educated guess as to what the first mode's displacement should be and compare the results of using your guess in Rayleigh's quotient with the

actual first natural frequency. What was your error? $k_1 = 100$ N/m, $k_2 = 100$ N/m, $k_3 = 100$ N/m, $k_4 = 1000$ N/m, $m_1 = 1$ kg, $m_2 = 1$ kg, and $m_3 = 1$ kg.

## Section 6.4

**10.** Say you are shown three vectors that satisfy orthogonality with respect to the mass and stiffness matrices of a discrete system. Explain how it would be possible that they are not eigenvectors of the system.

**11.** Your assistant has given you the following results of his Rayleigh-Ritz analysis of a 5-DOF structural problem and asked for your opinion as to its correctness. The actual system natural frequencies are indicated by the $\omega_i$ and his estimates are shown as circular dots on the graph. Are the results plausible? How many errors can you spot? What may have caused them? Four trial functions were used.

**12.** Consider the system shown here ($k_i$ and $m_i$ values shown on figure). Without actually solving the problem, can you tell which is a better set of trial vectors for the illustrated problem?

$$\text{Set A}: \begin{Bmatrix} 1 \\ 1 \\ 1 \\ 1 \end{Bmatrix}, \begin{Bmatrix} 1 \\ 1.1 \\ 1.1 \\ 1 \end{Bmatrix}, \begin{Bmatrix} 1 \\ .9 \\ .8 \\ 1 \end{Bmatrix}$$

or

$$\text{Set B}: \begin{Bmatrix} 1 \\ 2 \\ 2 \\ 1 \end{Bmatrix}, \begin{Bmatrix} 1 \\ 1 \\ -1 \\ -1 \end{Bmatrix}, \begin{Bmatrix} 1 \\ -.4 \\ -.4 \\ 1 \end{Bmatrix}$$

**13.** Explain why the Rayleigh-Ritz procedure will break down if you use a trial vector that is a linear combination of previous trial vectors. You can consider a 3-DOF system for simplicity. (Hint: Look at the $[\mathcal{M}]$ and $[\mathcal{K}]$ matrices and consider the consequences of the third trial vector being a linear combination of the first two.)

**14.** Are

$$\left\{\begin{matrix}1\\2\\1\end{matrix}\right\}, \quad \left\{\begin{matrix}1\\-1\\1\end{matrix}\right\}, \quad \left\{\begin{matrix}1\\0\\1\end{matrix}\right\}$$

good trial vectors to use in a Rayleigh-Ritz analysis?

**15.** If two $10 \times 1$ vectors are orthogonal with respect to $[M]$ and $[K]$ are they necessarily eigenvectors? ($[\mathcal{M}]$ and $[\mathcal{K}]$ are $10 \times 10$). What if ten $10 \times 1$ vectors were considered?

## Sections 6.5 and 6.6

**16.** Consider the set of trial functions

$$\phi_n(x) = \sin(n\pi x)$$

Would these be reasonable trial functions to use if you were interested in using the Rayleigh-Ritz method for the fixed-fixed tensioned string of length 1 m? Note that the tension and density of the string are not constant.

**17.** Repeat Problem 16 for the system shown in Problem 10, Chapter 5. The tension and density are variable. $l = 1$ m.

**18.** Consider the system of Problem 10, Chapter 5. Let the string properties be constant. For $k = 0$, the first four natural frequencies are equal to 1 rad/s, 2 rad/s, 3 rad/s, and 4 rad/s. You then calculate the natural frequencies for $k \neq 0$ and come up with a new set of natural frequencies: $\omega_1 = 2.5$ rad/s, $\omega_2 = 5$ rad/s, $\omega_3 = 7.5$ rad/s, and $\omega_4 = 10$ rad/s. Do you have confidence that the analysis for the $k \neq 0$ case was done correctly?

**19.** Set up the appropriate equations from which we can solve for the free-vibration solution of the illustrated system using a Rayleigh-Ritz approach. $\rho$ and $T$ are constant and $m$ is a lumped mass attached to the string.

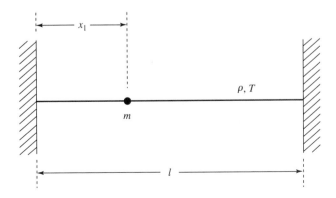

**20.** Set up the appropriate equations from which we can solve the free vibration of the illustrated system using a Rayleigh-Ritz approach. $\rho$ and $T$ are constant.

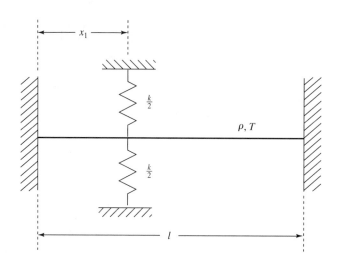

**21.** Consider the system of Problem 19. Let the lumped mass be denoted as $m_1$. How much will the frequencies of the first four modes change if we look at two cases: $m_1 = 1$ kg and $m_1 = 50$ kg? Use $\phi_n(x) = \sin\left(\frac{n\pi x}{l}\right)$ as your trial functions. Eight trial functions will be a good number to use. Solve the problem for the two values of $m_i$. Do the results make sense? $\rho = 1$ kg/m, $T = 1000$ N, $l = 2$ m, and $x_1 = .5l$.

**22.** Consider a tensioned string with constant tension, but variable mass. $T = 1000$ N and $l = 1$ m. Use trial functions $\phi_n(x) = \sin(\frac{n\pi x}{l})$ to determine the system's natural frequencies. Plot the approximation to the first eigenfunction. Use three mass distributions:

  **1.** $\rho(x) = \rho_0$; $\rho_0 = 1$ kg/m
  **2.** $\rho(x) = \rho_0(1 + .5 \sin \frac{\pi x}{l})$; $\rho_0 = 1$ kg/m
  **3.** $\rho(x) = \rho_0(1 + 5 \sin \frac{\pi x}{l})$; $\rho_0 = 1$ kg/m

You should be able to verify analytically the correctness of case 1, since it is simply the case of uniform stiffness and density.

**23.** Consider the system of Problem 10, Chapter 5. Calculate the approximate natural frequencies and mode shapes for the system. Use $\phi_n(x) = \sin \frac{n\pi x}{2l}$ as trial functions, $n = 1, 2, \ldots, 8$. Plot the first two mode shapes. The first two natural frequencies are equal to 88.68 rad/s and 179.2 rad/s. $\rho = 1$ kg/m, $T = 1000$ N, $k = 8000$ N/m, and $l = 1$ m.

**24.** This question is an extension of Problem 23. In this problem you'll be analyzing a tensioned string with constant tension and variable mass. The endpoints of the string are fixed. Use trial functions $\phi_n = \sin\left(\frac{n\pi x}{l}\right)$ in a Rayleigh-Ritz analysis to determine the system's natural frequencies. Plot the approximation to the first eigenfunction. Use the mass distribution:

$$\rho = \left[ 1000\delta\left(x - \frac{l}{2}\right) + 1 \right] \text{kg/m}$$

This is equivalent to a lumped mass of 1000 kg placed at midspan along with a small constant density. Try using an eight trial-function approximation. When you have your results, compare your estimate for $\omega_1$ with the SDOF approximation (in which the string simply acts as an equivalent spring).

25. This problem builds upon the results of Problem 24. Instead of using $\phi_8(x) = \sin\left(\frac{8\pi x}{l}\right)$, as was done in the previous problem, use

$$\phi_8(x) = x; \qquad 0 \le x < \frac{l}{2}$$

$$\phi_8(x) = l - x; \qquad \frac{l}{2} \le x \le l$$

that is, a linear increase followed by a linear decrease (*tent* mode). We know that this is very close to the actual response shape for the first mode, when a large mass is placed in the string. We know intuitively that this is very close to the actual response shape for the first mode when a large mass is placed in the center of the string. Compare the two Rayleigh-Ritz results from this and Problem 24, both for the natural frequencies as well as the model deflections. Go ahead and compare $\omega_1$ and the actual model deflection to those found from an exact analysis as well. Explain why using the tent mode improves the estimate for $\omega_1$ so dramatically. What's the problem with the Rayleigh-Ritz approach and $\phi_n(x) = \sin(n\pi \frac{x}{l})$ trial functions?

26. Repeat Problem 24, but with a much more regular mass distribution of

$$\rho = \left[\delta\left(x - \frac{l}{2}\right) + 1\right] \text{kg/m}$$

Does the Rayleigh-Ritz method do better for this case when $\phi_n = \sin\left(\frac{n\pi x}{l}\right)$ than it did in Problem 24? Compare your results with the exact solution and also see how replacing $\phi_8$ with the "tent" mode of Problem 24 helps the analysis.

27. In Problem 23 we computed the natural frequencies and eigenmodes for the system shown in Problem 10, Chapter 5, using $\phi_n = \sin\left(\frac{n\pi x}{2l}\right)$. This time use $\phi_n = \sin\left(\frac{(2n-1)\pi x}{2l}\right)$. Plot the first and second mode along with the exact results. Why aren't the approximate results as good as you might hope? $\rho = 1$ kg/m, $T = 1000$ N, $k = 8000$ N/m, and $l = 1$ m. The first two natural frequencies are equal to 88.68 rad/s and 179.2 rad/s.

28. In Problem 27 we solved a spring restrained, tensioned string problem using $\phi_n(x) = \sin\left(\frac{(2n-1)\pi x}{2l}\right)$ as trial functions and in Problem 23 we used $\phi_n(x) = \sin\left(\frac{n\pi x}{2l}\right)$. These trial functions correspond to the eigenfunctions of the problem when the spring stiffness approaches zero and infinity. Let's see what happens if we use trial functions that have nothing particular to do with either limiting case. Choose $\phi_n(x) = \sin(\sqrt{2}n\pi)$ and see how well you can approximate the solution. The first six natural frequencies are equal to 88.68 rad/s, 179.2 rad/s, 272.05 rad/s, 366.8 rad/s, 262.9 rad/s, and 559.8 rad/s. $\rho = 1$ kg/m, $T = 1000$ N, $k = 8000$ N/m, $l = 1$ m.

29. Show via a Rayleigh-Ritz analysis that the $\beta$ values for the illustrated system approach the pinned-pinned case for $k_\theta \to 0$ and the pinned-clamped case for $k_\theta \to \infty$. Show that the $\beta_i$ lie between these two extremes when $k_\theta$ is finite. Let the overall numerical ratio of $EI$ and $k_\theta$ be $\frac{k_\theta}{EI} = 8$ (all units used

consistently). Use $\phi_n(x) = \sin \dfrac{(n\pi x)}{l}$. Eight trial functions will be sufficient. $EI$ and $\rho$ are constant $k_\theta$ is a torsional spring attached to the right end of the beam.

30. Show by a Rayleigh-Ritz analysis that the first three natural frequencies for the illustrated system match the 3-DOF lumped approximation shown here. (You'll have to determine $k_1 - k_4$ from a static analysis). Use $\phi_n(x) = \sin(n\pi \frac{x}{l})$, $n = 1, 2, \ldots, 8$. $m_1 = 10$ kg, $m_2 = 20$ kg, $m_3 = 30$ kg, $T = 1000$ N, $\rho = .3$ kg/m, $l = 1$ m.

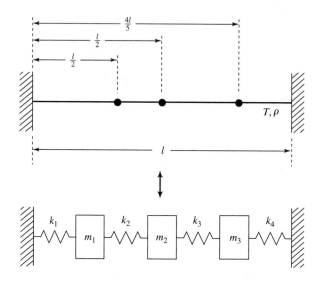

31. Assume that you have completed a Rayleigh-Ritz analysis of a continuous system using four trial functions and came up with the four vectors shown below as the resultant eigenvectors of the Rayleigh-Ritz formulation. These aren't the displacement eigenvectors; they are the eigenvectors that indicate trial-function participation as calculated from the Rayleigh-Ritz procedure. What do they tell you about your original trial functions?

$$\left[[\mathcal{K}] - \omega_i^2[\mathcal{M}]\right] A_i = 0$$

$$A_1 = \begin{Bmatrix} 0 \\ 1 \\ 0 \\ 0 \end{Bmatrix} \; ; \quad A_2 = \begin{Bmatrix} 1 \\ 0 \\ 0 \\ 0 \end{Bmatrix} \; ; \quad A_3 = \begin{Bmatrix} 0 \\ 0 \\ 0 \\ 1 \end{Bmatrix} \; ; \quad A_4 = \begin{Bmatrix} 1 \\ 1 \\ 1 \\ 1 \end{Bmatrix}$$

# 7

## Seat-of-the-Pants Engineering

Who cares exactly how tall it is?! It's big enough to have eaten my convertible; is that accurate enough for you?

## 7.1 INTRODUCTION

The next few pages will examine explicitly what we've already seen in some of the homework problems, the question of how to get a *good enough* answer instead of an exact one. Although this chapter and these problems could have been broken up and distributed among the previous chapters, the approach seems to be distinctive enough to merit a more unified presentation. To a large extent the ability to do good seat-of-the-pants engineering is an acquired skill, one that comes after having gone through the tedium of producing exact results for a great many problems. The difficulty for the student is that the whole idea of seat-of-the-pants analysis is one that is very seldom referred to in any text, even though it underlies much of the successful engineering of this world. Really good engineers rarely start by sitting down and precisely analyzing a problem. Rather they consider many options, get quick approximations, redesign, get some more approximate solutions, and then finally zero in on a final solution. Only after a particular problem has been adequately pinned down do really serious analysis tools come into play. This chapter attempts to give the student some sense of what can be accomplished by stepping back and thinking about a particular problem in a general way before bringing out all the analytical tools that the preceding chapters have given us.

## 7.2 GETTING APPROXIMATE RESULTS

*Seat-of-the-pants engineering* refers to the ability to analyze a system reasonably well, *but not exactly*, in a short time. Obviously, you can analyze a given problem in great depth and detail if you're given the necessary time and resources. Nevertheless, it's often the case that you won't be given all the time you might like and you must produce an answer anyway. In addition to situations like these, you might want to generate an approximate answer quickly because you're considering different design options and you don't want to spend too much development time on an idea that's not going to pan out anyway. Thus there's a good deal of value in having the ability to get a good enough answer fairly quickly.

As an example of how we can cut quickly to an answer, let's revisit our spring-mass problem of Section 1.2. You'll recall that in this problem we were trying to model the system as a lumped spring and mass and to determine the combined natural frequency of the system. What made the analysis approximate was the fact that the spring had mass of its own, mass that wasn't accounted for in the analysis. This same complication will appear more strongly when considering bars or beams.

To illustrate how we can reasonably account for this added mass, we'll consider longitudinal motions of a bar with a lumped mass at one end, as shown in Figure 7.1.

If the bar actually was a pure spring, then the deflected position of the bar under an applied load at the end would vary linearly, from zero at the fixed end to a maximum value at the free end. In Section 1.2 we found the relation between the deflection at the end and the applied force (1.2.57),

$$\xi(l) = \frac{fl}{EA} \tag{7.2.1}$$

If our simplifying assumptions hold, then a linearly varying displacement along the bar implies that the velocities will vary linearly also, since finding the velocities simply involves differentiation in time. No spatial operations are involved, so the spatial relations stay the same.

We've already seen that the natural frequencies of a system can be found by taking ratios of the system energies (Rayleigh's quotient). Since we know what the velocity field looks like within the bar (if it behaves according to our assumptions), then we can modify the kinetic energy part of Rayleigh's quotient and solve for the modified natural frequency. Originally, the kinetic energy was given by

$$k_e = \tfrac{1}{2} m_1 \dot{\xi}^2(l) \tag{7.2.2}$$

that is, the only component of the kinetic energy was that due to the lumped mass at the right end.

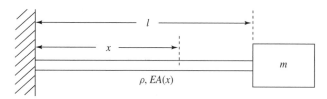

**Figure 7.1**   Uniform bar with a lumped mass at the right end

Now we're going to add the energy of the bar. If the velocity distribution were truly linear then we'd have

$$\dot{\xi}(x) = \tfrac{x}{l}\dot{\xi}(l) \tag{7.2.3}$$

In reality, the actual distribution won't be precisely linear. For large end masses, the first mode will be very close to a linear distribution and the accuracy of this approximation will degrade as the end mass is reduced with respect to the bar's mass.

Taking the bar's mass into account, our new kinetic energy becomes

$$k_e' = \frac{1}{2}m_1\dot{\xi}^2(l) + \frac{1}{2}\int_0^l \rho\dot{\xi}^2(x)dx \tag{7.2.4}$$

or, using (7.2.3),

$$k_e' = \frac{1}{2}m_1\dot{\xi}^2(l) + \frac{1}{2}\int_0^l \rho\left(\frac{x}{l}\right)^2\dot{\xi}^2(l)dx \tag{7.2.5}$$

We can factor out a couple of terms to leave us with

$$k_e' = \frac{1}{2}\dot{\xi}^2(l)\left(m_1 + \int_0^l \rho\left(\frac{x}{l}\right)^2 dx\right) \tag{7.2.6}$$

If we now evaluate the integral, we'll be left with

$$k_e' = \frac{1}{2}\left(m_1 + \frac{\rho l}{3}\right)\dot{\xi}^2(l) \tag{7.2.7}$$

Comparing (7.2.2) and (7.2.7), we see that we have an equivalent mass. Whereas previously our mass was just equal to $m_1$, now it's equal to $m_1 + \frac{\rho l}{3}$. Thus we can go ahead and use our old $\omega_n = \sqrt{\frac{k}{m}}$ formulation (with $k = \frac{EA}{l}$), but this time we'll use the modified mass rather than just the lumped mass. Therefore

$$\omega_n = \sqrt{\frac{EA}{l(m_1 + \frac{\rho l}{3})}} \tag{7.2.8}$$

If we calculate the natural frequency for the case of $EA = 10$ N, $l = 1$ m, $m_1 = 10$ kg, and $\rho = 1$ kg, we'll find that $\omega_n = .9837$ rad/s. Without the correction for $\rho$, our estimate for the natural frequency is 1.0 rad/s, giving us an error of about 2 percent from the exact result of .9836 rad/s (found from solving the actual PDE). By modifying our mass term, we've reduced the error in the estimate to about one one-hundredth of a percent.

Let's next go ahead and tackle the tougher problem of $\rho = 10$ kg. For this case we previously had a 16 percent error (Section 1.2), due to the fact that the lumped mass isn't as it was when we compared with the mass of the bar. Our modified estimate for the natural frequency (using the new modified mass) is equal to .8660 rad/s. If we compare this with the exact value of .8603 rad/s, we see that we've cut the error from 16 percent to about .5 percent.

These are major improvements! Just by using a little insight and realizing that the motions of the bar will contribute to the system's kinetic energy, we've improved our frequency estimates tremendously. Of course, if you really want the exact solution you can always find it using the techniques of Chapter 5. But what we have instead is a very accurate estimation with very little effort.

Okay, we've now seen how we can improve on our frequency estimate for an isolated single-degree-of-freedom problem. Let's move beyond that and into an MDOF estimation problem. For this example, we'll assume that you work for a major automobile manufacturer and your job involves suspension design. A colleague has come up with a very nice model of a vehicle, one that includes seven degrees of freedom. The precise model isn't important because we won't be doing an exact analysis, but Figure 7.2 shows the basic setup. The tires are modeled as spring elements, as are the suspension pieces between the axles and the body. The front and rear axles are included as rigid bars and the body mass is supported by the suspension springs. Since the model has seven degrees of freedom, it also has seven eigenvectors, eigenvectors that can be associated with specific types of responses. For instance, the lowest natural frequency mode is associated with what automotive designers call *bounce*. This is the same mode that airplane designers call *heave* and corresponds to purely up-and-down motions of the body. Although the physical mode that's called bounce doesn't exactly correspond to pure up-and-down motions (there's some small amount of pitch, etc.), the approximation found by ignoring everything but the up-and-down motions is actually quite good. There's also a *roll* mode, which corresponds to rotations about the longitudinal axis of the car. Pitching refers to rotations about a lateral axis through the car (the sort of motion you'd get from strongly applying the brakes). These low-frequency modes are *rigid-body* modes since they reflect motions of the car body itself, oscillating on its suspension. The remaining modes are higher-frequency modes that involve motions of smaller subsystems within the vehicle.

The data to be used in this example are taken from an actual car analysis discussed in [7]. In this model, there are two kinds of mass elements. The car body is called a *sprung* mass because it's supported by the suspension. The mass elements that aren't directly attached to the body, like the rear axle, are called *unsprung* mass elements. Because these masses aren't directly connected to the body, they can experience their own, higher frequency modes. One of these is called *wheel*

**Figure 7.2**  Simplified automobile vibration model

**TABLE 7.1**  PARAMETER VALUES FOR
1975 COMPACT TWO-DOOR

| Parameters | Parametric values |
|---|---|
| Tire stiffness | 184 kN/m |
| Front suspension stiffness | 18.7 kN/m |
| Rear suspension stiffness | 26.1 kN/m |
| Vehicle mass | 1065 kg |
| Total unsprung mass | 175 kg |
| Front unsprung mass | 73 kg |
| Rear unsprung mass | 102 kg |
| Rear axle | 52 kg |
| Body roll inertia | $190 \ kg \cdot m^2$ |
| Body pitch inertia | $1080 \ kg \cdot m^2$ |
| Rear axle rotational inertia | $8 \ kg \cdot m^2$ |
| Wheelbase | 2.42 m |
| Front wheels to CG | 1.2 m |
| Rear wheels to CG | 1.22 m |
| Ground to CG | .46 m |
| Front track | 1.38 m |
| Rear track | 1.35 m |

*hop*, which occurs when the front or back wheels bounce up and down in unison. If a set of wheels is moving in an antisymmetric way (one up and the other down), then the mode is called *tramp*. The various physical parameters of the vehicle model are given in Table 7.1.

In order to analyze the system and to determine what the frequencies and mode shapes are, you could construct the equations of motion for the complete system, code it, and analyze it. But now we're going to show how you can get a quick and reasonable estimate of the responses without going to all this trouble. Just for comparison, Table 7.2 lists the exactly calculated natural frequencies for the modes we'll be looking at.

Let's start with the bounce mode. We know that this mode ideally consists of the body moving up and down. If you ignore the mass between the tire springs and suspension springs, you can

**TABLE 7.2**  NATURAL FREQUENCIES FOR
SOME CAR VIBRATION MODES

| Vibration mode | Frequency of oscillation |
|---|---|
| Bounce (sprung mass) | 1.33 Hz |
| Pitch (sprung mass) | 1.71 Hz |
| Roll (sprung mass) | 2.27 Hz |
| Front wheel hop | 11.87 Hz |
| Front tramp | 11.87 Hz |
| Rear wheel hop | 10.19 Hz |
| Rear wheel tramp | 12.63 Hz |

model them as two springs in series. Thus the equivalent spring at each corner of the car is equal to

$$k_f' = \frac{k_{ft}k_{fs}}{k_{ft} + k_{fs}} = 16,975 \text{ N/m} \tag{7.2.9}$$

and

$$k_r' = \frac{k_{rt}k_{rs}}{k_{rt} + k_{rs}} = 22,858 \text{ N/m} \tag{7.2.10}$$

The subscripts $f$ and $r$ indicate the front or rear springs and the subscripts $s$ and $t$ indicate the suspension spring or the tire. Thus $k_{ft}$ is referring to the stiffness due to the tire at the front of the car.

Remember that we're after a single mode, the one corresponding to bounce. If we call $x_b$ the coordinate of the bounce mode, then we can write

$$1065\ddot{x}_b + 2(16,975 + 22,858)x_b = 0 \tag{7.2.11}$$

Note that 1065 is the total mass (in kilograms) and $2(16,975+22,858)$ is the total spring constant. (We have spring forces from both the left and right side, thus giving us the factor of 2.)

Solving for the natural frequency gives us

$$\omega_n = \sqrt{\frac{79,666}{1065}} = 8.65 \text{ rad/s} = 1.38 \text{ Hz} \tag{7.2.12}$$

This can be compared to the exact value found in Table 7.2, which is 1.33 Hz. We see that our quick and dirty solution gave us an answer that is within 4 percent of the actual answer.

Given this success, let's go crazy and try to replicate another answer. The exact solution for the roll mode gives a natural frequency of 2.27 rad/s for the sprung mass. The moment of inertia for the sprung mass about the longitudinal axis is equal to 190 kg·m$^2$. All that we need do is determine the angular stiffness. Figure 7.3 shows what we actually must deal with. When the car rolls, the equivalent springs at the front and back are deformed. As Figure 7.3 shows, the equivalent torsional spring for each spring is equal to $k_{eq}l^2$, where $l$ is the distance from the roll center to the spring and $k_{eq}$ corresponds to the $k_f'$ and $k_r'$ defined above. Note that the distances from the roll center to the equivalent spring is given by the front and rear track data in Table 7.1. Thus our equation of motion for the roll mode is given by

$$190\ddot{x}_r + 2\left(16,975\left(\frac{1.38}{2}\right)^2 + 22,858\left(\frac{1.35}{2}\right)^2\right)x_r = 0 \tag{7.2.13}$$

For this case, the natural frequency turns out to be

$$\omega_n = \sqrt{\frac{34,598}{190}} = 14.0 \text{ rad/s} = 2.22 \text{ Hz} \tag{7.2.14}$$

a result that's only off by 2 percent from the exact solution.

Finally, let's look at one more mode, that of front-wheel hop. This is an interesting mode because we can choose from among different approximation levels. The most basic observation to

**Figure 7.3**  Roll mode

make is that we have a relatively large mass (890 kg) connected to a relatively small one (73 kg) by way of two suspension springs, each with a stiffness equal to 18,700 N/m. The 890-kg mass corresponds to the sprung mass, while the 73-kg mass represents the front unsprung mass (the mass between the tire spring and the suspension). The unsprung mass is sprung to ground by means of the two tire springs, each having a stiffness equal to 184,000 N/m. If we completely ignore the sprung mass and suspension springs, we're left with the unsprung mass supported by two tire springs

$$73\ddot{x}_h + 2(184,000)x_h = 0 \tag{7.2.15}$$

This gives us a natural frequency estimate of 11.3 Hz, which is only 5 percent off the exact value of 11.87 Hz. But wait, there's more! We observe that this frequency is much higher than the bounce frequency. Now remember what happened in your physical experiment with the mass and spring when you moved your hand at different speeds. When you vibrated your hand quickly, the mass didn't move; it was isolated by means of the spring. Well, just think about our car. According to our estimates, if the rest of the car isn't moving, then the front-tire hop occurs at a frequency of 11.3 Hz. Let's make believe that the tires are bouncing at this frequency. Will the large mass of the car itself respond? No way! The total sprung vehicle mass is equal to 890 kg. The springs between this mass and the wheels have a combined spring constant of $2(18,700) = 37,400$ N/m. Thus the natural frequency of this mass on these springs is equal to

$$\omega_n = \sqrt{\frac{37,400}{890}} = 6.48 \text{ rad/s} = 1.03 \text{ Hz} \tag{7.2.16}$$

This is more than 10 times lower than the excitation frequency. Thus we have exactly the same situation we had when we vibrated our hand at a much higher frequency than the system's natural frequency. End result? No motion of the big mass. Thus we can improve our approximation by saying that the sprung mass of the vehicle stays fixed, and we're left with the wheels being restrained by two sets of springs, namely the suspension and the tire springs. Since these are acting in parallel on the sprung mass we'll have

$$73\ddot{x}_h + 2(184,000 + 18,700)x_h = 0 \tag{7.2.17}$$

which gives us a natural frequency of

$$\omega_n = \sqrt{\frac{405,400}{73}} = 74.5 \text{ rad/s} = 11.86 \text{ Hz} \tag{7.2.18}$$

The result? An amazingly good approximation, one in which the error is less than a tenth of a percent.

Now that you've seen the power of this kind of analysis, let's go over what we're doing more carefully to see when it will work and when it will not. First of all, the approach depended upon the existance of distinct and *identifiable* modes of vibration in our system. For a general system, there might be no easily identified modes of the sort we have in the car problem. The first mode might involve both front wheels moving up and down in unison, while the rear ones moved antisymmetrically. Or it might have involved an equal amount of bounce and roll happening simultaneously. If this had been the case, then our assumptions of pure bounce or pure wheel hop would have been wrong and the frequency estimates associated with them would have been equally wrong. This is where engineering intuition first comes in. Whether from analyzing different cars in the past or from an examination of the model, you have to suspect that the bounce, roll, and hop modes will show up. Only then can you solve for their frequencies of vibration.

After you've decided that certain physical modes should predominate in the response, you still must determine the essential factors that govern its motion. For the bounce mode, the key fact was in deciding that the two springs (tire and suspension) could be modeled as a series of springs, with a resultant equivalent single spring. For the wheel-hop mode, we had to use a bit more insight and realize that certain elements of the vehicle wouldn't be involved in the motions (the sprung mass) because they were isolated from the high-frequency vibrations by their relatively soft springs. If the suspension had been a very hard one, one in which the suspension spring constants were as large as the tire spring constants, then the sprung mass *wouldn't* have been isolated and would have moved along with the wheels. This would have complicated our analysis.

**Example 7.1**

> **Problem**  In order to illustrate the concepts we've just learned, let's analyze the system shown in Figure 7.4. We want to determine the modes of vibration and their natural frequencies without solving the entire coupled problem.
>
> **Solution**  As you can see, we have three masses and six springs. The first item of business is to determine how many modes there will be. $m_1$ can move only vertically, as can $m_2$. However, $m_3$ can translate vertically and can also rotate. Thus this mass supports two degrees of freedom. This

**Figure 7.4**  4-DOF system

means that our system has a total of four degrees of freedom and we expect to find four modes of oscillation.

First we'll note the relative masses. $m_1$ and $m_3$ are tiny in comparison to $m_2$. It seems very likely that for the lowest frequency mode we'll have $m_2$ ponderously going up and down, dragging everything else along with it. To find this mode, we must determine the equivalent spring that's restraining $m_2$. In total, we have two $k_2$ springs, plus an equivalent spring that's made up of the $k_3$ spring in series with two $k_4$ springs. Thus the overall spring constant is

$$k_{\text{equiv}} = 2k_2 + \frac{k_3(2k_4)}{k_3 + 2k_4} = 24,878 \text{ N/m} \tag{7.2.19}$$

using the values given in Figure 7.4. The simple (and accurate) answer for the oscillations of this mass is to set it equal to $1500 + 40$, which is the total of $m_1$ and $m_2$. A more involved approach is to realize that $m_3$ is basically along for the ride, adding inertia but not doing much else in this

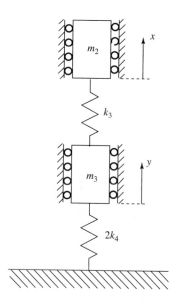

**Figure 7.5**  Static analysis to find equivalent mass contribution of $m_3$

low-frequency mode. Since we know what the spring constants are surrounding this mass, we can determine how much it moves for a given motion of the main mass $m_2$. We'll just do a static analysis, like that shown in Figure 7.5. If we move the $m_2$ by $x$, then we must do a force balance to decide how much $m_3$ moves. The force balance gives us

$$200,000y = 5000(x - y) \tag{7.2.20}$$

or

$$y = \frac{5000}{205,000}x \tag{7.2.21}$$

Thus the effect of $m_3$ will be to add $\left(\frac{5000}{205,000}\right)$ (48 kg) = 1.17 kg to the overall mass. This reflects the fact that $m_3$ doesn't move nearly as much as $m_2$, so we shouldn't include all of its inertia.

Using our equivalent mass and spring, we can now solve for the natural frequency

$$\omega_n = \sqrt{\frac{24,878}{1541}} = 4.018 \text{ rad/s} \tag{7.2.22}$$

We can also find the approximate mode itself. If we use the coordinates

$$X^T = \{x_1 \quad x_2 \quad x_3 \quad \theta\} \tag{7.2.23}$$

(as shown in Figure 7.4), then we see that the mode is equal to

$$\tilde{X}_1^T = \{1.000 \quad 1.000 \quad .024 \quad 0\} \tag{7.2.24}$$

where the tilde shows that this is only our estimate, not the actual eigenvector. This eigenvector was constructed by setting the response amplitude of $x_2$ to 1.0. Since $m_1$ is modeled as being rigidly attached to $m_2$, it also has amplitude 1.0. We just saw that the motion of $m_3$ is $\frac{5}{205}$ the size of $m_2$, which in this case equals .024. Finally, the rotational mode isn't excited in this motion. Therefore it has an amplitude of zero.

Solving the fully coupled problem exactly results in a frequency of 4.019 rad/s and a mode shape of

$$X_1^T = \{1.000 \ .987 \ .024 \ 0\} \tag{7.2.25}$$

You can't get much better than this. There's hardly any difference at all between the approximate and exact results.

Now how about the other modes? We have already alluded to one. If $m_3$ is given an initial rotation, it will experience a rotary vibration. All the other motions in the system are purely translational, and there's no coupling between rotation and translation for this system. Thus we can analyze this mode immediately by setting all translations to zero and examining the rotational mode that's left. Note that this would definitely not be the case if the problem wasn't as symmetric as it is. For our system, $m_3$ is uniform, both the springs from $m_3$ to ground are identical, and the attachment of the $k_4$ and $k_3$ springs are symmetric. This is what lets us state that the rotations will not couple with the translational motion.

Taking the rotational part by itself, we have

$$4\ddot{\theta} + 2(100,000)\left(\frac{1}{2}\right)^2 \theta = 0 \tag{7.2.26}$$

where the factor of 4 comes from the rotational inertia of $m_3$ and the rotational-spring term comes from determining the restoring moment due to the $k_4$ springs. This supports a natural frequency of

$$\omega_n = 111.8 \ \text{rad/s} \tag{7.2.27}$$

and an eigenmode estimate of

$$\tilde{X}_2^T = \{0 \ \ 0 \ \ 0 \ \ 1\} \tag{7.2.28}$$

Since there really are no other interactions for this mode, the exact solution and the approximate one are identical.

For our next approximate solution let's examine the small mass $m_1$ that's attached to the very large mass $m_2$. Common sense tells us that the vibrations of a very small mass aren't likely to do much to affect the motion of a huge one. Thus we can approximate our next mode as being one for which $m_1$ oscillates and $m_2$ remains fixed. This leaves us with the SDOF problem

$$40\ddot{x}_1 + 50,000x_1 = 0 \tag{7.2.29}$$

The associated natural frequency is equal to 35.36 rad/s and the approximate mode is given by

$$\tilde{X}_3^T = \{1 \ \ 0 \ \ 0 \ \ 0\} \tag{7.2.30}$$

That was certainly quick. Does it agree with the exact answer? Well, the exact solution will produce a natural frequency of 35.83 rad/s and an eigenvector of

$$X_3^T = \{1.000 \ -.027 \ -.001 \ 0\} \tag{7.2.31}$$

You'd have to be a very stubborn individual not to agree that these two results are just about identical.

Finally, we must address what's happening with the last mode. Applying the same insight we used in the car problem, we can check to see if there's a mode that oscillates at a much higher frequency than the large mass would want to oscillate, thus isolating it. The first choice is to look at the motion that hasn't had any play yet, namely translational motions of $m_3$. If we rearrange our system as shown in Figure 7.6, we can view it as two SDOF systems joined by a central spring. (We've neglected the $m_1$-$k_1$ combination as being too small to worry about.) As shown in Figure 7.6, $m_2$ is sprung to ground through the two $k_2$ springs and attached to $m_3$ through $k_3$. We've already

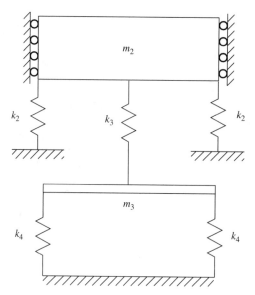

**Figure 7.6**   Reduced system ignoring $m_1$ and $k_1$

seen that the mode involving $m_2$ occurs at 4.018 rad/s. If we examine the $m_3$-$k_3$-$k_4$ combination, keeping $m_2$ stationary, we have a natural frequency of $\omega_n = \sqrt{\frac{205,000}{48}} = 65.35$ rad/s. Thus $m_3$ would like to oscillate at a much higher frequency than the low frequency mode of $m_2$. Therefore $m_2$ is isolated from the vibrations and we can confidently approximate it as being stationary.

If $m_2$ is taken as stationary, then we know from the previous paragraph that the natural frequency associated with $m_3$'s motion is 65.35 rad/s. The mode shape would be

$$\tilde{X}_4^T = \{0 \quad 0 \quad 1 \quad 0\} \tag{7.2.32}$$

The actual eigenvector is

$$X_4^T = \{.000 \quad -.001 \quad 1.000 \quad .000\} \tag{7.2.33}$$

and the natural frequency occurs at 65.35 rad/s. Not much difference, eh?

It wasn't an accident that the approximation results were so close to the exact results; the system was designed so that this would happen. The point is that if you hadn't taken the time to think about what the system represented and consider ways to get approximate solutions, you'd have been forced to solve the full-blown, 4-DOF problem in order to get any results—and this isn't always the most intelligent approach. Real systems invariably will be complex and solving them will require time and effort. If all you really want is an estimate of a couple of eigensolutions, it's worth your while to think a bit about intelligent approximations before jumping in.

## 7.3 LIMITING CASES

Up to now, our examples always had finite values for the mass and spring stiffness. Although this is very reasonable, it is also of use to consider what happens when one or both of these terms goes to zero or infinity. This is true for a variety of reasons. For one thing, we may wish to suggest design changes to a structure that involve large reductions of mass or increases in spring stiffness. In these cases, knowing the limiting value response for zero mass or infinite spring stiffness will let us know the greatest effect that such an alteration could have on our system. In addition, an easy way to alter the system we're considering, if it is coded on the computer, is to have the computer make certain mass or stiffness elements very small or very large (approximating zero and infinity, respectively). This will serve to lock out or release degrees of freedom. Again, we would want to know what the analytical effect of this will be ahead of time in order to correctly interpret the computer's output.

This last technique, that of driving some terms to zero or infinity, is perhaps most useful in debugging programs. When the program is complex, it's hard to determine if the output is really reasonable or not. You can often push the system into a simpler configuration, one that's easier to check, if you set certain parameters to their limiting values.

In Chapter 1 we've already considered the case of zero spring stiffness. Equation (1.2.2) showed the response of a free mass, which is the same as a mass-spring system for which the spring constant has been set to zero. As Equation (1.2.2) indicated, the mass can have an initial displacement and may also translate at a constant velocity if no forces are applied. This behavior is also indicated from our solution of a finite spring-mass system. Since the natural frequency of such a system is equal to the square root of $\frac{k}{m}$, we see that the consequence of $k$ going to zero is that the natural frequency also goes to zero. And a zero frequency implies that the period of the oscillation is infinity. Thus the mass will move off and not come back. This is a very important point to remember: zero frequency implies translation (or rotation) with no oscillations. It is the free-body motion referred to earlier. Note that increasing the mass a great deal, while holding $k$ fixed, will give us the same result; it drives the natural frequency to zero.

This observation also has a more compelling intuitive feel. It is easy to see that, given a huge mass and a tiny spring, once the mass is started in motion, the spring is going to have a very hard time reversing that motion. Since the force generated in the spring is directly proportional to its stretch, it's also clear that the little spring will have to stretch a great deal in order to generate sufficient force to bring the mass back, that is, the mass will translate for a long time before the spring can affect it appreciably.

Next we'll look at the case of the mass going to zero (or equivalently, the stiffness going to infinity). In this case, the natural frequency will go to infinity since either $m$ going to zero or $k$ going to infinity will cause $\frac{k}{m}$ to go to infinity. This makes sense if you consider a fixed mass attached to a spring that gets stiffer and stiffer. As the stiffness gets huge, the amount of motion needed to counteract any inertial forces is quite small. This is exactly the opposite of the previous case. For small stiffnesses, the system became free. Now we see that large stiffnesses lock the system up.

These single-degreee-of-freedom examples also hold when we look at multiple degrees of freedom. For instance, let's look at the system shown in Figure 7.7. This system consists of three masses and four springs. We'll fix the values of $m_1$–$m_3$ and $k_1$–$k_3$, but let $k_4$ vary from zero to 12 N/m. As we'll see, this range will show all the effects that we'd see by varying the stiffness from zero to infinity.

Since there are three masses, the system will display three distinct natural frequencies for each value of $k_4$. The variation of $\omega_i^2$ is shown in Figure 7.8 for all $k_4$ values between zero and 12 N/m.

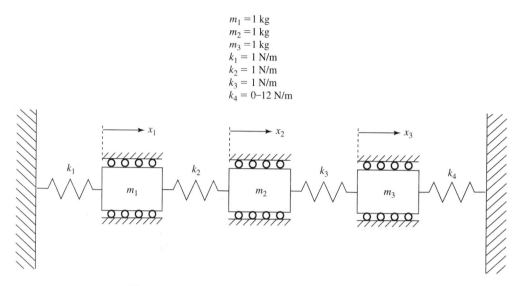

**Figure 7.7**   Spring-mass system with variable end spring

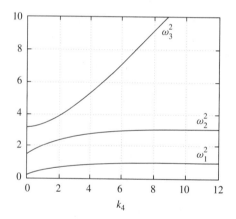

**Figure 7.8**   Variation of natural frequencies as a function of spring constant $k_4$

We already know that we should expect all the natural frequencies to increase as $k_4$ is increased, since we're adding stiffness to the system. We can also predict what the final frequencies will be. When $k_4$ becomes huge, it locks down the motions of $m_3$. Thus we're left with a 2-DOF system. Analytically, we can solve for the natural frequencies and find that they are equal to 1 rad/s and $\sqrt{3}$ rad/s. Since Figure 7.8 plots the square of the natural frequencies (the eigenvalues of the problem), we should see results of 1 and 3 for large $k_4$, which we do. In fact, you'll notice that essentially all variation of the first two natural frequencies has occurred by the time $k_4$ has reached 8 N/m. This is quite typical. As soon as the effect of a particular element has become sufficiently large, the effect on the overall system response is as if it were equal to infinity.

We can also see that the third natural frequency does not level off but, in fact, climbs steeply as $k_4$ is increased. This frequency is associated with the reduced system of $m_3$, $k_3$, and $k_4$. For high values of $k_4$, the other masses are unaffected; the vibration is too high for them to respond. Thus $m_1$ and $m_2$ are essentially stationary, leaving the mass $m_3$ to vibrate, restrained by $k_3$ and $k_4$. The natural frequency squared of this mode is given by

$$\omega_3^2 = \frac{1 + k_4}{1} \tag{7.3.1}$$

(i.e., $\dfrac{k_3 + k_4}{m_3}$ with the numerical values of $m_3$ and $k_3$ equal to 1.) Therefore we would expect it to rise linearly as soon as $k_4$ becomes substantially greater than 1, which it does.

Next, let's consider small values of $k_4$. Setting $k_4$ equal to zero still leaves us with a 3-DOF problem. If you go ahead and solve for the exact natural frequencies for this case, you'll see that they're equal to those given in Figure 7.8 for the case of $k_4 = 0$.

You can also let damping and/or mass elements vary from small values to very high values and obtain similar results. For instance, letting a mass element go to infinity will mean that you'll have a natural frequency that goes to zero (due to the huge mass), while the others go to the values they'd have if the mass were fixed in space. Damping can also lock out degrees of freedom. When the damping goes to infinity, it's as if the mass is moving in incredibly thick tar. Obviously, it won't move much in this case.

These insights can and should be used in your engineering work. Let's say you want to increase a natural frequency in a system because one of the current natural frequencies is near an external excitation frequency. You may want to move it up by 20 percent and you may think you can do it by increasing a certain spring stiffness in the system. By setting this stiffness to infinity, then analyzing the resulting vibrations, you can determine what the maximum frequency changes are going to be. If they're not sufficient, you know not to waste time trying to alter that particular spring; it's not going to be sufficient to do the job.

## 7.4  VERIFYING YOUR ANALYSES

One of the most important topics that you'll need to master is how to verify your analyses. You may have developed a nice analytical or approximate model for a system, and you're ready to produce some results. But how do you know if your model is any good? Or perhaps you've included some typos in your computer code, typos that don't keep the code from running but that do give you incorrect answers. How can you tell?

The first secret is a simple one. Go slow. Go REALLY SLOOOOWWWW. This probably sounds very silly, but it isn't. It's incredibly easy to make an error when flying through a derivation. Although it's extremely difficult, you should try to develop the discipline to go a line at a time and *check each line*. The time saved (by all the errors you'll catch) is well worth the time it takes.

Okay, let's say you went slow and did what you think is a good job. What now? Well, one thing you can do with your written equations is to check to see that the limiting cases work out. For instance, do your equations give you the correct static result? If you go through your equations and cross off all the time-dependent terms, you'll be left with a static equation, one that's easy to solve. It's usually pretty straightforward to determine what the physical response should be for a static situation. Thus you can check that at least part of your equation is correct. The same reasoning also applies to the other extreme. If you kill all but the highest-frequency terms, you'll be left with equations that describe the upper-frequency limit for your problem, something that you can then verify.

If you're attempting a modal approximation, you could include only a single mode to start. If your system is restrained to ground, this could be a rigid-body mode. Your approximate analysis should then match the results you'd get from letting the distributed system behave rigidly. If it doesn't, you've got an error somewhere.

Most of the time, you'll end up with computer code that's supposed to accurately represent your equations, which in turn represents some physical reality. By looking at limiting cases for all of these, and cross checking, you'll go a long way toward verifying that your work is correct.

## 7.5 HOMEWORK PROBLEMS

### Section 7.2

1. Figure 2.3 shows a seismically excited system. For an input displacement $y = \bar{y}(\cos \omega t)$, what will the phase lag of the output $x$ to input $y$ equal? An answer that's within 10 percent is good enough. $m = 10$ kg, $k = 9000$ N/m, $c = 60$ N/m/s, $\bar{y} = .1$ cm, $\omega = 600$ rad/s.

2. Your research assistant has given you three vectors that she says represent the three eigenvectors of a particular system. You put these together into the model matrix $[U]$ and form the matrices $[U]^T[M][U]$ and $[U]^T[K][U]$, which turn out to be

$$[U]^T[M][U] = \begin{bmatrix} 1 & 0 & 0 \\ 0 & 1 & 0 \\ 0 & 0 & 1 \end{bmatrix}, \quad [U]^T[K][U] = \begin{bmatrix} 6 & 0 & 0 \\ 0 & 4 & 1 \\ 0 & 1 & 3 \end{bmatrix}$$

What can you say about the three vectors?

3. The following figure shows two snapshots of a tensioned string under free vibration for two instants in time. Explain how such a situation could arise if the system modes were proportionally damped. Explain how it could also occur under nonproportional damping.

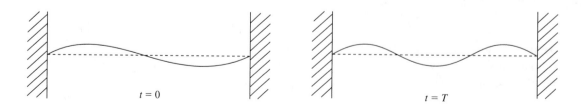

$t = 0$                    $t = T$

**4.** Consider the system made up of a torsional rod with a heavy mass at one end. For what frequency range (in terms of $\rho$, $G$, and $J$) would it be appropriate to model this system as a SDOF system (that is, the lumped inertia of the mass at the end and the rod stiffness act together as a SDOF spring-mass system)?

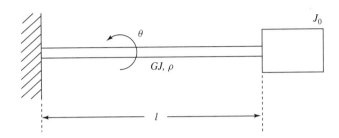

**5.** Can you think of what response variable would produce the following amplitude/phase outputs? (Hint: Consider the types of system we've seen—mass excited, rotating imbalance, seismically excited—and what could be measured $x, \dot{x}, \ddot{x}$, etc.) Note that the magnitude of the output goes to zero as the forcing frequency goes to infinity.

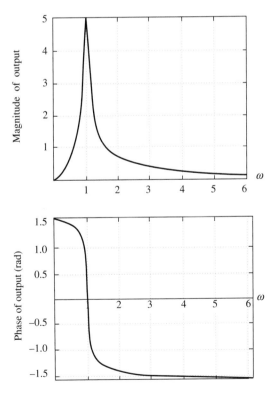

**6.** Consider the following tensioned string with a string-mass system suspended from it. The string can be forced at a variety of frequencies. If the forcing is equal to the frequency of the first string mode (ignoring the spring-mass), what should $k/m$ be in order to minimize the response? What if the forcing was at the second natural frequency? Can the spring-mass system be placed anywhere along the string in both these cases or are there restrictions as to where it should be located? Note that this is *not* a Rayleigh-Ritz problem. $\rho = 1$ kg/m, $l = 1$ m, $T = 1$ N.

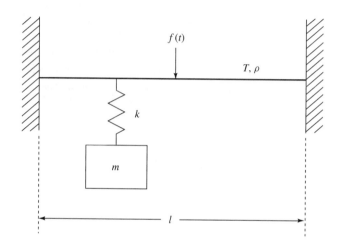

**7.** Find the eigenvectors and natural frequencies for the illustrated system. Describe the modes that seem to involve only part of the structure and explain why the whole structure (all $x_i's$) isn't involved. $k_0 = 1000$ N/m, $k_1 = 10$ N/m, $k_2 = 50$ N/m, $k_3 = 75$ N/m, $k_4 = 2000$ N/m, $k_5 = 1$ N/m, $k_6 = 2,000$ N/m, $m_1 = 1$ kg, $m_2 = 10$ kg, $m_3 = 50$ kg, $m_4 = 1$ kg, and $m_5 = 200$ kg.)

**8.** What's wrong with the amplitude versus frequency response illustrated below?

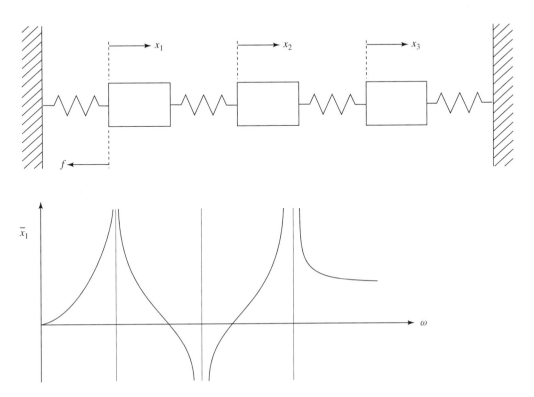

**9.** Show how you would find the *approximate* natural frequencies and eigenvectors for the illustrated system. Don't solve it exactly, show how you'd approach it in an approximate sense. Then solve it exactly and compare the results. $k_i = 10$ N/m, $m_1 = 1$ kg, $m_2 = 100$ kg, $m_3 = 2$ kg.

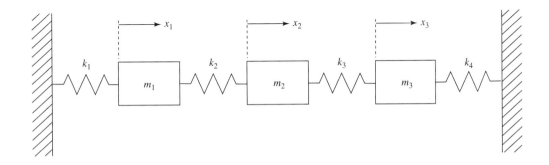

**10.** In this problem we'll try to gain some insight as to when analyzing simpler subsystems is a valid approach. Four different configurations of the relative masses and stiffnesses of Figure 4.4 will be considered. Solve the problems exactly and comment on which aspects of the solution are predictable from an analysis of a simpler subsystem. For all the problems, $f_i = 0$ and $k_3 = 0$.

a)  $m_1 = 100$ kg
    $m_2 = 1$ kg         Base system
    $k_1 = 1000$ N/m
    $k_2 = 10$ N/m

b)  $m_1 = 100$ kg
    $m_2 = 1$ kg
    $k_1 = 1000$ N/m
    $k_2 = 1000$ N/m

c)  $m_1 = 100$ kg
    $m_2 = 100$ kg
    $k_1 = 1000$ N/m
    $k_2 = 1000$ N/m

d)  $m_1 = 100$ kg
    $m_2 = 100$ kg
    $k_1 = 1000$ N/m
    $k_2 = 10$ N/m

11. Let's spend some more time with the example from Section 7.2. A revised version of the system is shown below. The change is the fact that $k_3$ is no longer centered, but is offset by an amount $e$. Determine how much of an effect such a misalignment can have. Do the changes in the frequencies and eigenvectors make physical sense?

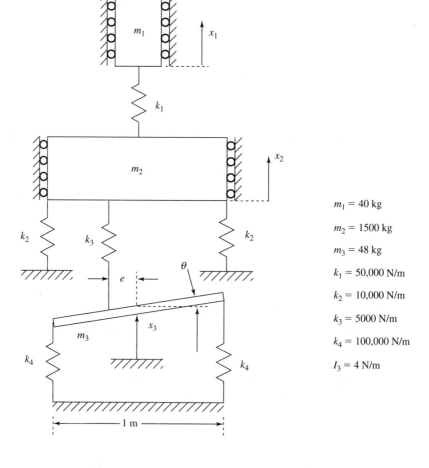

$m_1 = 40$ kg

$m_2 = 1500$ kg

$m_3 = 48$ kg

$k_1 = 50,000$ N/m

$k_2 = 10,000$ N/m

$k_3 = 5000$ N/m

$k_4 = 100,000$ N/m

$I_3 = 4$ N/m

12. In this problem we'll again consider the example of Section 7.2. This time you should vary the mass $m_1$. When it was small, it decoupled nicely from the rest of the system. As its mass increases, the assumption of decoupling will become invalid. At what mass do the errors in the frequency estimates of Section 7.2 exceed 10 percent when compared to exact results?

13. If you want to analyze the system modes that involve lateral vibrations of Beam 2, do you need to treat the boundary conditions as spring restrained or is it reasonable to approximate them as fixed? All beams have the same parameter values. $E = 2.1 \times 10^{11}$ N/m, $\rho = 7.8 \times 10^3$ kg/m$^3$, $\ell = 1$ m, $A = .0001$ m$^2$, $\bar{I} = 8.\bar{3} \times 10^{-10}$ m$^4$.

## Section 7.3

14. The illustrated system is composed of a lumped mass $m_1$ that's connected to ground by means of two tensioned strings. The length of each string is 1 m. The linear density of the strings is 2 kg/m, the lumped mass is variable, and the tension is equal to 1800 N. The lumped approximation gives the frequency of the first mode as $\sqrt{\frac{2T}{m_1}}$ (from $m\ddot{x}_1 = -2T\theta$, where $x$ is the displacement of the mass). How small can $m_1$ be before the actual first natural frequency differs from this estimate by 10 percent?

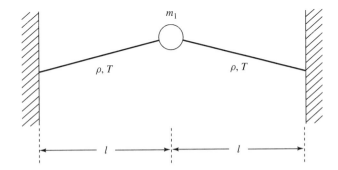

15. Shown below is a uniform bar, spring restrained to ground, with a lumped mass attached to it. The lumped mass can be placed anywhere along the bar. Analyze the system for different placements of the

mass (from midspan to the end) and comment on how badly the symmetry of the solution (when the mass is at midspan) is affected by its movement off from the center. $\ell = 2$ m, $m_2 = 3$ kg, $m_1 = 1$ kg, $k = 10,000$ N/m.

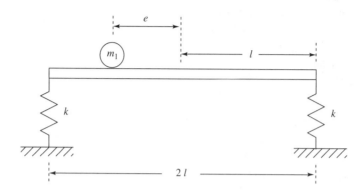

16. Shown below is a uniform bar that is spring restrained to ground. Analyze the system for the case of equal spring constants and then allow one of the springs to become stiffer. How do the solutions alter as the spring constant is increased? How much can it change and still leave the symmetric solutions (associated with equal spring constants) relatively unchanged? $m_1 = 1$ kg, $k_1 = 1000$ N/m, $\ell = 1$ m.

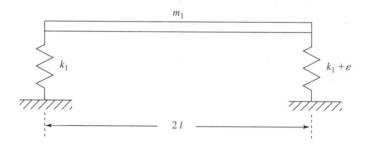

17. The figure below shows a tensioned string with an attached mass $\epsilon$. The mass can move only vertically. If the mass is large, the first vibratory mode can be analyzed as if the system is a SDOF system, with the string tension supplying the restoring force and the lumped mass acting as the dominant mass in the system. However, as the mass reduced, the other modes of the string problem become more important and, eventually, the SDOF approximation isn't an accurate approximation. At what value of $\epsilon$ does the SDOF approximation for the lowest natural frequency exceed 5 percent error when compared with the actual solution of the coupled string-mass problem? The tension in the string is equal to 100,000 N. Neglect gravity. $\rho = 1$ kg/m, $\ell = 1$ m.

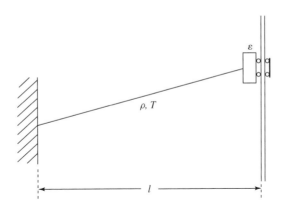

**18.** Figure 4.4 shows a 2-DOF system. Let $f_1 = f_2 = 0$ and let $k_3 = 0$. If $k_1$ is reduced sufficiently, the system can be viewed as supporting two decoupled modes. One involves motion of both $m_1$ and $m_2$, which moves like a rigid body. The other involves the two masses moving in opposition to each other, with $k_1$ contributing little to the response (a breathing mode). Both these modes are shown below. For what values of $k_1$ is the assumption of modes like those shown in the figure a good one? (Note: You'll have to define an error measure for when the modes look reasonable.) $m_1 = m_2 = 1$ kg, $k_2 = 900$ N/m.

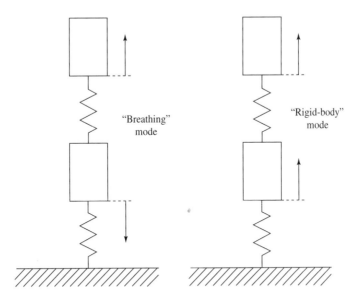

**19.** Consider a tensioned string with a lumped mass and spring at the right end (shown below). Solve for the actual natural frequencies. Next, use your insight as to how the first mode will differ qualitatively from the higher ones and show how you can obtain an approximate first natural frequency without solving the entire continuous problem. Don't forget to include the equivalent lumped mass and stiffness provided by the string. $\ell = 1$ m, $m_1 = 10$ kg, $k = 9000$ N/m, $\rho = .1$ kg/m, $T = 50$ N.

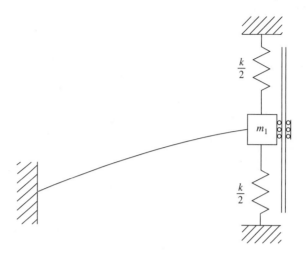

**20.** Assume that you've solved for the responses of the vibration absorber problem shown in Figure 4.11$b$. $m_1 = 1000$ kg, $k_1 = 36,000$ N/m, $m_2 = 2$ kg, $k_2 = 72$ N/m, $\omega = 6.03$ rad/s, and the forcing amplitude is .7 N. You determine that $x_1 = -.00783$ m and $x_2 = .781$ m. Verify that these results make sense by analyzing the SDOF system made up of $m_2$ and $k_2$, presuming that the system is seismically excited by $m_1$.

# 8

## Experimental Methods and Real World Behavior

## 8.1 INTRODUCTION

Up to now we have been concerning ourselves with deterministic vibrations, that is, periodically excited systems. The nice thing about periodic forcings is that once you know one period of the input, you know them all. Nothing else can happen; the input is completely determined. Thus we call the excitation deterministic. Although we could continue to restrict ourselves in this way, it turns out that expanding the scope of our excitations to include random vibrations will be very beneficial. For one thing, many systems are subjected to some form of random loading. The wind loads on an airliner, the forces acting on an off-shore oil rig, and the accelerations felt by an object in a delivery truck are all well described by a random signal, not a periodically varying one. Furthermore, many of the algorithms that modal analyzers use to determine a system's natural frequencies and mode shapes are based on a random-vibrations viewpoint. This chapter will present some of the important information that you will need if you ever plan to carry out modal analyses, but please remember that this is just an introduction. The field of random vibrations is quite large and there is a great deal of material that we won't be considering. Those interested in studying random vibration and modal analysis in more depth can consult [3] and [9].

We'll start the chapter by laying out some fundamentals of random vibrations and finish by introducing some of the concepts of modal analysis.

## 8.2 SIGNAL DESCRIPTIONS

Let's begin this section by asking what a random signal is in the first place. Actually, you probably already have a feel for what a random signal is, even if you haven't thought about it in a technical sense. Random signals look irregular, with no discernable pattern, such as the example shown in Figure 8.1. At a simple level, we can think of random signals as waveforms that never repeat themselves. Thus a sinusoid isn't random because it repeats every $T$ seconds, where $T$ is the period of the oscillation. This is, however, not an adequate definition. Obviously, we wouldn't want to call $x(t) = at$ a random signal, even though it doesn't repeat itself. It's intuitively clear that $x(t) = at$ isn't random; it's just a constantly increasing signal. This is an example of a nonstationary situation. By stationary we mean that one or more averaged quantities remain fixed. For instance, in Figure 8.1 we see that the average value of $x$ is zero; the signal seems to be centered about $x = 0$. Thus this signal exhibits stationarity. Signals that aren't stationary will drift with time as $at$ does. Generally we will be concerned with stationary processes.

Even if the signal is stationary, nonperiodicity isn't enough. For example,

$$f(t) = \sin(t) + \sin(\sqrt{2}t)$$

isn't periodic, but it certainly isn't random, as Figure 8.2 shows. Adding two sine waves whose frequencies aren't rationally related will always produce a nonperiodic result. We know that each individual sinusoidal signal isn't random, and thus adding two sinusoidal signals together shouldn't

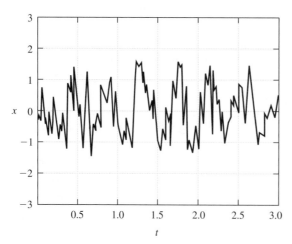

**Figure 8.1**　Random time response

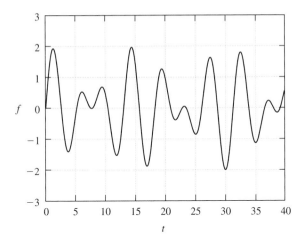

**Figure 8.2**  Response of two sinusoids with nonrationally related frequencies

be considered random since the function's value at a time $t$ is completely predictable, as we know it is simply composed of two sinusoids. This observation leads to the idea that random signals cannot be broken down into purely harmonic components. Thus we won't be able to find a Fourier series representation of a truly random signal.

Although people like to talk about random signals as if they truly exist, as far as our computations are concerned, they don't. Even if we had a steady-state, truly random signal, one that continually and unpredictably changes as time goes on, we'd have to cut off the data stream eventually. At this point our analyses can go one of two ways. We can view the actual signal as being zero from $t = -\infty$ to 0, equal to our data from $t = 0$ to $t = t_1$ and equal to zero for time greater than $t = t_1$, as shown in Figure 8.3. In this case the actual signal we're concerned with is a transient

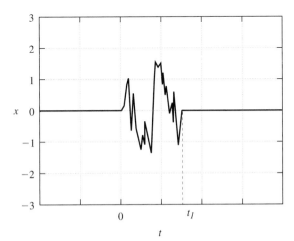

**Figure 8.3**  Truncated random time response

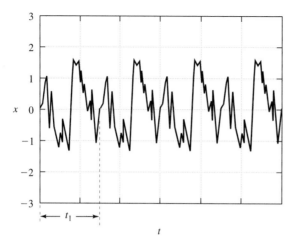

$x$

$t$

**Figure 8.4**   "Periodic" response

one, not a continuously varying, aperiodic one, and we obtain it by simply truncating the actual signal.

The other option is to impose a periodicity on the signal by assuming that the segment we captured repeats endlessly, with a period equal to $t_1$ (Figure 8.4). Just as in the previous case, this viewpoint is not completely accurate. Whereas previously we truncated the signal (making it into a transient), now we've made the signal periodic. Thus we can analyze either transient or periodic signals. The trick is to realize that as we increase the time $t_1$, we should be *approaching* the actual random signal. If we're lucky, these approximations will asymptotically approach the true random signal and we'll be left with only a small error for large enough $t_1$. It is important for the reader to keep these points in mind as we introduce the notions of random signals later in this chapter. Just as it was mathematically convenient to deal with viscous damping rather than the more realistic hysteretic or dry-friction damping models, so it is more convenient to assume truly random signals, even when we know our analyses will actually be based on nonrandom ones.

You may have noted that both *random process* and *random signal* were used in the foregoing paragraphs. It is now time to ask what is meant by these words. Technically, there can be only random processes. It really makes no sense to call an acquired signal random since, once it has been acquired, it is completely determined. There isn't any randomness at all because the value of the signal for each point in time has been recorded, and there is no surprise about what value might occur for a given moment of time. Having said this, it's clear that this is a pretty pedantic viewpoint. If you have a signal such as that shown in Figure 8.1, and covered over the graph for $t$ greater than 1 s, you would be hard pressed to reproduce the graph from a knowlege of the signal from zero to 1 s. That's what we really mean by random; that you can't predict what the signal will be at a later time, even if you know exactly what it was equal to in the past. This is quite different from the case of a sinusoid, for which we know what all future values will be once we know the period and amplitude of the signal.

When we say random what we're really concerned with are random *processes*, processes that generate an output that is unpredictable. Luckily, even if a random process produces an output that isn't predictable in the fine details, the outputs are nevertheless quite predictable in an averaged sense. For instance, if a skylight was opened during a rainstorm, you could pretty well predict that the floor beneath the skylight will get wet, while the floor away from it will stay dry. Furthermore, you would expect the amount of water to be uniform, on average, over the wetted surface (this is ignoring drips from the sides of the skylight, of course). You wouldn't be able to predict exactly where the next drop of water will fall, but you would be able to predict their general location and you would also be able to predict the average amount of rain per unit time, given your observations of how much has been falling over the preceding minutes.

If the average amount of water per unit time stays constant, then the process is stationary. If the average amount of water per unit time increases (the rainfall moves from a gentle sprinkle to a thunderstorm), then the process is nonstationary. Obviously the property of stationarity depends upon our viewpoint. The rain will eventually stop, therefore it can't be strictly stationary. Thus stationarity depends on the time span we're interested in. If the average rainfall stays constant over a five-minute span and we're not interested in times greater than this, then for our purposes we can consider the process to be stationary.

This kind of view, looking at averages rather than fine details, is the key to describing random processes. In the next few pages we'll be looking at some types of averages and see how they can be used to get a handle on random processes.

The most common signal descriptor we'll deal with is the *mean value* of a signal, given by

$$E(y) = \frac{1}{n} \sum_{i=1}^{n} y_i \tag{8.2.1}$$

Note that $E(y)$ is used to indicate the expected value (also called the *mean value*) of a variable. If we were concerned with a continuous process $y(t)$, we would have defined the mean as

$$E(y(t)) = \frac{1}{t_1} \int_0^{t_1} y(t) dt \tag{8.2.2}$$

where our data sample has a duration of $t_1$. The only problem with such a representation is that we will almost never be dealing with continuous signals. All real signals must be sampled and the samples processed. Granted, a continuous data stream is more familiar and is sometimes easier to work with mathematically (just as viscous damping is easier than hysteretic), but the sad fact is that all our processing will necessarily be digital. Thus we'll often be concerned with a digital outlook.

Certainly the mean of a signal gives us important information. If we know the dynamic pressure loading on an airplane wing, then the mean of this loading gives us the static pressure that the wing must be designed to withstand. The mean, however, isn't usually enough. Consider Figure 8.5, in which two different force loadings are shown, both with mean equal to 10. Clearly these loadings are very different, even though that difference doesn't show up in the mean value. The signal in Figure 8.5a has a much lower amplitude, as well as a lower frequency, than that

(a)

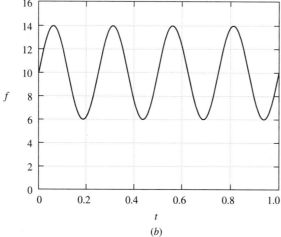

(b)

**Figure 8.5**  Same averaged behavior, but different dynamic behavior

of Figure 8.5*b*. Thus, although the mean value is important, we can see that more information is needed.

The next descriptor we need to construct is one that will capture how much the signal deviates from its mean value. One choice would be to calculate

$$\sum_{i=1}^{n} \left[ y_i - E(y) \right] \tag{8.2.3}$$

Although this is perhaps the most obvious choice, it is actually quite a poor one. For instance, this calculation would yield zero for both signals shown in Figure 8.5. The reason for this is that we've allowed both positive ($y_i > E(y)$) and negative ($y_i < E(y)$) values. Thus they can act to cancel each other out. What we actually care about is how far the signal deviates from the mean value, and we aren't really concerned about whether the deviation is positive or negative. So it is sensible to ensure that all deviations give a positive contribution to our descriptor. Although we could come up with a variety of different descriptors having this property, the one we'll introduce is the one used in common practice, called the *variance*:

$$\text{var}(y) = \frac{1}{n} \sum_{i=1}^{n} \left[ y_i - E(y) \right]^2 \tag{8.2.4}$$

You'll note that this could also be written as

$$\text{var}(y) = E\left[ y - E(y) \right] \tag{8.2.5}$$

The variance is also called the mean-square value.

Often we'll talk about the *standard deviation*, $\sigma$ of the signal,

$$\sigma = \sqrt{\text{var}(y)} \tag{8.2.6}$$

which gives us a sense of exactly how far our signal is likely to be from the mean value (in an averaged way). This is also called the root-mean-square value (RMS) of the signal. If our signals are continuous in time rather than discrete, then the variance is found from

$$\text{var}(y) = \frac{1}{t_1} \int_0^{t_1} \left[ y - E(y) \right]^2 \tag{8.2.7}$$

Although $t_1$ is always finite in any real analysis, we'll often set it to $\infty$ for the purposes of our mathematical analyses.

**Example 8.1**

**Problem**   Consider two different discrete signals. The first ($s_a$) is given by

$$s_a = 1, \ 1, \ 1, \ 1, \ 1, \ 13 \tag{8.2.8}$$

and the second ($s_b$) is given by

$$s_b = 1, \ 5, \ 1, \ 5, \ 1, \ 5 \tag{8.2.9}$$

Determine the expected value and variance of these signals.

**Solution**   The expected value for both these discrete signals is equal to 3. However the variances are quite different. For $s_a$ we have

$$\text{var}(s_a) = 20 \tag{8.2.10}$$

while the variance of $s_b$ is

$$\text{var}(s_b) = 4 \tag{8.2.11}$$

The variance for $s_b$ is quite a bit smaller than that of $s_a$. This is because $s_b$ varies only moderately from the mean value for all the entries, while $s_a$ has a huge variation in its last entry. Thus we see that the variance allows us to distinguish between two signals as a function of how scattered the data is.

The foregoing descriptions have dealt with averaged quantities. It is possible to come up with many more averaged descriptors, but for our purposes the mean and variance are sufficient. What we'll next want to look at is a way of describing signals that depends on the particular value of the signal at particular times. Before doing this, however, we'll need to introduce the Fourier transform. The next section will describe this transform, and then we'll get back on track, discussing both correlation functions and spectral density functions, both of which are important components of *spectral analyses*.

## 8.3 FOURIER TRANSFORM ANALYSIS

The basic building block of spectral analyses is Fourier transform analysis. As you'll recall from Section 3.2, we can take any signal of a fixed duration and determine the frequency components within it by calculating the Fourier series corresponding to the signal. The definition of a Fourier series was given by

$$x(t) = \sum_{n=0}^{\infty} a_n \cos(n\omega_0 t) + \sum_{n=1}^{\infty} b_n \sin(n\omega_0 t) \tag{8.3.1}$$

for which

$$a_0 = \frac{\omega_0}{2\pi} \int_0^{\frac{2\pi}{\omega_0}} x(t)dt \tag{8.3.2}$$

$$a_n = \frac{\omega_0}{\pi} \int_0^{\frac{2\pi}{\omega_0}} x(t) \cos(n\omega_0 t)dt \tag{8.3.3}$$

and

$$b_n = \frac{\omega_0}{\pi} \int_0^{\frac{2\pi}{\omega_0}} x(t) \sin(n\omega_0 t)dt \tag{8.3.4}$$

Although this approach is fine for periodic signals, random signals are decidedly *not* periodic. Indeed, as we've already mentioned, a working definition of a random signal is that it's a signal that is not periodic and yet doesn't die down to zero (obviously, a signal that goes to zero isn't periodic, but isn't necessarily random).

If the signal isn't periodic, then we can't identify a fundamental frequency $\omega_0$ and can't use (8.3.1–8.3.4). So what do we do? One approach is to view the signal as we did in Figure 8.4. By artificially making the signal look periodic, we can apply the Fourier series approach and find its spectral components. It's clear that this approach will give more accurate approximations as the period we're using increases. To see where this approach ultimately leads us, we'll mathematically let the period go to infinity. This means that $\omega_0$ goes to zero and the difference between the various harmonics also goes to zero. As this happens, our summation mutates into an integration. This isn't surprising when you realize that what we do on a computer to simulate an integration is to represent it as a finite summation. For more accuracy, we use more terms in the summation. Conceptually we're running this in reverse here, going from a finite summation to an integration. The difference between harmonics ($\omega_0$) will become $d\omega$ (the differential frequency), and our individual harmonics won't be indicated by $n\omega_0$ anymore, but by $\omega$. If you go through the details of this process you'll obtain

$$x(t) = \int_{-\infty}^{\infty} a(\omega)\cos(\omega t)d\omega + \int_{-\infty}^{\infty} b(\omega)\sin(\omega t)d\omega \qquad (8.3.5)$$

for which

$$a(\omega) = \frac{1}{2\pi} \int_{-\infty}^{\infty} x(t)\cos(\omega t)dt \qquad (8.3.6)$$

and

$$b(\omega) = \frac{1}{2\pi} \int_{-\infty}^{\infty} x(t)\sin(\omega t)dt \qquad (8.3.7)$$

Equations (8.3.5–8.3.7) are one way to define the Fourier transform of a signal $x(t)$.

The other way of representing a Fourier transform is to use complex notation, just as we did when dealing with deterministic forced vibrations. If we let $X(\omega)$ represent the Fourier transform of $x(t)$, we'll have

$$x(t) = \int_{-\infty}^{\infty} X(\omega)e^{i\omega t}d\omega \qquad (8.3.8)$$

and

$$X(\omega) = \frac{1}{2\pi} \int_{-\infty}^{\infty} x(t)e^{-i\omega t}dt \qquad (8.3.9)$$

Both complex and real forms are equally valid. For the complex case $X(\omega)$ is complex and $x(t)$ is real, while in the real form $x(t)$, $a(\omega)$, and $b(\omega)$ are real.

**Example 8.2**

**Problem**   Determine the Fourier transform of the signal shown in Figure 8.6. Consider both the real and complex form of the transform equations.

**Solution**   Starting with the real form, we have

$$a(\omega) = \frac{1}{2\pi} \int_{0}^{a} \cos(\omega t)dt = \frac{1}{2\pi\omega}\sin(\omega a) \qquad (8.3.10)$$

and

$$b(\omega) = \frac{1}{2\pi} \int_{0}^{a} \sin(\omega t)dt = \frac{-1}{2\pi\omega}(\cos(\omega a) - 1) \qquad (8.3.11)$$

$x(t)$

**Figure 8.6**   Rectangular pulse

Now let's consider the complex form. Equation (8.3.9) tells us that

$$X(\omega) = \frac{1}{2\pi} \int_0^a e^{i\omega t} dt = \frac{-1}{2\pi i \omega} \left( e^{-i\omega a} - 1 \right) \tag{8.3.12}$$

Expanding the complex exponential in terms of sine and cosine components gives us

$$X(\omega) = \frac{-1}{2\pi \omega} \Big[ \sin(\omega a) + i(\cos(\omega a) - 1) \Big] \tag{8.3.13}$$

From this it seems like the real part of $X(\omega)$ is simply equal to the $a(\omega)$ we calculated from (8.3.6) and the imaginary part is equal to negative $b(\omega)$, found from (8.3.7).

To see that this wasn't just a coincidence, we can rewrite (8.3.9) as

$$X(\omega) = \frac{1}{2\pi} \int_{-\infty}^{\infty} x(t) \Big[ \cos(\omega t) - i \sin(\omega t) \Big] dt \tag{8.3.14}$$

Comparing (8.3.14) with (8.3.6) and (8.3.7) we see that the real part of the integral gives us $a(\omega)$, while the imaginary part yields $-b(\omega)$, thus supporting the validity of our calculated results.

When Fourier transforms are first encountered they often seem advanced and difficult to understand. We've now seen, however, that they're easily viewed as limiting cases of a Fourier series, something that you're probably quite comfortable with. In addition, you'll rarely have to worry about actually computing the Fourier transform in any real-world applications; such calculations are carried out as internal functions in the spectrum analyzers you'll be using.

## 8.4 SPECTRAL ANALYSES

Now that we've seen what Fourier transform analysis is all about, we can get back to finding useful descriptors of our signals. The next approach we'll examine is not one that looks at an overall average, but rather asks if there's any correlation between a signal at two points in time.

Clearly, a signal is always perfectly correlated with itself. But what if there's a correlation between a signal at time $t$ and at $t + \tau$? If the signal is a sine wave with period $\tau$, then we know there should be an excellent correlation, since the response at these two times will be identical. A purely random signal should be uncorrelated for any $\tau$'s besides zero since the response changes in a completely unpredictable way. The autocorrelation function, $R_{xx}(\tau)$, deals with just this sort of question and is defined by

$$R_{xx}(\tau) \equiv \lim_{T \to \infty} \frac{1}{T} \int_{\frac{-T}{2}}^{\frac{T}{2}} x(t)x(t + \tau)dt \tag{8.4.1}$$

We've already mentioned that infinite time traces aren't really practical, thus the limit in the above formula really applies to $T$ getting large, not truly infinite. Also, if our signal $x(t)$ has been truncated, then the integral will have finite limits defined by where $x(t)$ goes to zero.

If $\tau = 0$, then (8.4.1) becomes

$$R_{xx}(0) \equiv \lim_{T \to \infty} \frac{1}{T} \int_{\frac{-T}{2}}^{\frac{T}{2}} x^2(t)dt \tag{8.4.2}$$

This is simply the mean-square value of $x(t)$. Thus the autocorrelation of a signal is equal to the signal's mean-square value when the delay $\tau = 0$.

A similar correlation function can be found by asking what the correlation is between two different signals, again with varying time delay $\tau$:

$$R_{xy}(\tau) \equiv \lim_{T \to \infty} \frac{1}{T} \int_{\frac{-T}{2}}^{\frac{T}{2}} x(t)y(t + \tau)dt \tag{8.4.3}$$

This form, called the cross-correlation form, is more general since we can always let $y = x$ if we wanted the autocorrelation.

Cross-correlation forms would be useful if you were trying to determine whether a vibration at one point of a structure was being influenced by vibrations at some other point. If the cross correlation is high for some value of $\tau$, then you could deduce that there is a relation between the two vibrations and that the transit time for the motions to be transmitted from one point to the other is equal to $\tau$. Cross correlations are also used in radar applications. By sending out a signal and looking for the same signal delayed in time (delayed because it is bounced off a plane being tracked) the radar unit determines how far away the plane is by using the detected time delay and knowing the signal's transmission speed.

These correlation functions are useful for our purposes due to the functions that are defined from them. The first of these functions is known as the *power spectral density*, $S_{xx}(\omega)$, and is found from

$$S_{xx}(\omega) = \frac{1}{2\pi} \int_{-\infty}^{\infty} R_{xx}(\tau)e^{-i\omega\tau}d\tau \tag{8.4.4}$$

If you look back at (8.3.9) you'll note that the form shown above suggests that the power spectral density is the Fourier transform of the autocorrelation function. Logically, we can therefore deduce that $R_{xx}$ can be written as the inverse transform of $S_{xx}(\omega)$,

$$R_{xx}(\tau) = \int_{-\infty}^{\infty} S_{xx}(\omega)e^{i\omega\tau} d\omega \qquad (8.4.5)$$

The reason this is highly relevant is that the power spectral density does what its name implies. It tells us the density level of power that's present in a signal for any frequency of interest. If we want to know the total power between any two frequencies, then we need to integrate $S_{xx}(\omega)$ between these frequencies, taking into account both positive and negative $\omega$'s.

This particular descriptor is one of the cornerstones of spectral analysis. The outlook here is purely frequency based. Rather than worrying about the specific details of a signal in *time*, we're concerned with what's happening with respect to *frequency*. This can be made clear by an example.

**Example 8.3**

**Problem**  To see how the power spectral density gives us a frequency-based insight into our analyses, determine $S_{xx}(\omega)$ by means of $R_{xx}(t)$ for

$$x(t) = a\sin(\omega_0 t) \qquad (8.4.6)$$

**Solution**  The autocorrelation is found from

$$R_{xx}(\tau) \equiv \lim_{T\to\infty} \frac{1}{T} \int_{\frac{-T}{2}}^{\frac{T}{2}} a\sin(\omega_0 t)a\sin(\omega_0(t+\tau))\, dt \qquad (8.4.7)$$

Expanding out the second sine term yields

$$\begin{aligned}
R_{xx}(\tau) = &\lim_{T\to\infty} \frac{a^2}{T} \int_{\frac{-T}{2}}^{\frac{T}{2}} \sin(\omega_0 t)\sin(\omega_0 t)\cos(\omega_0\tau)\, dt \\
&+ \lim_{T\to\infty} \frac{a^2}{T} \int_{\frac{-T}{2}}^{\frac{T}{2}} \sin(\omega_0 t)\cos(\omega_0 t)\sin(\omega_0\tau)\, dt
\end{aligned} \qquad (8.4.8)$$

The second integral goes to zero as $T$ goes to infinity, while the first will leave us with

$$R_{xx}(\tau) = \frac{a^2\cos(\omega_0\tau)}{2} \qquad (8.4.9)$$

This result makes sense as it tells us that the sine is maximally correlated with itself for delays of $\frac{2\pi}{\omega_0}$ (the times when $\cos(\omega_0\tau)$ equals 1.0) and is not correlated at all for a time shift of $\frac{\pi}{2\omega_0}$. Since a time shift of $\frac{\pi}{2\omega_0}$ alters a sine into a cosine (which we know is an independent function), we'd expect zero correlation.

Equation (8.4.9) predicts that the mean-squared value $R_{xx}(0)$ is equal to $\frac{a^2}{2}$. We can verify that this is so analytically. If $x(t) = a\sin(\omega_0 t)$, then $x^2 = a^2\sin^2(\omega_0 t) = a^2\left(\frac{1}{2} - \frac{\cos(2\omega_0 t)}{2}\right)$. Thus, since the $\cos(2\omega_0 t)$ will average out to zero, we're left with the mean-square value being equal to $\frac{a^2}{2}$, just as predicted.

Now that we have the autocorrelation function, we can compute $S_{xx}(\omega)$:

$$S_{xx}(\omega) = \lim_{n\to\infty} \frac{1}{2\pi} \int_{\frac{-n\pi}{\omega_0}}^{\frac{n\pi}{\omega_0}} \frac{a^2\cos(\omega_0\tau)}{2} e^{-i\omega\tau} d\tau \qquad (8.4.10)$$

You'll notice that this is a bit different from (8.4.4) in that it involves a limit. If we immediately tried to use the infinite limits, we'd be unable to evaluate the integral. By taking the time integral to be integer multiples of the autocorrelation's period (using $\frac{n\pi}{\omega_0}$ instead of $\infty$), and letting the number of multiples go to infinity, we'll get the correct final solution and also be able to see the way in which we approach this solution.

Carrying out the integrations shown above will leave us with

$$S_{xx}(\omega) = \lim_{n\to\infty} \frac{1}{4\pi} \left[ \frac{a^2}{\omega_0 - \omega} \sin\left(n\pi - \frac{n\pi\omega}{\omega_0}\right) + \frac{a^2}{\omega_0 + \omega} \sin\left(n\pi + \frac{n\pi\omega}{\omega_0}\right) \right] \qquad (8.4.11)$$

Figure 8.7 shows what this function looks like for different values of $n$. Only $\omega$ values greater than zero are shown. Since $S_{xx}(\omega)$ is an even function (which we haven't proved, but is true nonetheless), there's no need to plot it for negative $\omega$ values as the behavior is identical to that for positive $\omega$'s. For $n = 4$ (Figure 8.7a) we see that a spike is developing at $\omega = 5$ rad/s (for this plot $\omega_0$ was chosen to be 5). If we increase the integration duration by a factor of 20 (bringing $n$ to 80), we get Figure 8.7b. What you can see from these two plots is that as $n$ is increasing, the spike at $\omega = 5$ is also increasing. We can determine this analytically in a straightforward manner. Let $\omega = \omega_0 - \epsilon$ in (8.4.11). The second term isn't going to be important because the first is going to completely dominate the response. The first term becomes

$$S_{xx}(\omega) = \frac{a^2}{2\pi} \lim_{n\to\infty} \frac{\sin\left(\frac{n\pi\epsilon}{\omega_0}\right)}{\epsilon} \qquad (8.4.12)$$

As we look at smaller and smaller terms we obtain an indeterminant fraction. Expanding the sine in terms of its trigonometric representation (or using L'Hôpital's rule) gives us

$$S_{xx}(\omega_0) = \lim_{n\to\infty} \frac{na^2}{2\omega_0} \qquad (8.4.13)$$

Clearly, as $n$ goes to infinity, so does $S_{xx}(\omega)$.

Everything away from $\omega = \omega_0$ or $\omega = -\omega_0$ goes to zero and we're left with two Dirac delta functions, one located at $\omega = \omega_0$ and the other at $\omega = -\omega_0$. The value of each delta function turns out to be .25, that is, in the limit of $n \to \infty$ we have (Figure 8.7c)

$$S_{xx}(\omega) = \frac{a^2}{4}\delta(\omega + \omega_0) + \frac{a^2}{4}\delta(\omega - \omega_0) \qquad (8.4.14)$$

This is a very relevant finding. It tells us that all the power in the sine wave is concentrated at one frequency. Thus, when we run modal tests and find sharp power spectral density peaks, we'll know

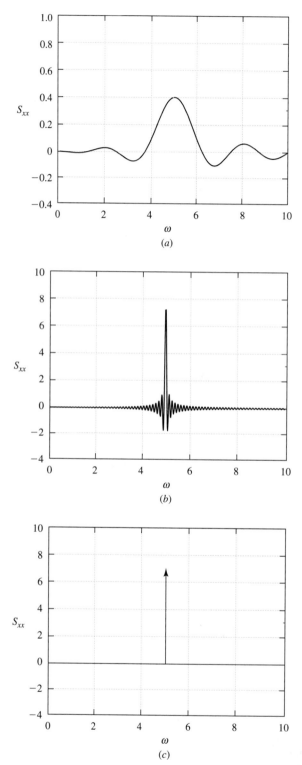

**Figure 8.7** Power spectral density of a 5 rad/s sine wave with increasing numbers of cycles included in the calculation

it means that there are distinct harmonics in the signal under analysis and that the frequencies of the harmonics correspond to the frequencies of the peaks. Note that we got this frequency-based information from an analysis of a time-based function (the autocorrelation function). The whole field of spectral analysis revolves around this viewpoint, which is based in the frequency domain rather than in time.

A tremendously useful result (which we won't derive, but shall simply present) is the following:

$$S_{xx}(\omega) = |g(\omega)|^2 S_{ff}(\omega) \qquad (8.4.15)$$

$g(\omega)$ is the transfer function of the system under consideration, $f(t)$ is the input to the system, and $x(t)$ is the system's output. $S_{xx}(\omega)$ is the power spectral density of the output and $S_{ff}(\omega)$ is the power spectral density of the input.

For example, assume that our equation of motion was given by

$$m\ddot{x} + kx = f(t) \qquad (8.4.16)$$

in which case the transfer function is given by

$$g(t) = \frac{1}{k - m\omega^2} \qquad (8.4.17)$$

If we know what the power spectral density of the forcing is, then we can immediately calculate the power spectral density of the output from (8.4.15) as

$$S_{xx}(\omega) = \left| \frac{1}{k - m\omega^2} \right|^2 S_{ff}(\omega) \qquad (8.4.18)$$

Equation (8.4.15) can be used in a number of ways. The most obvious was just discussed, namely finding the response power spectral density from knowing the system transfer function and the forcing power spectral density. But think about how else we can use it. One use would be to determine the input excitation to a system by measuring the output power spectral density and dividing it by the square of the transfer function. This might be useful if we can't easily determine what the input is directly, but have access to the system output and know the system characteristics.

The most common way in which this equation is used is as a means of actually determining the system characteristics simply by measuring the input and output signals. Dividing (8.4.15) by $S_{ff}(\omega)$ gives us

$$|g(\omega)|^2 = \frac{S_{xx}(\omega)}{S_{ff}(\omega)} \qquad (8.4.19)$$

Thus

$$|g(\omega)| = \sqrt{\frac{S_{xx}(\omega)}{S_{ff}(\omega)}} \qquad (8.4.20)$$

Obtaining the power spectra of the input and output signals is usually easy. Common transducers for acquiring data of this sort would be an impact hammer to excite the system and produce an input signal and an accelerometer to produce output data. Both signals would be fed into a modal analyzer, which would calculate the system's transfer function in a manner similar to that shown above. Of course, in a real experimental set-up there will be noise in the measurements and the analyzer wouldn't carry through exactly as shown in (8.4.20). The general approach, however, is valid and the differences between the analyzer's algorithms and (8.4.20) would simply be modifications introduced to minimize computational errors due to experimental error sources. Further information on this subject can be found in [3] and also (briefly) later in this section. One interesting point is that (8.4.15) doesn't include any phase information for the system's transfer function. This information can be found with a slightly different approach. Again, details are in [3].

**Example 8.4**

**Problem**    As an example of how the foregoing approach is used, we'll look at a problem that we already know how to solve, namely the response of a spring-mass under sinusoidal forcing. Consider the system shown in Figure 2.10. Let the input force be equal to $2\sin(\omega t)$ newtons, $m = 2$ kg, and $k = 100$ N/m. For this example we'll let $\omega = 10$ rad/s. Determine the mean-squared value of the response using two approaches and compare the results. First, determine the power spectral density of the output, find the autocorrelation function from this, and evaluate the autocorrelation at $\tau = 0$ (giving the mean-square response). Next, solve the problem exactly and simply calculate the mean-square response from the known solution. The two methods should yield the same answer.

**Solution**    The transfer function for this problem is given by

$$g(\omega) = \frac{1}{k - m\omega^2} \tag{8.4.21}$$

The power spectral density of the input is given by (8.4.14)

$$S_{ff}(\omega) = \frac{2^2}{4}\left(\delta(\omega + 10) + \delta(\omega - 10)\right) \tag{8.4.22}$$

Using (8.4.15) we find

$$S_{xx}(\omega) = \left|\frac{1}{k - m\omega^2}\right|^2 \left(\delta(\omega - 10) + \delta(\omega + 10)\right) \tag{8.4.23}$$

Substituting the values for $m$, $k$, and $\omega$ yields

$$S_{xx}(\omega) = \frac{1}{10,000}\left(\delta(\omega - 10) + \delta(\omega + 10)\right) \tag{8.4.24}$$

We can next take the inverse transform of this to find the autocorrelation $R_{xx}(\tau)$,

$$R_{xx}(\tau) = \frac{1}{10,000} \int_{-\infty}^{\infty} \left(\delta(\omega - 10) + \delta(\omega + 10)\right) e^{i\omega\tau} \, d\omega \tag{8.4.25}$$

which yields

$$R_{xx}(\tau) = \frac{1}{5000} \cos(10\tau) \qquad (8.4.26)$$

Evaluating this at $\tau = 0$ gives us

$$R_{xx}(0) = \frac{1}{5000} \qquad (8.4.27)$$

Now let's see if this matches the result we'd get from calculating the actual forced response and evaluating the mean-square value from it. We know from earlier in the chapter that the amplitude response of the system at $\omega = 10$ rad/s is given by

$$\bar{x} = \frac{\bar{f}}{k - m\omega^2} \qquad (8.4.28)$$

where $\bar{x}$ is the response magnitude and $\bar{f}$ is the forcing magnitude. Substituting the values for this example yields

$$\bar{x} = \frac{2}{100 - 2(100)} = -\frac{1}{50} \qquad (8.4.29)$$

Thus the total response is given by

$$x(t) = -\frac{1}{50} \sin(10t) \qquad (8.4.30)$$

The mean-square value of this is given by

$$E(x^2) = \left(\frac{1}{50^2}\right)\left(\frac{1}{2}\right) = \frac{1}{5000} \qquad (8.4.31)$$

Note that this agrees with (8.4.27). We got the same answer by directly determining the mean-squared value from a knowledge of the actual response, as well as by deducing it from a knowledge of the response's power spectral density.

We've now seen that the transfer function of a system relates the input power spectral density and the output power spectral density:

$$S_{xx}(\omega) = |g(\omega)|^2 S_{ff}(\omega) \qquad (8.4.32)$$

Two other formulas that are useful in modal analysis (discussed previously) are

$$S_{fx}(\omega) = g(\omega)S_{ff}(\omega) \qquad (8.4.33)$$

and

$$S_{xx}(\omega) = g(\omega)S_{xf}(\omega) \qquad (8.4.34)$$

where the $xf$ and $fx$ indicate that these are power spectral densities found from the cross-correlation functions between the input and output. We can estimate the system transfer function in two more ways, now. The first estimate, found from (8.4.33), is

$$g_1(\omega) = \frac{S_{fx}(\omega)}{S_{ff}(\omega)} \tag{8.4.35}$$

and the second, found from (8.4.34), is

$$g_2(\omega) = \frac{S_{xx}(\omega)}{S_{xf}(\omega)} \tag{8.4.36}$$

Noise in the measurements enters into these expressions in different ways, and they can be used together in order to find the best overall transfer-function estimate. Normally, due to experimental problems, the estimates are not exactly the same. If we use the two estimates shown above, we can form their ratio

$$\gamma^2 = \frac{g_1(\omega)}{g_2(\omega)} \tag{8.4.37}$$

$\gamma^2$ is called the coherence function. This frequency-dependent function is a widely used measure of how believable our measurements are. We won't derive it here, but it can be shown that $\gamma^2$ always lies between 0 and 1. If the coherence is near zero, it means that our measurements should be viewed with great suspicion, whereas a coherence near 1.0 indicates that the both $g_1(\omega)$ and $g_2(\omega)$ are in very close agreement, which implies that the measurements are believable and that noise and nonlinear effects aren't significant. Most all analyzers will display the coherence whenever they compute a system transfer function, allowing the user to decide whether or not to believe the computed results.

## 8.5 NOISE

Now that we've seen how to analyze general random inputs, it is time to examine the most common random signal around, namely *white noise*. Figure 8.1 showed what a white-noise signal might look like when plotted versus time. As you can see, it is a very jagged function, not at all regular. It is so irregular, in fact, that it is completely unpredictable from one instant to the next. This means that its autocorrelation function is simpy a delta function. For $\tau = 0$ the signal is completely correlated and for any other $\tau$ it is completely uncorrelated. If we denote the magnitude of the delta by $w_0$, then our signal's autocorrelation is given by

$$R_{xx}(\tau) = w_0 \delta(\tau) \tag{8.5.1}$$

The power spectral density for this function is given by

$$S_{xx}(\omega) = \frac{w_0}{2\pi} \tag{8.5.2}$$

As you can see in Figure 8.8, the power spectral density is a constant, regardless of frequency. Thus the density of power is the same at all frequencies.

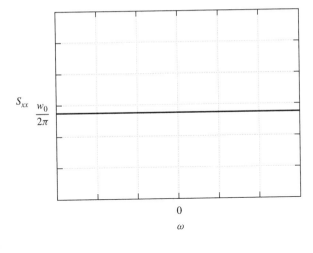

**Figure 8.8**  Constant power spectral density

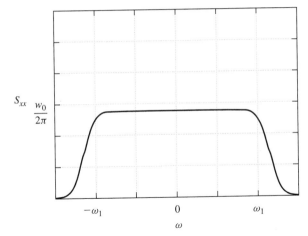

**Figure 8.9**  Power spectral density of band-limited white noise (gradual frequency cutoff)

Does this make sense from a physical point of view? Well, if the output is a constant at all frequencies, then we can integrate over all frequencies to find the total power of the signal. Since the power spectral density has the same value everywhere and extends out to infinity, the integral is infinite. Thus the signal must have infinite energy. This is a problem. It's pretty obvious that we can't produce a signal that has an *infinite* amount of energy; all real signals must be finite.

What's the conclusion? Simply that our assumption of a constant power spectral density over all frequencies must be erroneous. At some point, the power spectral density must drop off, as shown in Figure 8.9. This figure shows what all real-world, white signals really look like. They're reasonably constant over some range, and then the energy drops off to zero. An idealization of this situation is shown in Figure 8.10, in which we assume a constant level to $\omega_1$ and then have an immediate cut-off to zero. Such signals are called *band-limited* noise.

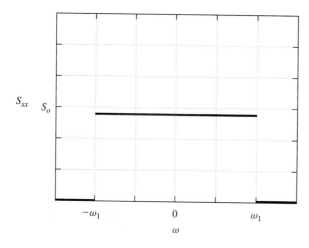

**Figure 8.10** Power spectral density of band-limited white noise (sharp frequency cutoff)

Luckily, the difference between ideal white noise and actual white noise doesn't really affect our analyses. As we've already seen, the transfer function of our mechanical system will drop to zero as we pass its highest natural frequency. Thus, even if we have an ideal white-noise input, the response will drop to zero at high frequencies due to the transfer function itself going to zero. To use a white-noise assumption we simply have to make sure that the white-noise signal is pretty much constant up to and a little beyond the highest natural frequency of the system that we're concerned with.

Another way of seeing how ideal white noise differs from what we see in real life is to realize that the autocorrelation won't really be a delta function. If the output in time was really changing in a completely random manner, then the implication is that the acceleration experienced in the system is infinite since only an infinite acceleration could move the output from one state to another in an infinitesimal time span. We realize that real systems have some limitation on how fast they can change and know that an actual signal at $t = \tau$ will be close to its value at $t = 0$, for small $\tau$. An *infinitely jagged* function (such as ideal white noise will have) requires energy at ultrahigh frequencies. This high-frequency content is what lets the signal "wiggle" so quickly.

We can compute the autocorrelation corresponding to a more realistic, band-limited signal to see how cutting off the spectral energy at some finite frequency affects matters. Figure 8.10 shows our power spectral density. It is constant at a level $S_0$ from $-\omega_1$ to $\omega_1$ and zero everywhere else. Calculating the autocorrelation for this case is straightforward. Recalling that for the general problem we have

$$R_{xx}(\tau) = \int_{-\infty}^{\infty} S_{xx}(\omega)e^{i\omega\tau}d\tau \tag{8.5.3}$$

for our particular problem we have

$$R_{xx}(\tau) = S_0 \int_{-\omega_1}^{\omega_1} e^{i\omega\tau}d\tau = 2S_0\frac{\sin(\omega_1\tau)}{\tau} \tag{8.5.4}$$

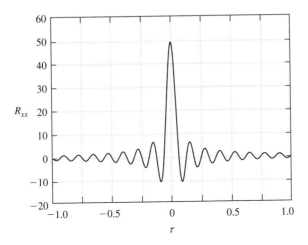

**Figure 8.11**  Autocorrelation of band-limited white noise

This autocorrelation function is plotted in Figure 8.11. For this plot the value of $S_0$ was set equal to .5 and $\omega_1$ was equal to 50 rad/s. As you can see, the correlation drops off quickly, but not instantly. As $\tau$ increases, the autocorrelation goes to zero. Finally, the magnitude at $\tau = 0$ is finite, not infinite as would be the case for a delta function.

Figures 8.12–8.15 give an overview of the most basic responses we might expect to see. Figure 8.12 shows the relationship between white noise (a function of time), its autocorrelation (a function of time delay $\tau$), and its power spectral density (a function of frequency). For white noise, the autocorrelation $R_{xx}(\tau)$ is a pure delta function and the power spectral density $S_{xx}(\omega)$ is a constant.

Real-world effects mean that the closest we'll ever get to Figure 8.12 is the situation shown in Figure 8.13. The time function looks similar to that of the previous case. The only difference is that if it was examined in fine detail we'd see that the response of Figure 8.13$a$ is continuous, while that of Figure 8.12$a$ is discontinuous (implying finite changes of output for infinitesimal changes in time). The autocorrelation is, therefore, not a pure delta function, and the power spectral density drops off at high frequencies.

Figure 8.14 shows what we'll see if the signal is a pure sinusoid. The autocorrelation for this case is also sinusoidal (telling us that the signal is periodic), and the power spectral density is composed of deltas at the frequency of the input signal. As we've already seen (Figure 8.7), this precise response won't ever be seen because we'd have to observe the input sinusoid for an infinitely long time in order to get a power spectral density that's actually a delta. Furthermore, most real signals will have some noise superimposed on them, as shown in Figure 8.15$a$. The autocorrelation of this signal is shown in Figure 8.15$b$. The effect of the noise is to create a large response at $\tau = 0$, whereas the rest of the response is dominated by the sinusoidal signal (and thus gives us largely periodic behavior). Finally, Figure 8.15$c$ shows the power spectral density. Since the noise is independent of the sinusoidal signal, what we end up with are the delta functions associated with sinusoids, along with a low level of constant energy density (associated with the noise).

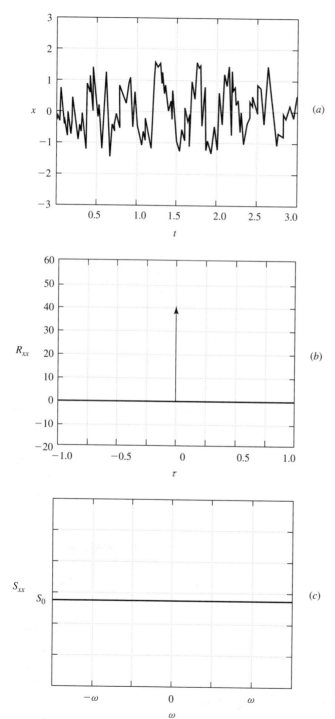

**Figure 8.12** Time trace, autocorrelation, and power spectral density for white noise

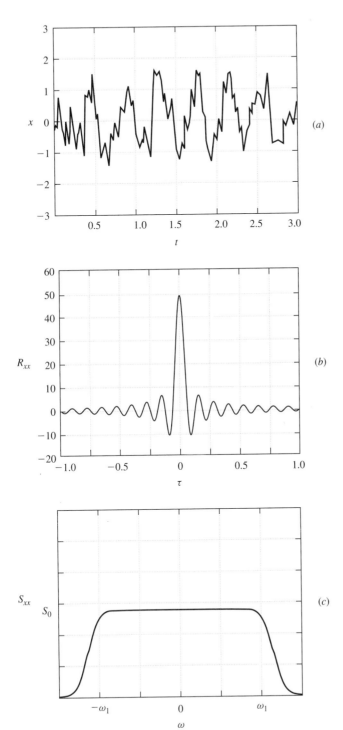

**Figure 8.13** Time trace, autocorrelation, and power spectral density for band-limited white noise

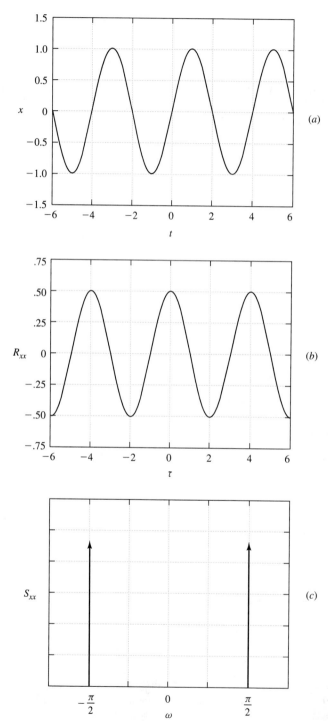

**Figure 8.14** Time trace, autocorrelation, and power spectral density for $\sin\left(\frac{\pi}{2}t\right)$

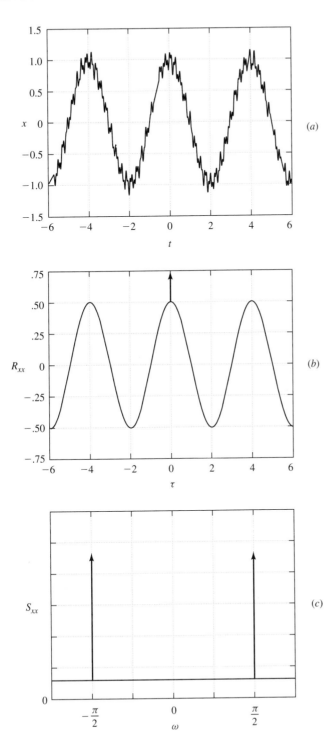

(a)

(b)

(c)

**Figure 8.15**  Time trace, autocorrelation and power spectral density for a combination of white noise and a sinusoidal signal

## 8.6 SENSORS AND ACTUATORS

If you're going to run a modal test, there are a few basic ingredients that must be present. You need some way of exciting the system, you need a way to sense the system's response, and you need a device that can use the input and output information in order to deduce some of the system properties. Let's start with the question of excitation. The two most widely used excitations are the *impulse hammer* and the *shaker* (electrohydraulic or electromechanical).

The impulse hammer (Figure 8.16) looks very much like an ordinary hammer and is used in a similar way. You hold the impulse hammer in much the same way as a regular hammer and quickly strike the structure being tested. Some care is necessary to make sure that the hammer hits the structure only once. (It is possible for the head to contact the structure two or more times in quick succession if the hammer isn't held correctly, an occurrence that will ruin the data acquisition). We've mentioned in Section 3.4 that loadings that occur quickly look and act like delta functions. We've also seen that delta functions have a constant power spectral energy. This characteristic is great for our analyses. Recall that the absolute value of the transfer function can be found from (8.4.20),

$$|g(\omega)| = \sqrt{\frac{S_{xx}(\omega)}{S_{ff}(\omega)}} \tag{8.6.1}$$

If the input power spectral density is a constant (call it $S_0$), then we have

$$|g(\omega)| = \frac{\sqrt{S_{xx}(\omega)}}{\sqrt{S_0}} \tag{8.6.2}$$

Thus we see that instead of having to take the ratio of two frequency-dependent functions, we need only divide the output's power spectral density by a constant. Actually, our analyzers will

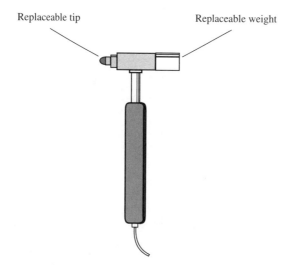

Replaceable tip                    Replaceable weight

**Figure 8.16**  Impact hammer

still calculate the ratio of the two power spectral densities, but since the input is essentially flat, calculation errors are minimized.

You'll notice from the picture that the impulse hammer has a wire coming out of it. This wire is connected to the head of the hammer. Just behind the tip of the head (the part striking the structure) is a piezoelectric sensor that generates a signal proportion to the impact force. By itself, this element is known as a *force transducer*. The signal from this transducer is fed into the modal analyzer. The analyzer can then determine what the precise force profile versus time was during the strike and from that can calculate the power spectral density of the input force.

Impulse hammers come with weights that can be added to the head and different tips. The weights allow different force levels to be input to the structure and the tip hardness determines how long the tip stays in contact with the structure (which affects the spectral energy input). Using a hard tip gives a wider frequency input, but lowers the overall energy that's transferred to the structure. Thus the hard tips allow you to analyze higher frequency responses, but might not supply enough excitation energy for accurate calculations. Of course, softening the tip lets you put more energy in at the expense of a restricted excitation bandwidth.

The advantage of impulse hammers is that they are fast to use and relatively inexpensive (on the order of $1000). In addition, they aren't attached to the structure and therefore do not alter its dynamic characteristics. They also have the advantage of being able to input energy at any structural point you can get a clean hit on. One disadvantage is that you don't know precisely where you're hitting the structure and thus you're bringing some uncertainty into your measurements. In addition, if you hit the structure too hard you could damage it, but if you don't hit it hard enough you might not be able to input enough energy to adequately excite the system.

The second type of excitation device is known as a *shaker*. These devices move the test structures mechanically (electromechanical shaker) or hydraulically (electrohydraulic shaker). The most common is the electromechanical shaker (Figure 8.17). If you're familiar with a loudspeaker then you're also familiar with electromechanical shakers; there's really little fundamental difference between the two. The speaker cone moves because a voice coil vibrates due to an alternating current. This is precisely what happens in an electromechanical shaker. The only difference is in the power and accuracy involved. Shakers must be capable of producing a significant time-varying force with which to excite the test structure (unlike loudspeakers, which merely move the speaker cone and a bit of air).

The shaker is physically connected to the test structure by means of a thin rod of plastic or metal, known as a stinger. The stinger is much stiffer along its length than it is laterally. Thus, if it is placed in line with the shaker's axis of movement, it effectively transmits the shaker forces, but doesn't cause any unwanted twisting of the structure.

Because we are able to control the electrical signal reaching the shaker, we can produce any input waveform we like. The most obvious signal would be a sinusoidal one, allowing us to determine the structure's response at a single frequency. We could also input a random signal (band-limited white noise) that would let us calculate the entire frequency response of the system. Another common input is called a swept sine. This consists of a sinusoidal excitation that starts

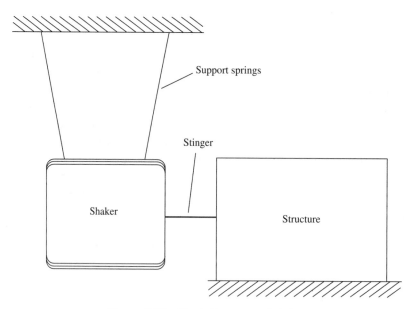

**Figure 8.17**   Elastically supported shaker

at low frequencies and ramps upward. This signal also allows us to determine the structure's complete frequency response. Transient inputs are also possible and are used to mimic shock loadings.

The advantages of a shaker are the variety of excitations that can be used, the range of force levels, and the accuracy with which the inputs can be applied. The disadvantages are cost (small shakers and associated power amps can run $2000, with larger ones hitting $10,000–$20,000 without difficulty) and the fact that the shakers will modify the structures that they are attached to. The stinger and the moving piece of the shaker that the stinger is attached to have their own mass and stiffness and therefore affect the structure. The shakers are also heavy so that it is often difficult to attach them to the structure where we might wish to. Often they are suspended near the structure by elastic cords, which serve to dynamically isolate them, as well as position them. Shakers can come in a wide range of sizes. The smallest ones in general use are about the size of a baseball. The largest are called shaker tables, and the entire structure is placed on on the shaker in order to seismically excite it. Custom-made shaker assemblies can be big enough to shake a small building (used in earthquake studies, not surprisingly).

Of course, it's not enough to just excite a structure, we also must measure the response. The most common devices used to measure a structure's response are *accelerometers* (previously discussed in Section 2.11. These devices are usually piezoelectric in nature and allow us to determine the structure's acceleration as long as we make sure that the natural frequency of the accelerometer is greater than the highest frequency we're interested in measuring. Accelerometers come in a variety of masses, from .5 to 500 g. The mass is clearly of concern if the test structure is light.

Since the accelerometer is physically attached to the structure, the mass is increased and must be analytically accounted for if the test results are to be accurate.

It always seems strange to someone new to modal testing, but the most common means of attaching an accelerometer is to simply stick it to the structure with a thin layer of beeswax. Alternatively, we can mount it semipermanently by screwing it to the structure.

A fairly new class of measurement devices is laser based. A laser beam is aimed at a structure, reflected back to the apparatus (called a *laser vibrometer*), and compared to the output beam. Using this information the laser vibrometer determines what the structure's vibrational motions are like at the target point. Since there is no loading on the structure, this method is an excellent way of examining very lightweight test objects (for which an accelerometer would represent too much of an additional mass loading for accurate results). The major drawbacks to this approach are cost (they start at \$15,000–\$20,000) and the difficulty of getting the laser signal reflected back to the vibrometer.

An old sensor that is still widely used is the *strain gauge*. A strain gauge is composed of a material that experiences a resistance change when subjected to strain. By measuring the resistance variation, we can deduce the local strain of the test specimen. The sensors are very cheap, but are also relatively noisy and have a limited dynamic range.

Another new way of sensing and actuating a structure is by means of a distributed piezoelectric piece of material. In addition to piezoelectric crystals (used in force transducers and accelerometers), it is possible to make large flat pieces of piezoactive materials. These can be flexible or rather brittle. Since the material is distributed over a wide area, we can use it to directly sense particular modes of the structure, or conversely, excite particular modes. Another variant on this approach is to combine the sensing and actuation into a single piece of piezoelectric material, resulting in a single device that both excites a structure and measures the structure's vibrations. Since the sensor/actuator pair is collocated for this device, it is attractive for controls work.

After a signal is sensed it must be amplified through an appropriate *signal conditioner* (charge amplifier or voltage amplifier) and then fed to an *analyzer*. The range of analyzers (known as vibration analyzers, modal analyzers, spectrum analyzers, etc.) is very wide nowadays. You can purchase simple, two channel units (one input and one output), as well as sophisticated multichannel ones capable of handling 64 channels of data or more. A few years ago all the analyzers were stand-alone pieces of equipment, but it is becoming more common to see analyzers that work along with a separate computer. In this way you can mix and match the amount of analyzer function and computational speed that you desire. If you already have the computer, then you can add the analyzer section for a relatively low cost and buy only as many channels of input and output capability as you wish.

All these devices attempt to determine the transfer functions between the input and output signals. For instance, the output might be a sinusoidal signal that is fed to a shaker. The input would be a signal proportional to the structure's acceleration, as measured by an accelerometer. If the test was done at a single frequency, then the analyzer would determine the amplitude and phase response of the structure, for this frequency of excitation. If the input was a random signal, then the analyzer would use the spectral analysis we've gone over to determine the complete transfer function over a range of frequencies. By analyzing the peak responses, the half-power points, the zeros, etc., the analyzer could also construct estimates of the natural frequencies and damping of the system.

By getting responses from multiple points along the structure, we can build up a data base of response information that will allow the analyzer to compute the actual mode shapes of the structure. Analyzers that do this are called *modal analyzers*. Sophisticated analyzers will allow the user to create a model of the system from experimental data and then modify the system in order to see how changes will affect the overall vibrational response.

## 8.7 NONLINEAR EFFECTS

A central assumption in all of the work so far has been linearity. At the risk of being obvious, this means that all the physical quantities we measure vary in a linear manner. Thus the force in a spring is linearly proportional to the stretch; double the extension and the force doubles. The same is true for the force necessary to accelerate a mass; double the force and mass's acceleration doubles (if there are no additional forces acting on it). Thus a nonlinearity in the system implies something isn't behaving in a linear manner. One obvious example is a coil spring, like those found in automobile suspensions. For small oscillations this spring will behave linearly. What happens, however, if the load in the car becomes so great that the spring is completely compressed and all the coils touch each other? In this case it won't act anything like a linear spring. Additional forces now serve to compress the actual metal of the spring, rather than simply bending a coil. Thus the effective spring constant will be immense. This behavior is shown in Figure 8.18. The deflection versus force shows a normal linear relationship up until the critical deflection at which the coils become fully compressed. After this the force versus deflection curve still behaves linearly, but at a much increased spring constant. Physical elements that display relationships, such as that shown in the previous example, are called piecewise linear. The overall behavior is nonlinear, but in fixed regimes the behavior can be well approximated as linear.

Sources of nonlinearity are common. One example occurs whenever two surfaces rub against each other. The friction between a sliding surface and a fixed surface is given by $\mu N$, where $\mu$ is

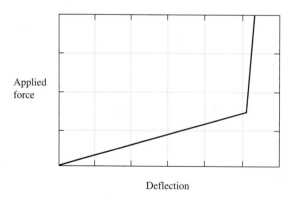

**Figure 8.18** Force versus deflection for a coil spring

the coefficient of friction and N is the normal force. Certainly this isn't linear; it shows no variation at all with velocity, let alone a linearly proportional one. This kind of friction (called Coulomb friction or dry friction) will always be present in the joints and connections of any structure.

What happens to our linear analyses when an unmodeled nonlinearity is present? If the nonlinearities are small, there's no great effect. However, as they become more significant they can substantially alter the modal analyzer's calculations (not unexpectedly). One of the difficulties facing the analyst is trying to determine whether problems with the analyses are due to nonlinearities in the system or noise. To illustrate how nonlinearities can alter a system's response, look at Figures 8.19–8.23. The system is a simple vibrating beam, but with a twist (Figure 8.19). A metal block is placed near the beam's midspan. When the block is far from the beam, the vibrating beam never touches it, resulting in a linear system. As the restraint is moved closer, it becomes more and more common for the beam to strike the restraint as it is vibrating, a strongly nonlinear effect. Four cases are examined, corresponding to linear, weakly nonlinear, moderately nonlinear, and strongly nonlinear responses.

Figure 8.20 shows the fully linear response (case 1). Two peaks are present, at approximately 30 and 175 rad/s. A zero is also indicated at about 105 rad/s. This simply means that the output is close to zero when the system is excited at that frequency. The phase response is similarly clear. The phase angle versus frequency is very well defined and shows that the system is purely in phase or out of phase, depending upon the frequency. The phase switches around each resonance peak as well as at the zero. Finally, the coherence is excellent. Everywhere away from the two peaks and the zero we see a coherence of 1.0. This result is typical of a good modal test. You shouldn't expect a good coherence around resonances or zeros. At resonance, the amplification effect magnifies any noise in the excitation device, affecting $S_{ff}(\omega)$. When at a zero, the actual response is very small. In this case the measured response will be dominated by sensor noise and $S_{xx}(\omega)$ will be adversely affected. The fact that the coherence is excellent everywhere else is an assurance that the test results are believable.

In Figure 8.21 (case 2) we show the results of having just a little bit of contact with the block. Because of this contact, we've got a bit of nonlinear contamination. Of course the linear algorithms in the analyzer don't know this and so they go ahead as if the system is linear. The phase and amplitude responses have changed a bit. The most obvious change is the apparent reduction in damping (shown by the flatter resonance peak) and the noise in the phase response. The real indicator of a problem is the coherence results. As you can see, the coherence has degraded quite a bit, especially around the zero (105 rad/s).

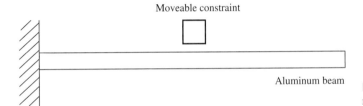

Moveable constraint

Aluminum beam

**Figure 8.19**  Clamped-free beam with rigid constraint

Magnitude and Phase

Coherence

**Figure 8.20**   Linear beam response (case 1)

Magnitude and phase

Coherence

**Figure 8.21**   Nonlinear beam response (case 2)

Magnitude and phase

Coherence

**Figure 8.22**  Nonlinear beam response (case 3)

Magnitude and phase

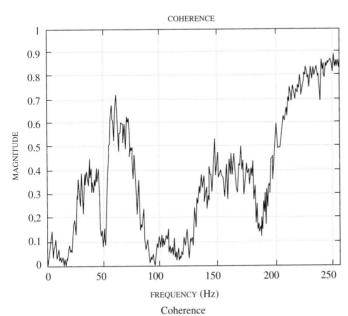

Coherence

**Figure 8.23**  Nonlinear beam response (case 4)

**TABLE 8.1**  FREQUENCY/DAMPING ESTIMATES FOR
NONLINEAR BEAM VIBRATION

|        | Mode 1 | | Mode 2 | |
| --- | --- | --- | --- | --- |
|        | Frequency | Damping | Frequency | Damping |
| Case 1 | 31.2 | 0.009 | 174 | 0.005 |
| Case 2 | 32.2 | 0.011 | 174 | 0.005 |
| Case 3 | 34.8 | 0.003 | 173 | 0.020 |
| Case 4 | 56.0 | 0.110 | 174 | 0.051 |

A moderate nonlinearity (Figure 8.22) causes further degradation. The amplitude response is now very irregular, as is the phase response. Hardly any frequencies show a good coherence response. Finally, Figure 8.23 (case 4) shows the results for a severe nonlinearity. For this case the block is being struck for almost every oscillation of the beam. The results are clearly a mess. The phase response is completely garbled and we've lost any sense of separate resonance peaks in the amplitude response. As expected, the coherence is terrible, indicating that we shouldn't have any faith at all in these responses.

Table 8.1 shows what the first two calculated natural frequencies were for these tests, as well as the calculated linear damping factor approximation. Since the block provides a constraint, and constraints typically stiffen systems, we would expect it to cause the calculated first natural frequency to increase, as indeed happens. Note also the wide variation in the damping estimates. This simply reflects the fact that the measurements are contaminated by the nonlinearity and thus the damping algorithms work badly, trying to find the linear damping response for a system that isn't truly linear.

## 8.8 HOMEWORK PROBLEMS

### Section 8.2

1. Assume that

$$x(t) = \sin(t)$$

is the signal you're concerned with. Generate a discrete time series by recording $n$ values of $x(t)$ for

$$t = \frac{2\pi i}{n} \qquad i = 1 \cdots n$$

Let $n = 10$, 20, and 100. Does the average of these values equal 0, which is the correct average value for the continuous signal $x(t)$?

2. Repeat Problem 1, but let

$$t = \frac{\pi}{5} + \frac{2\pi i}{n} \qquad i = 1 \cdots n$$

that is, include a phase shift. Does the average come out to zero? Why?

**3.** Repeat Problem 1, but take the average of a 10 s interval, that is, average

$$x(t) = \sin(t)$$

over $0 \le t \le 10$. Use a 10 point discretization. Does the average equal zero? Why? How does this impact a real-world data analysis?

**4.** Repeat Problem 1, but vary the averaging period. Take the average over 10, 20, and 100 s. To keep the resolution constant, use 10 data points for the 10-s average, 20 for the 20-s average and 100 for the 100-s average. Does the average equal zero? How do longer averaging times affect the average? Comment on the implications for data analysis.

**5.** Can you find a finite averaging time for which a discrete sampling of

$$x(t) = \sin(t) + \sin(\sqrt{2}t)$$

will average to zero?

**6.** Calculate the variance of

$$x(t) = 3\sin(2\pi t) + 4\cos(4\pi t)$$

over the time interval $0 \le t \le 1$. Demonstrate numerically that the variance of the sum of the two sine waves is equal to the sum of the variance of the individual components.

**7.** Problem 6 showed that the sum of the variance is equal to the variance of the sum when several signals are being added together. Examine the same function,

$$x(t) = 3\sin(2\pi t) + 4\cos(4\pi t)$$

and compute the variance for 20 data points, 200 data points, and 600 data points. Use a sampling interval of .0415 s. Explain the results.

## Section 8.3

**8.** Compute the complex Fourier transform of the $x(t)$ shown below.

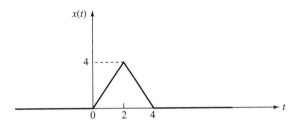

**9.** Compute the complex Fourier transform of the $y(t)$ shown below. How does the answer differ from that of Problem 8? Explain how you can obtain the answer to Problem 9 if you already know the answer to Problem 8 (i.e., without explicitly calculating the Fourier transform).

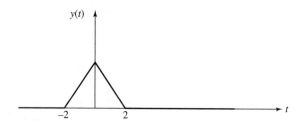

10. Show that the Fourier transform of a signal $y(t)$, where $y(t)$ is simply a delayed version of a signal $x(t)$, is equal to the Fourier transform of $x(t)$ multiplied by $e^{-i\tau\omega}$, where $\tau$ is the time delay.

11. Show that the complex Fourier transform is purely imaginary if the time function is odd and the transform is purely real for even functions. (Hint: Consider what the Fourier series representation of odd and even functions looks like.)

12. Calculate the Fourier transform of $x(t)$ and $y(t)$. How does the delay in time appear in the Fourier transform?

## Section 8.4

13. Find the solution to $\ddot{x} + .2\dot{x} + 4x = \delta(t - 1)$; $x(0) = 0$, $\dot{x}(0) = 0$.

14. Find the solution to $\ddot{x} + .2\dot{x} + 4x = \delta(t - 1)$; $x(0) = .01$, $\dot{x}(0) = 0$.

15. Find the solution to $\ddot{x} + .2\dot{x} + 4x = \delta(t)$; $x(0) = 0$, $\dot{x}(0) = 2$.

16. What does the impulse response look like for an overdamped system?

17. Calculate and plot $R_{xx}(\tau)$ for the illustrated power spectral density. Let $\omega_1 = 1.6$ rad/s and $\omega_2 = 2.0$ rad/s.

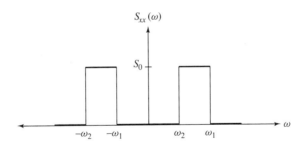

**18.** How does the answer to Problem 17 change as $\omega_1$ goes towards $\omega_2$? (Keep the total energy constant.)

**19.** Consider the system shown below. Let the forcing be described by band-limited white noise. Such an excitation has a power spectral density like that shown below. The input is unpredictable and has energy from $\omega = 10$ rad/s to $\omega = 100$ rad/s. Determine the mean-square value of the system's response. $m = 1$ kg and $k = 90$ N/m.

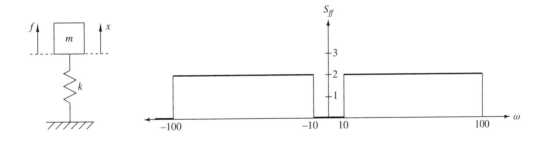

**20.** Verify your answer to Problem 19. Break the forcing spectra into segments of width $\Delta\omega$ as shown below. Using the information in the text, determine what a delta function, of magnitude equal to $2\Delta\omega$ and centered at $10 + \frac{\Delta\omega}{2} + i\Delta$ (that is, the center points of $s_1$, $s_2$, $s_3$, etc.) corresponds to in the time domain. (Hint: It corresponds to a sine wave of frequency $10 + \frac{\Delta\omega}{2} + i\Delta$ with a magnitude you have to determine.) Calculate the mean-square response of the system to each sine and sum them. This should give you the same result as you found from Problem 19, since we're just approaching the problem from the time domain rather than the frequency domain.

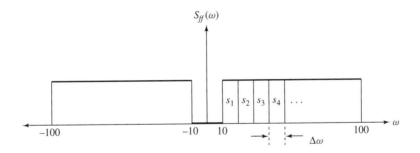

**21.** In this problem we'll show an alternative way to numerically evaluate the answer to Problem 19. As in Problem 20, break $S_{ff}$ into segments and approximate the input by a string of $\delta$ functions.

$$\omega_1 = 10 + \frac{\Delta\omega}{2}$$

$$\omega_2 = 10 + \frac{3\Delta\omega}{2}$$

$$\omega_3 = 10 + \frac{5\Delta\omega}{2}$$

$$\vdots$$

Evaluate the spectral integral relating $S_{xx}(\omega)$, $\dfrac{1}{k - m\omega^2}$, and $S_{ff}(\omega)$ by using the $\delta$ approximation and show that you obtain the same answer as in Problems 19 and 20.

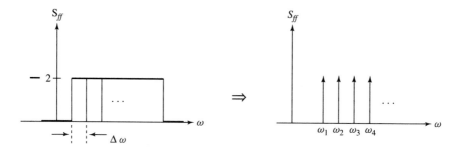

**22.** Determine the autocorrelation function for the $x(t)$ shown below.

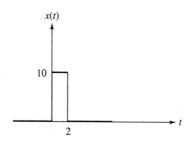

**23.** How does the autocorrelation change from that of Problem 22 if the rectangular function is repeated as shown below?

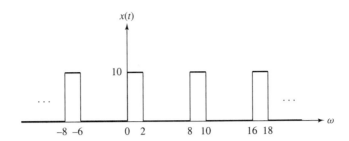

24. Determine the autocorrelation function for the function having the illustrated power spectral density function. The magnitude of each delta function is 20. What kind of signal does this correspond to?

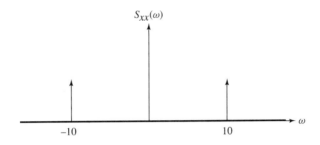

25. Determine the autocorrelation function for the function having the following power spectral density function:

$$S_{xx}(\omega) = 20\delta(\omega - 10) + 20\delta(\omega + 10) + 40\delta(\omega)$$

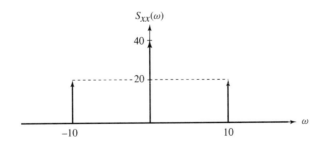

26. What is the mean-square response of a system having a transfer function

$$g(\omega) = \frac{2}{6 + i\omega}$$

for an ideal white-noise input, that is, the input has a power spectral density equal to

$$S_{ff}(\omega) = S_0?$$

**27.** How does the answer to Problem 26 change if the input is band limited so that

$$S_{ff}(\omega) = S_0 ; \quad -7 \leq \omega \leq 7$$

$$S_{ff}(\omega) = 0 ; \quad |\omega| > 7$$

## Section 8.7

**28.** If you're concerned about nonlinearities in your system you might use an impact hammer and tap the structure lightly, identify the system parameters by means of a modal analyzer, and then repeat the procedure with a more forceful hammer strike. How could you use the resulting information to determine if the structure is acting nonlinearly or not?

**29.** Numerically integrate

$$\ddot{x} + x + x^3 = 0; \quad \dot{x}(0) = 0$$

and compare the frequency of the resultant oscillations for $x(0) = .1$, $x(0) = 0.2$, and $x(0) = 0.6$. Does the nonlinearity act to *stiffen* the system (raise the natural frequency) or *soften* it (lower the natural frequency)? Plot the force versus deflection curve for the equivalent spring

$$F = x + \epsilon x^3$$

and try to explain why the system behaves as it does.

#

# Four Continuous Systems

This appendix lists four simple continuous systems that have qualitatively identical equations of motion, as well as the equation of motion governing transverse vibrations of an Euler-Bernoulli beam.

$$(EA\,\xi_x)_x - m\,\ddot{\xi} = 0$$

$E =$ Young's modulus
$m =$ mass density per unit length
$\xi =$ longitudinal deflections of the bar

**Figure A.1**  Equation of motion for a bar

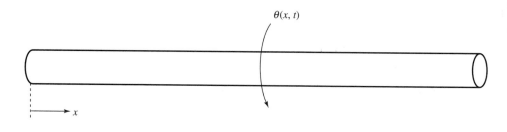

$$(G\theta_x)_x - \rho\ddot{\theta} = 0$$

$G =$ shear modulus
$\rho =$ mass density
$\theta =$ angular rotation of of the rod

**Figure A.2**   Equation of motion for a circular rod

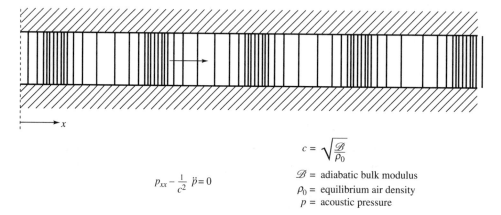

$$p_{xx} - \frac{1}{c^2}\ddot{p} = 0$$

$$c = \sqrt{\frac{\mathscr{B}}{\rho_0}}$$

$\mathscr{B} =$ adiabatic bulk modulus
$\rho_0 =$ equilibrium air density
$p =$ acoustic pressure

**Figure A.3**   Equation of motion for one-dimensional duct acoustics

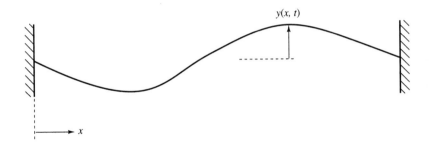

$$(Ty_x)_x - \rho\ddot{y} = 0$$

$T$ = string tension
$\rho$ = mass per unit length
$y$ = lateral string deflection

**Figure A.4**   Equation of motion for a tensioned string

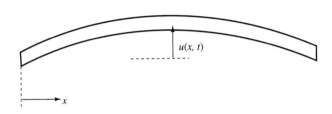

$$(EIu_{xx})_{xx} + m\ddot{u} = 0$$

$E$ = Young's modulus
$m$ = mass density per unit length
$I$ = area moment of inertia
$A$ = cross-sectional area
$u$ = beam deflection

**Figure A.5**   Equation of motion for an Euler-Bernoulli beam

# B

## Lumped Spring Constants

This appendix lists the SDOF, lumped spring constant for several common situations.

$$k = \frac{EA}{l}$$

**Figure B.1**  Longitudinal bar

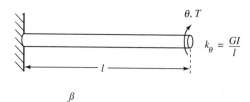

$$k_\theta = \frac{GI}{l}$$

**Figure B.2**  Torsional rod

$x, f$

$l$

$$k = \frac{12EI}{l^3}$$

**Figure B.3**  Lateral motion of cantilevered beam (both ends constrained to remain vertical)

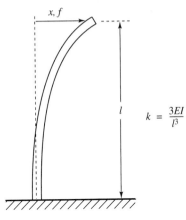

$x, f$

$l$

$$k = \frac{3EI}{l^3}$$

**Figure B.4**  Lateral motion of cantilevered beam (free end allowed to rotate)

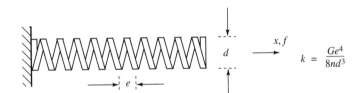

$d$

$x, f$

$e$

$$k = \frac{Ge^4}{8nd^3}$$

**Figure B.5**  Extensional motion of a coil spring ($n$ equals number of turns in coil)

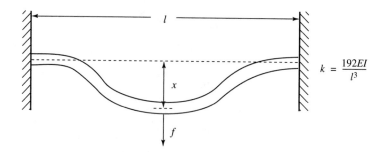

**Figure B.6**   Lateral motion of clamped-clamped beam

$$k = \frac{192EI}{l^3}$$

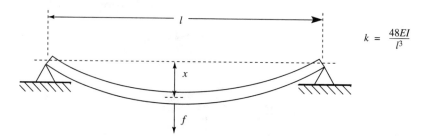

**Figure B.7**   Lateral motion of pinned-pinned beam

$$k = \frac{48EI}{l^3}$$

# C

# Assorted Material Constants

This appendix lists some material constants for several common materials.

**TABLE C.1** DAMPING FACTORS OF COMMON MATERIALS

| Material | Damping factor |
|---|---|
| Steel | $(5-8) \times 10^{-4}$ |
| Aluminum | $(2-3) \times 10^{-4}$ |
| Concrete | $(2-3) \times 10^{-2}$ |
| Wood | $(2-4) \times 10^{-3}$ |
| Riveted steel truss | $(2-4) \times 10^{-2}$ |
| Soft rubber | $(3-5) \times 10^{-2}$ |

**TABLE C.2** MATERIAL CONSTANTS

| Material | Density $\rho$ (kg/m$^3$) | Young's modulus $E$ (N/m$^2$) | Shear modulus $G$ (N/m$^2$) |
|---|---|---|---|
| Steel | $7.8 \times 10^3$ | $2.1 \times 10^{11}$ | $7.9 \times 10^{10}$ |
| Aluminum | $2.7 \times 10^3$ | $7.1 \times 10^{10}$ | $2.6 \times 10^{10}$ |
| Concrete | $1.1 \times 10^3$ | $4.2 \times 10^9$ | — |
| Beryllium copper | $8.2 \times 10^3$ | $1.2 \times 10^{11}$ | $4.8 \times 10^{10}$ |
| Soft rubber | $.95 \times 10^3$ | $2.0 \times 10^9$ | $7.8 \times 10^8$ |

# D

# Elementary Matrix Relations

This appendix is by no means meant to be a substitute for a good course in linear algebra. All we will be be doing here is reminding you of a few matrix facts and operations that you should have already seen in a previous course.

To start, a matrix is an ordered rectangular array of numbers, arranged in rows and columns. The simplest matrix we can have is a $1 \times 1$ matrix, that is 1 row and 1 column. Of course, this is just a scalar. Matrices that have only one row or one column of numbers are the next jump in complexity. We usually call such a one-dimensional array a *vector*. Obviously, a vector in the form of a column is called a column vector and a vector in the form of a row is termed a row vector. If $X$ is a $4 \times 1$ column vector made up of 1's, for example, then

$$X = \begin{bmatrix} 1 \\ 1 \\ 1 \\ 1 \end{bmatrix} \qquad (D.0.1)$$

Finally, an honest-to-goodness matrix consists of a number of columns and rows, such as the matrix illustrated below:

$$[P] \equiv \begin{bmatrix} p_{1,1} & p_{1,2} & \cdots & & p_{1,n} \\ p_{2,1} & p_{2,2} & \cdots & & \vdots \\ \vdots & \cdots & p_{m-1,n-1} & p_{m-1,n} \\ p_{m,1} & \cdots & p_{m,n-1} & p_{m,n} \end{bmatrix} \qquad (D.0.2)$$

Note that we've used the same notation here as in the main text. A matrix is always indicated by a capital letter, surrounded by square brackets. A vector is indicated by a capital letter without any brackets and a scalar is indicated by a lowercase letter. If we're considering a particular matrix, say $[H]$, then we'll always indicate the entries of the matrix by a lowercase $h$, for example $h_{2,3}$. Writing $[h_{i,j}]$ is equivalent to writing $[H]$; they both represent the matrix composed of the elements $h_{i,j}$. We indicate the dimension of a matrix by the number of rows and columns it has. Thus the matrix $[P]$ illustrated above is an $m \times n$ matrix.

Taking the transpose of a matrix simply means interchanging the rows and columns. The transpose of an $m \times n$ matrix is therefore an $n \times m$ matrix. We indicate the transpose by the superscript $T$. Thus if $[B]$ is the transpose of $[A]$, we'd write

$$[B] = [A]^T \tag{D.0.3}$$

In this case we'd have

$$b_{i,j} = a_{j,i} \tag{D.0.4}$$

Matrices $[P]$ for which $p_{i,j} = p_{j,i}$ are called symmetric. Such matrices must be square (that is, the same number of rows as columns). The elements $p_{i,i}$ of a square matrix are known as the main diagonal elements (and lie on the main diagonal), while the other elements ($p_{i,j}$ with $i \neq j$) are known as the off-diagonal elements of the matrix.

For two matrices to be equal, all elements must be equal and in the same relative position, that is, for $[P]$ to equal $[Q]$ we'd require $p_{i,j}$ to equal $q_{i,j}$ for all $i$ and $j$. Only if two matrices are the same size can they be added or subtracted. If $[C] = [A] \pm [B]$, then $c_{i,j} = a_{i,j} \pm b_{i,j}$.

Multiplication is a bit more tricky. If we're multiplying a matrix by a scalar, then we simply need multiply each element in the matrix by the scalar. Thus if $[B] = \alpha[A]$, where $\alpha$ is some scalar constant, then $b_{i,j} = \alpha a_{i,j}$. This isn't too hard. It's when we're multiplying two matrices together that we'll have difficulties. If we want to multiply two matrices together, such as $[P] = [Q][R]$, then

$$p_{i,j} = q_{i,1}r_{1,j} + q_{i,2}r_{2,j} + q_{i,3}r_{3,j} + \cdots = \sum_{k=1}^{o} q_{i,k}r_{k,j} \tag{D.0.5}$$

In the above operation, $[Q]$ is $m \times o$, $[R]$ is $o \times n$, and the resultant matrix $[P]$ is $m \times n$. It's pretty clear that, in general,

$$[Q][R] \neq [R][Q] \tag{D.0.6}$$

In fact, only if $[Q]$ and $[R]$ are square matrices can we form both $[Q][R]$ and $[R][Q]$ and, even then, it's usually not the case that both multiplications lead to the same result.

Division by matrices will be defined soon. First, let's see what is meant by matrix division. With scalars, division is straightforward. If

$$c = \frac{a}{b} \tag{D.0.7}$$

then we know that $a = cb$ and also $a = bc$.

When we talk about matrix division we don't write $\frac{[Q]}{[R]}$. Rather we'd write $[Q][R]^{-1}$ or $[Q]^{-1}[R]$. $[R]^{-1}$ is the inverse of $[R]$ and is the matrix equivalent of $\frac{1}{r}$. Note that the meanings of the two "divisions" just mentioned are quite different. If $[P] = [Q][R]^{-1}$ then $[P][R] = [Q]$. If $[P] = [Q]^{-1}[R]$ then $[Q][P] = [R]$. You can see that the important fact is what side the inverse is on. Why? If you look at the intermediate steps it'll be clear. We started with

$$[P] = [Q][R]^{-1} \tag{D.0.8}$$

If we postmultiply both sides by $[R]$, we'll have

$$[P][R] = [Q][R]^{-1}[R] \tag{D.0.9}$$

One of the facts about matrix inverses is, if a matrix has an inverse, then the matrix times its inverse is the identity matrix, that is,

$$[R][R]^{-1} = [R]^{-1}[R] = [I] \tag{D.0.10}$$

Thus we can simplify (D.0.9) to

$$[P][R] = [Q] \tag{D.0.11}$$

which is what we wanted to show. The same holds for $[P] = [Q]^{-1}[R]$, except in this case you'd premultiply both sides by $[Q]$. Note that a matrix can't have an inverse unless it is square. Furthermore, its determinant must be nonzero. And now, having mentioned and used both determinants and inverses, let's define them.

The determinant of a matrix $[P]$ is represented by $\det[P]$ and by $\|[P]\|$ and is defined by

$$\|[P]\| = \sum_{j=1}^{n} p_{i,j}\pi_{i,j} \tag{D.0.12}$$

where

$$\pi_{i,j} = (-1)^{i+j}\left|[\tilde{P}_{i,j}]\right| \tag{D.0.13}$$

$[\tilde{P}_{i,j}]$ stands for the matrix (not matrix entry) that's composed of what's left over when the $i$th row and $j$th column of $[P]$ have been removed. This smaller matrix is called a *minor* of $[P]$. $\pi_{i,j}$ is called the *cofactor* of $[P]$.

It's easy to calculate low-order determinants. The determinant of a scalar is just the scalar itself. If the matrix $[P]$ is $2 \times 2$, then its determinant is given by $p_{1,1}p_{2,2} - p_{1,2}p_{2,1}$. For higher order determinants you must rely on (D.0.12).

Using minors and cofactors, we can define a matrix inverse. First, we need to construct the adjoint of $[P]$. The adjoint is simply the transpose of the matrix composed of the cofactors of $[P]$,

$$\text{adj}[P] = [\pi_{i,j}]^{T} \tag{D.0.14}$$

One nice fact about the adjoint is that

$$[P]\text{adj}[P] = \|[P]\|[I] = \|[P]\| \tag{D.0.15}$$

Since

$$[P]^{-1}[P] = [P][P]^{-1} = [I] \tag{D.0.16}$$

we can use the foregoing two relations to find

$$[P]^{-1} = \frac{\text{adj}[P]}{|[P]|} \tag{D.0.17}$$

Notice that if the determinant of $[P]$ is equal to zero, then we won't get an inverse. In this case we'd say that $[P]$ is *singular*. Clearly, matrices that have nonzero determinants are nonsingular.

You should realize that you'll rarely be performing these operations by hand; it's far more efficient to use software such as MATLAB. It is, however, nice to know where the algorithms come from, especially if you run into some tricky matrices that give the software problems.

**TABLE D.1**  ASSORTED MATRIX
RELATIONS

| |
|---|
| If $[P] = [M] + [N]$  then  $p_{i,j} = m_{i,j} + n_{i,j}$ |
| $[M] + [N] = [N] + [M]$ |
| $[M][N] \neq [N][M]$   in general |
| $[M]([N][P]) = ([M][N])[P]$ |
| $([M] + [N])^T = [M]^T + [N]^T$ |
| $([M][N])^T = [N]^T[M]^T$ |
| $([M][N][P])^T = [P]^T[N]^T[M]^T$ |
| $([M][N][P])^{-1} = [P]^{-1}[N]^{-1}[M]^{-1}$ |

# References

[1] Boyce, W. E. and DiPrima, R. C. (1977) *Elementary Differential Equations and Boundary Value Problems* (3d ed.). John Wiley and Sons, New York.

[2] Dowell, E. H. (1984) "On Asymptotic Approximations to Beam Mode Shapes." *Journal of Applied Mechanics* 51:439.

[3] Ewins, D. J. (1984) *Modal Testing: Theory and Practice*. Research Studies Press, Letchworth, England.

[4] Ginsberg, S. H. (1995) *Advanced Engineering Dynamics* (2d ed.). Cambridge University Press, Cambridge, England.

[5] Greenwood, D. T. (1988) *Principles of Dynamics* (2d ed.). Prentice Hall, Englewood Cliffs, New Jersey.

[6] Inman, D. J. (1994) *Engineering Vibration*. Prentice Hall, Englewood Cliffs, New Jersey.

[7] Kamal, M. M. and Wolf, J. A., Jr., ed. (1982) *Modern Automotive Structural Analysis*. Van Nostrand Reinhold Co., New York.

[8] Karnopp, D. and Rosenberg, R. (1975) *System Dynamics: A Unified Approach*. John Wiley and Sons, New York.

[9] Newland, D. E. (1984) *An Introduction to Random Vibrations and Spectral Analysis* (2d ed.). Longman Scientific and Technical, New York.

[10] Rayleigh, J. W. S. (1945) *The Theory of Sound, Volume Two*. Dover Publications, Inc., New York.

[11] Synge, J. L. and Griffith, B. A. (1959) *Principles of Mechanics*. McGraw Hill, New York.

[12] Wahl, A. M. (1949) *Mechanical Springs*. Penton Publishing Co., Cleveland, Ohio.

# Selected Readings

There are many vibrations text out there that can and should be perused by the interested student. Some are listed below.

*Structural Dynamics: An Introduction to Computer Methods*, R. R. Craig, Jr., John Wiley and Sons, New York, 1981.

*Mechanical Vibration* (4th ed.), J. P. Den Hartog, McGraw-Hill, New York, 1956.

*Vibration for Engineers*, A. D. Dimarogonas and S. Haddad, Prentice Hall, Englewood Cliffs, New Jersey, 1992.

*Engineering Vibration*, D. J. Inman, Prentice Hall, Englewood Cliffs, New Jersey, 1994.

*Elements of Vibration Analysis* (2d ed.), L. Meirovitch, McGraw-Hill, New York, 1986.

*Analytical Methods in Vibration*, L. Meirovitch, Macmillan, New York, 1967.

*Fundamentals of Vibration Analysis*, N. O. Myklestad, The Maple Press Co., York, Pennsylvania, 1956.

*Mechanical Vibrations* (2d ed.), S. S. Rao, Addison-Wesley, Reading, Massachusetts, 1990.

*The Theory of Sound, Volume Two*, J. W. S. Rayleigh, Dover Publications, Inc., New York, 1945.

*An Introduction to Mechanical Vibrations* (3d ed.), R. F. Steidel, John Wiley and Sons, New York, 1989.

*Theory of Vibration with Applications* (4th ed.), W. T. Thomson, Prentice Hall, Englewood Cliffs, New Jersey, 1993.

# Answers to Selected Problems

Answers to most of the problems are given below. Note that only the final result is given, not the derivations and discussion that a complete solution must, of course, contain. If the answer is "yes," you'd better be able to back that "yes" up with some reasons.

**Chapter 1**

1. $a_1 = \dfrac{2\omega - 3i}{2\omega}$, $a_2 = \dfrac{2\omega + 3i}{2\omega}$

3. $a = 4.143$, $\phi = -.761$ rad

4. $x(8) = 101.\bar{3}$ m

7. Amplitude $= 4$ cm, period $= 4$ s, frequency$= 1.57$ rad/s

9. $\omega_n^2 = \dfrac{2k_3 + \frac{k_1 k_2}{k_1 + k_2}}{m}$

11. Not possible.

13. Compressed: $\omega_n = 38.7$ rad/s; uncompressed: $\omega_n = 38.7$ rad/s

15. $t^* = .479$ s

17. $y(t) = -.01 \cos(7.91t)$ m

21. 50 percent

22. $m' = \frac{m}{8}$

25. $\omega_n = 28.3$ rad/s

27. $\omega_n = 12.6$ rad/s

29. $\omega_n = 31.62$ rad/s

31. $k = 5.48$ N/m

**34.** $\omega_n = 2.59$ rad/s

**35.** $\omega_n = \sqrt{\dfrac{g\rho_{\text{water}}}{l\rho_{\text{wood}}}}$

**37.** $T = \dfrac{2\pi}{\omega_n}$, $\omega_n = \sqrt{\dfrac{2g}{27r_1}}$

**40.** $\omega_n = \sqrt{\dfrac{k_\theta - mgl\cos(\theta_{\text{eq}})}{ml^2}}$

**42.** $\omega_n = .367$ rad/s

**44.** $\omega_n = \sqrt{\dfrac{k_\theta}{m_2l_2^2 + m_1l_1^2}}$

**46.** $\ddot{\theta} - \dfrac{g}{l}\theta = 0$

**48.** $\omega_n = 4.35$ rad/s

**50.** $\omega_n = 6.85$ rad/s

**52.** $\omega_n = 8.38$ rad/s

**54.** $k_\theta = 6.25$ N·m/rad

**56.** $\omega_n = \sqrt{\dfrac{2g}{l}}$

**58.** $\omega_n = \sqrt{\dfrac{4Tr}{I_0 l}}$

**60.** The frequency will decrease.

**62.** Increasing $m$ will reduce the decay rate.

**64.** Yes

**66.** Frequency will be reduced by a factor of 2.

**68.** Decay rate is unaffected by $k$.

**70.** $\zeta = .05$

**72.** Damping isn't linear.

**74.** $\zeta = .2$

**75.** $m\ddot{x} + (k_1 + k_2)x = 0$

**77.** $(m_2l_2^2 + m_1l_1^2)\ddot{\theta} + g(m_2l_2 - m_1l_1)\theta = 0$

**79.** $(m_1 + m_2)\ddot{x} + m_2l\ddot{\theta}\cos(\theta) - m_2l\dot{\theta}^2\sin(\theta) + (k_1 + k_2)x = 0$
$l^2\ddot{\theta} + l\ddot{x}\cos(\theta) - gl\sin\theta + \dfrac{k_\theta}{m_2}\theta = 0$

**80.** $l\ddot{\theta} - g\theta = 0$

**82.** $\ddot{x} = 0$

**85.** $16\ddot{\theta}_1 + 6\ddot{\theta}_2 + 78.48\theta_1 = 0$
$6\ddot{\theta}_1 + 3\ddot{\theta}_2 + 29.43\theta_2 = 0$

## Chapter 2

**2.** 3 m/s

**4.** Spring 1–Spring 2

**6.** $\phi(t) = \dfrac{a}{\omega_n^2 - \omega^2}\cos(\omega t)$, with $\omega_n^2 = \dfrac{4T}{ml}$ and $a = \dfrac{4rT\theta_0}{ml(r+l)}$

**8.** $\sqrt{\frac{kh}{ma}}$

**10.** $15 \sin(\omega t)$ (newtons)

**12.** $\sqrt{\frac{k_1 l_3^2 + k_2 l_2^2}{m_1 l_1^2}}$

**14.** $a_1 \cos(\omega_n t) + a_2 \sin(\omega_n t) + \frac{\bar{f}}{2\omega_n} t \sin(\omega_n t)$

**16.** $(m_1 l_1^2 + m_2 l_2^2 + m_3 l_3^2)\ddot{\theta} + c(l_1 - l_4)^2 \dot{\theta} + k l_2^2 \theta = -\omega c (l_1 - l_4) \bar{y} \sin(\omega t)$

**18.** Increases indefinitely

**20.** $\frac{\bar{f}_f}{\bar{f}} = \frac{k}{k - m\omega^2}$

**22.** $\frac{\bar{f}_\omega}{\bar{y}} = \frac{k_1 k_2}{k_1 + k_2 - m\omega^2}$

**24.** No

**26.** $\omega_n = 19.07$ rad/s, $\zeta = .0179$

**28.** No

**30.** $\frac{\bar{f}_f}{\bar{y}} = \frac{m\omega^2(k + i\omega c)}{k - m\omega^2 + i\omega c}$

**32.** $v = 2.97$ m/s, $|\bar{x}| = 10.2$ cm

**34.** 5 m

**36.** No

**38.** $\bar{x} = \frac{(l_2 m_2 - l_1 m_1)\omega^2}{k_1 + k_2 - \omega^2(m_1 + m_2)}$

**40.** Force magnitude = 1620 N

**42.** 1.8 mm

**44.** $\zeta = .3$

**46.** $\zeta = .1$

**48.** $4\,\mu mga$

**50.** $f > \frac{4}{\pi}\mu mg$

**52.** $a = 9.8$ mm

**54.** 3.4 mm

**56.** $\mu = .32$

**58.** 7.8-degree phase lag

**60.** $.1\omega_n$

## Chapter 3

**3.** $x(t) = \displaystyle\sum_{n=1}^{\infty} \frac{.04\omega_n^2}{n\pi(\omega_n^2 - (n\pi)^2)} \sin(n\pi t)$

**4.** $a_0 = 1$, $\omega_0 = \pi$ rad/s, $b_i = 0$, $a_i$ are nonzero

**7.** $x_n = \dfrac{\frac{8a}{(n\pi)^2}\sin(\frac{n\pi}{2})}{4 - \left(\frac{2n\pi}{T}\right)^2}\sin\left(\frac{2n\pi t}{T}\right)$, $x = \displaystyle\sum_{n=1}^{\infty} x_n$

**9.** $x(1) = 3.54$ mm, $\dot{x}(1) = 4.55$ mm/s

**11.** $x(t) = -y_0 (\cos(\omega_n t) - 1)$

**13.** 198 m/s$^2$

**15.** $k = 200$ N/m

**17.** No

## Chapter 4

**4.** $\omega_1 = 2.08$ rad/s, $\omega_2 = 4.82$ rad/s

**5.** $\omega_1 = 2.07$ rad/s, $\omega_2 = 6.69$ rad/s

**7.** $\omega_1 = .688$ rad/s, $\omega_2 = 2.28$ rad/s

**9.** $\omega_1 = .931$ rad/s, $\omega_2 = 1.296$ rad/s

**11.** No

**13.** Yes

**17.** $m\ddot{x} + k_1 x = 0$, $m\ddot{y} + k_2 y = 0$

**19.** $m\ddot{x} - 2\Omega m\dot{y} + (k_1 - m\Omega^2)x = 0$, $m\ddot{y} + 2\Omega m\dot{x} + (k_2 - m\Omega^2)y = 0$

**20.** $\omega_1 = 172$ rad/s, $\omega_2 = 430$ rad/s

**23.** $\omega_1 = 125$ rad/s, $\omega_2 = 399$ rad/s

**25.** $\omega_1 = 31.6$ rad/s, $\omega_2 = 109.5$ rad/s

**27.** $\omega_1 = 1.14$ rad/s, $\omega_2 = 1.51$ rad/s

**31.** $\omega_1 = 6.82$ rad/s, $\omega_2 = 13.2$ rad/s

**33.** $v = 17.2$ mph

**35.** $\lambda = 4$ m

**37.** Yes

**41.** Relative motion $= .065 \sin(18.1t) - .026 \sin(45.7t)$

**42.** $k_2 = 100$ m

**45.** $k_1 = 62,000$ N/m

**46.** No

**48.** $\bar{x}_1 = -7.83$ mm, $\bar{x}_2 = 781$ mm

**51.** $\omega_1 = \sqrt{\dfrac{k_1 + k_2}{m_1}}$, $\omega_2 = \sqrt{\dfrac{k_4 + k_5}{m_4}}$

**53.** $\omega_1 = \sqrt{\dfrac{k_1 + k_2}{m_1}}$

**56.** $\dfrac{k_4 + k_5}{m_3} = 900$

**58.** $\omega = 17.9$

**68.** Magnitude of modal response is $\dfrac{5}{3\sqrt{2}}$

**70.** $-.15 \pm .99i$, $-.2 \pm 1.21i$, $-.7 \pm 2.45i$

**73.** $x_1 = -1.3497 \sin(5t) - .0985 \cos(5t)$, $x_2 = .1769 \sin(5t) + .0415 \cos(5t)$

**75.** $x_1 = -.4208 \cos(20t) + .0038 \sin(20t)$, $x_2 = .0617 \cos(20t)$, $x_3 = 2.754 \cos(20t) + .1212 \sin(20t)$

**77.** No

**81.** $x_1 = 1.054$ m, $x_2 = 1.284$ m

**83.** $f_1 = 102$ N, $f_2 = -56$ N, $f_3 = 2$ N

**85.** $\frac{f_1}{k_1+k_2} = \frac{-f_2}{k_2}$

## Chapter 5

**2.** $\omega_1 = 17.09$ rad/s, $\omega_2 = 111.5$ rad/s, $\omega_3 = 219.0$ rad/s

**6.** $\frac{1}{\sqrt{2}}$

**8.** $200\frac{dl}{l}$ percent

**9.** $\omega_n = \frac{n\pi}{l}\sqrt{\frac{T}{\rho}}$

**19.** $w(x, t) = \frac{l}{\pi}\sqrt{\frac{\rho}{T}}\sin\left(\frac{\pi x}{l}\right)\sin\left(\frac{\pi}{l}\sqrt{\frac{T}{\rho}}t\right)$

**24.** $\xi(x, t) = a\sin\left(\frac{\pi x}{l}\right)\cos(\omega_1 t) + \frac{b}{\omega_3}\sin\left(\frac{3\pi x}{l}\right)\sin(\omega_3 t)$, $\omega_n = \frac{n\pi}{l}\sqrt{\frac{EA}{m}}$

**26.** $T = 2323$ N

**27.** steel: $8150(2n-1)$ rad/s; aluminum: $8060(2n-1)$ rad/s; copper: $6010(2n-1)$ rad/s

**29.** $a_n = \frac{2a\sin(\frac{n\pi}{2})}{\omega_n}$, $\omega_n = \beta_n\sqrt{\frac{T}{\rho}}$

**30.** 2484 rad/s; 7451 rad/s; 12,418 rad/s; 17,386 rad/s

**32.** 29,086 rad/s

**34.** Torsion

**36.** No difference

**38.** $I_0 = 5.7 \times 10^{-6}$ kg·m$^2$

**40.** $m_1 = .0534$ kg

**42.** 1811 rad/s; 25,656 rad/s; 25,785 rad/s

**47.** Frequency doubling

**50.** 3.6646, 6.6874, 9.7516, 12.8396

**52.** $\omega_n = n^2\pi^2\sqrt{\frac{EI}{\rho}}$

**55.** $\xi(x, t) = \frac{\bar{f}}{m}\frac{1}{\frac{EA}{m}\left(\frac{\pi}{2l}\right)^2 - \omega^2}\cos(\omega t)\sin\left(\frac{n\pi}{2l}\right)$

**59.** $u(x, t) = \frac{11}{\rho}\frac{t^2}{2}$

## Chapter 6

**2.** $\omega_1 = 3.1153\sqrt{\frac{EA}{ml^2}}$, $\omega_2 = 6.074\sqrt{\frac{EA}{ml^2}}$

**4.** Approximate results: $\omega_1 = 2.993\sqrt{\frac{EA}{ml^2}}$, $\omega_2 = 5.992\sqrt{\frac{EA}{ml^2}}$

**6.** At 5 percent error $k_1 = 180$ N/m

**8.** $\alpha = 1.6$ kg

**14.** No

**16.** Yes

**17.** No

**21.** For $m_1 = 1$ kg, $\omega_1 = 34.3$ rad/s

**23.** 88.69 rad/s, 179.22 rad/s, 272.1 rad/s

## Chapter 7

**1.** 90 degrees

**4.** $\omega \ll \frac{\pi}{2l}\sqrt{\frac{G}{\rho}}$

**6.** There are restrictions as to the placement.

**13.** A fixed assumption is reasonable.

**14.** 7 kg

**17.** .14 kg

**20.** The initial analysis was correct.

## Chapter 8

**1.** Average equals zero.

**3.** Average doesn't equal zero.

**5.** No

**6.** 12.5

**8.** $\frac{1}{\pi\omega^2}\left(2e^{-2i\omega} - e^{-4i\omega} - 1\right)$

**9.** $\frac{2}{\pi\omega^2}\left(1 - \cos(2\omega)\right)$

**12.** Time delay corresponds to a phase shift in the Fourier transform.

**14.** $x(t) = .01e^{-\zeta\omega_n(t-1)}\cos(\omega_d(t-1)) + .5011e^{-\zeta\omega_n(t-1)}\sin(\omega_d(t-1))$ for $t \geq 1$; $\omega_n = 2$, $\zeta = .05$.

**15.** $x(t) = 1.502e^{-.1t}\sin(1.9975t)$

**19.** $E[x^2] = 0.179$

**24.** $10\cos(10\tau)$

**26.** $2.094S_0$

**29.** Stiffen

# Index